Electrical Estimating

Mike Holt's Illustrated Guide

Electrical Estimating

neccode.com
888-NEC® CODE

Mike Holt Enterprises, Inc.
7310 W. McNab Road • Suite 201 • Tamarac, Florida, 33321
1.888.NEC.CODE • NECcode.com • Info@NECcode.com

NOTICE TO THE READER

Publisher does not warrant or guarantee any of the products described herein or perform any independent analysis in connection with any of the product information contained herein. Publisher does not assume, and expressly disclaims, any obligation to obtain and include information other than that provided to it by the manufacturer.

The reader is expressly warned to consider and adopt all safety precautions that might be indicated by the activities herein and to avoid all potential hazards. By following the instructions contained herein, the reader willingly assumes all risks in connections with such instructions.

The publisher makes no representation or warranties of any kind, including but not limited to, the warranties of fitness for particular purpose or merchantability, nor are any such representations implied with respect to the material set forth herein, and the publisher takes no responsibility with respect to such material. The publisher shall not be liable for any special, consequential, or exemplary damages resulting, in whole or part, from the readers' use of, or reliance upon, this material.

Cover Design: Paul Wright - wright1156@home.com

November, 2001.

COPYRIGHT © 2001 Charles Michael Holt Sr.

Printed in the United States of America
For more information, contact:

Mike Holt Enterprises, Inc.
7310 West McNab Road
Suite 201
Tamarac, Florida 33321

All rights reserved. No part of this work covered by the copyright hereon may be reproduced or used in any form or by any means graphic, electronic, or mechanical, including photocopying, recording, taping, or information storage and retrieval systems without the written permission of the publisher. You can request permission to use material from this text, phone 1.888.NEC.CODE, Sales@NECcode.com, www.NECcode.com.

NEC, NFPA, and National Electrical Code are registered trademarks of National Fire Protection Association.

This logo is a registered trademark of Mike Holt Enterprises, Inc.

ISBN 0-9710307-8-2

*I dedicate this book to the Lord Jesus Christ,
my mentor and teacher*

Table Of Contents

PREFACE 1
- Electrical Estimating 1
- About This Book 1
- Chapters 1 Through 8 2
- Appendixes 3
- Estimating Software Available 3
- About The Author 3
- The Emotional Aspect Of Learning 3
- How To Get The Most From This Book ... 4
- Acknowledgment 5

CHAPTER 1
INTRODUCTION 7
- Introduction 7
- 1.01 Estimating Versus Bidding 7
- 1.02 A Good Estimating System 8
- 1.03 Objectives And Purpose Of An Electrical Contractor 8
- 1.04 Why Are So Many Electrical Contractors Unsuccessful? 8
- 1.05 The Estimate Must Provide Information For Project Management 8
- 1.06 Job Management Ultimately Controls The Cost Of A Project 8
- 1.07 Can I Be Competitive? 8
- 1.08 The Market 10
- Summary 11
- Review Questions 13

CHAPTER 2
ABOUT ESTIMATING 15
- 2.01 Qualities Of A Good Estimator 15
- 2.02 Duties And Responsibilities Of The Estimator 15
- 2.03 The Estimating Work Space And Tools . 16
- 2.04 Types Of Bids 18
- 2.05 What An Accurate Estimate Must Include .19
- 2.06 Improper Estimating Methods 19
- 2.07 Proper Estimating Methods 20
- 2.08 The Detailed Estimating Method 20
- 2.09 How Accurate Can An Estimate Be? ... 21
- 2.10 Manual Estimate, Computer Assisted, Or An Estimating Service? 22
- Summary 24
- Review Questions 27

CHAPTER 3
UNDERSTANDING LABOR-UNITS 29
- Introduction 29
- 3.01 What Exactly Is A Labor-Unit? 29
- 3.02 How Labor-Units Are Used 30
- 3.03 What's Included In The Labor-Unit ... 30
- 3.04 Labor-Units Do Not Include 33
- 3.05 Labor-Unit Manuals, Which Should You Use? 33
- 3.06 Your Labor-Units As Compared To Your Competitors' 33
- 3.07 Knowing Your Competitors' Labor-Units . 34
- 3.08 How To Develop Your Own Labor-Units . 35
- 3.09 Variables That Impact Labor-Units 35
- 3.10 Labor-Unit Summary 40
- Summary 40
- Review Questions 43

vii

CHAPTER 4
THE ESTIMATING PROCESS 45
Introduction 45
4.01 Job Selection 46
4.02 Understanding The Scope Of Work (Step 1) . 48
4.03 The Take-Off (Step 2) 50
4.04 Determining The Bill-Of-Material (Step 3) . . 56
4.05 Pricing And Laboring (Step 4) 56
4.06 Extensions And Totals (Step 5) 57
Summary 62
Review Questions 64

CHAPTER 5
THE ESTIMATE SUMMARY AND
BID PROCESS 67
Introduction 67
Part A – The Estimate Summary 68
5.01 Labor-Hours (Step A) 69
5.02 Labor Cost (Step B) 71
5.03 Labor Burden 72
5.04 Total Material Cost (Step C) 72
5.05 Direct Job Expenses (Step D) 73
5.06 Estimated Prime Cost (Step E) 75
5.07 Overhead 75
5.08 Overhead Calculation 76
5.09 Applying Overhead (Part F) 77
5.10 Understanding Break Even Cost (Part G) . . 77
Part B – The Bid Process 78
5.11 Profit 78
5.12 Other Final Costs 79
5.13 Bid Accuracy 79
5.14 Bid Analysis 80
5.15 Bid Proposal 80
Part C – Unit Pricing 83
5.16 Unit Price Example 83
Summary 85
Review Questions 88

CHAPTER 6
ESTIMATING RESIDENTIAL WIRING . . . 95
Introduction 95
Part A – Plans And Specifications 95
Part B – Manual Estimate And Bid 108
6.01 Understanding Scope Of Work 108
6.02 The Take-Off 112
6.03 Completed Residential Take-Off . . . 120
6.04 Determining The Bill-Of-Material . . . 123
6.05 Pricing, Labor, Extending, And Totals . . 132
6.06 Estimate and Bid Summary 139
6.07 Bid Accuracy And Analysis 142
6.08 The Proposal 142
Part C – Computer Assisted Estimate And Bid . 143
6.09 Bid Preparation And Take-Off 143
6.10 Input Take-Off Into Computer 143
6.11 Bill-Of-Material, Pricing, Laboring,
  Extending, And Totaling 143
6.12 Estimate And Bid Summary 143
6.13 Bid Accuracy And Analysis 143
6.14 The Proposal 177
Review Questions 179

CHAPTER 7
ESTIMATING COMMERCIAL WIRING . . . 189
Introduction 189
Part A – Plans And Specifications 189
7.01 Electrical Specifications 196
Part B – Manual Estimate And Bid 214
7.02 Preparing The Estimate 214
7.03 Understanding Scope Of Work 214
7.04 Estimate And Bid Notes 214
7.05 The Take-Off 214
7.06 Determining The Bill-Of-Material . . . 223
7.07 Pricing, Laboring, Extending, And Totals . 223
7.08 Estimate And Bid Summary 233
7.09 Bid Accuracy And Analysis 235
7.10 The Proposal 235

Part C – Computer Assisted Estimate and Bid 236
7.11 Bid Preparation and Take-Off 236
7.12 Input Take-Off Into Computer 236
7.13 Bill-Of-Material, Pricing, Laboring, Extending, And Totaling 236
7.14 Estimate And Bid Summary 236
7.15 Bid Accuracy And Analysis 236
Review Questions 238

CHAPTER 8
COMPUTER ESTIMATING 247
Introduction 247
8.01 Computerized Estimating 247
8.02 Frequently Asked Questions 248
8.03 Training And Support 249
8.05 Pricing Services 249
8.06 Software Selection 249
8.07 How Much Should It Cost? 250
8.08 Who Sells Estimating Software? . . . 250
Review Questions 251

APPENDIX – ESTIMATING FORMS AND WORKSHEETS 253

INDEX 265

Preface

ELECTRICAL ESTIMATING

Electrical estimating is important to different individuals involved in the electrical industry. The apprentice and electrician need to have an understanding of estimating so as to gain a perspective of the value of their work. The aspiring estimator needs to understand the nuts and bolts of estimating to properly perform a complete and accurate estimate. The electrical contractor must know how to estimate to determine the job's selling price and for proper job management.

The primary purpose of this book is to give the reader a birds-eye view of both estimating and the bid process. This book explains how to determine material cost, labor cost, the proper application of direct cost, overhead, and profit.

ABOUT THIS BOOK

This book, *Electrical Estimating*, was designed to help you understand how to properly estimate, either manually or with the assistance of a computer. It also explains the importance of using the estimate for proper project management to insure that the job is completed profitably.

This book contains step-by-step instructions with blueprints on how to estimate residential and commercial electrical wiring. Also included at the end of this book is the *Electrical Estimating Workbook* which contains the scale blueprints required for this book.

As you proceed through this book you will find some sections you won't initially understand. Please don't get frustrated, just highlight the confusing section of the book and discuss the issue with someone who can add some insight. After you have completed the book, take the time and review those areas that you had problems with. I'm sure that most of those sticky points will have been cleared up.

Author's Comment. A live video presentation by Mike Holt is available to be used in conjunction with this book. Call toll free 1-888-NEC CODE for details.

Note. The management and financial topics in this book are directed to the electrical contractor and might not apply to the apprentice, electrician, or estimator if a specific management or financial topic does not apply to you, simply skip over that section.

This book is divided into eight Chapters and one Appendix. Chapters 1 through 5 contain comprehensive explanations of the estimating and bid process with concise summaries and review exams which consist of both essay as well as multiple choice questions.

Chapter 6 and 7 contain the details of a complete estimate and bid for both a residential as well as a commercial building, and detailed computer generated analysis reports. These two chapters contain an estimate to be used with the chapter review questions.

Chapter 8 explains the advantages and benefits of computer estimating. It also contains a review exam. This book does not contain a glossary of terms, because each new term or concept is explained as you proceed through the book.

CHAPTERS 1 THROUGH 8

Chapter 1 - Introduction
This chapter explains the purpose, importance, and need for proper estimating.

Chapter 2- About Estimating
This chapter lets you know what estimating is all about, the good, the bad, and the ugly.

Chapter 3 - Understanding Labor-Units
You will learn the details about labor-units, what they are, how they're used, and how to develop and use your own.

Chapter 4- The Estimating Process
This chapter explains the details about the estimating process, starting with estimate preparation and ending with extending and totaling of material cost and labor hours.

Chapter 5 - The Estimate Summary and Bid Process
This chapter is divided into three parts, Part A -The Estimate Summary, Part B The Bid Process, and Part C - Unit Pricing.

Part A - The Estimate Summary. This part covers how to summarize the estimate to determine the job's estimated break even cost. You'll learn how to determine total labor-hours including labor adjustments and how to:
1. Determine the labor rate per man-hour
2. Total material cost, taking into consideration miscellaneous material, waste, and theft.
3. Accounting for small tools, handling and markup cost.
4. Proper pricing of quotes, including handling and markup.
5. How to calculate direct job cost and overhead with examples, suggestions, and recommendations.

Part B - The Bid Process. This part explains how to:
1. Determine profit margins and the job's selling price.
2. Avoid common estimating errors, and how to perform a bid analysis.
3. Produce a bid proposal.

Part C- Unit Pricing. This part explains the process and gives an example of how to determine unit prices.

Chapter 6- Estimating Residential Wiring
This chapter contains the details of a complete estimate of a 3,700 square foot residence. The estimate is based on blueprints from Ray Mullin's best selling book *Electrical Wiring - Residential*, published by Del-mar Publishers.

This chapter is divided into two parts:

Part A - Manual Estimate. This part contains the complete details of a manual estimate and bid of the residence.

Part B – Computer Assisted Estimate. This part contains the complete details of a computer assisted estimate and bid of the residence, with sample computer printouts.

In addition, Chapter 6 contains a residential blueprint for the student to complete. The student is expected to estimate the residence and determine a bid price. This includes the bid preparation, take-off, bill of material, pricing, laboring, extending, totaling, estimate summary and applying profit.

Chapter 7- Estimating Commercial Wiring
This chapter contains the details of a complete estimate of a 7,740 square foot commercial building. The estimate is based on blueprints from Ray Mullin's and Robert Smith's book *Electrical Wiring - Commercial*.

This chapter is divided into three parts:

Part A - Plans and Specs. This part contains the blueprint and specification details required to understand the estimate contained in Parts B and C of this chapter

Part B - Manual Estimate. This part contains the complete details of a manual estimate and bid of the building.

Part C- Computer Assisted Estimate. This part contains the complete details of a computer assisted estimate and bid of the building, with sample computer printouts.

> **Note**. This book does not contain any specifics on estimating industrial wiring. However once you understand the proper technique of estimating, you should have no problem estimating any job you have experience with. This would include industrial wiring, fire alarms, service stations, hospitals, restaurants, etc.

Chapter 8 - Computer Estimating

This chapter should help you understand the technical aspects of computer assisted estimates. It explains the hardware requirements and the options in selecting software. If you are considering purchasing estimating software in the future, review this chapter to help you with your decision.

APPENDIX

Estimating Forms And Worksheets

This appendix contains typical estimating forms and worksheets, and explains their use.

ELECTRICAL ESTIMATING WORKBOOK

Chapter 1 - Price And Labor-Unit Catalog

Chapter 1 contains a price and labor-unit catalog for approximately 1,500 common material items.

Chapter 2 - Blueprints

Chapter 2 contains the blueprints necessary for use in conjunction with the questions contained in Chapters 6 and 7.

ABOUT THE AUTHOR

Mike Holt of Groveland, Florida, is the author of: *Understanding tile National Electrical Code*, *Understanding NEC Calculations*, *Journeyman Electrician's Exam Preparation*, *Master Electrician's Exam Preparation*, *Electrical Formulas and Reference Book*, and related workbooks. He worked his way up through the electrical trade as an apprentice, journeyman, master electrician, electrical inspector, electrical contractor, electrical designer and developer of training programs and software for the electrical industry. Mike is a contributing editor to *Electrical Construction and Maintenance* magazine (EC&M) and *Electrical Design and Installation* magazine (EDI). He has been a contributing writer for Electrical Contractor magazine (EC). With a keen interest in continuing education, Mike Holt attended the University of Miami Masters in Business Administration Program (MBA) for Finance.

Mike Holt has provided custom in-house seminars for: IAEI, NECA, ICBO, IBM, AT&T, Motorola, Boeing Airplanes, and the U.S. Navy, to name a few. He has taught over 1,000 classes on over 30 different electrical-related subjects ranging from Business Management, Estimating, Power Quality, Grounding and Bonding, Exam Preparation, and other continuing education programs. Many of Mike Holt's seminars are available on video as well. He continues to develop additional courses, seminars, and workshops to meet today's changing needs for individuals, organizations, vocational, and apprenticeship training programs.

Since 1982 Mike Holt has been helping electrical contractors improve the management of their business by offering business and legal consulting, business management seminars, and computerized estimating and billing software.

Contact Mike Holt about an Estimating and Project Management Seminar in your area, 1-888-NEC CODE.

THE EMOTIONAL ASPECT OF LEARNING

To learn effectively, you must develop an attitude that learning is a process that will hell) you grow 1)0th personally and professionally. The learning process has an emotional as well as an intellectual component that we must recognize. To understand what affects our learning, consider the following:

Uniqueness

Each of us will understand the subject matter from different perspectives and we all have some unique learning problems and needs.

Resistance to Change

People tend to resist change and resist information that appears to threaten their comfort level of knowledge. However, we often support new ideas that support our existing beliefs.

Dependence and Independence

The dependent person is afraid of disapproval and often will not participate in class discussion and will tend to wrestle alone. The independent person spends too much time asserting differences and too little time trying to understand others views.

Fearful

Most of us feel insecure and afraid with learning, until we understand what is going to happen and what our role will be. We fear that our performance will not match the standard set by us or others.

Egocentric

Our ego tendency is to prove someone is wrong, with a victorious surge of pride. Learning together without a win/lose attitude can be an exhilarating learning experience.

Emotional

It is difficult to discard our cherished ideas in the face of contrary facts when overpowered by the logic of others.

HOW TO GET THE MOST FROM THIS BOOK

Studies have concluded that for students to get the most out of any subject, they must learn to get the most from their natural abilities. It's not how long you study or how high your IQ is, it's what you do and how you study that counts. To get the most from this book, you must make a decision to practice as many of the following techniques as possible.

Attitude

Maintaining a positive attitude is important. It helps keep you going and helps keep you from getting discouraged.

Clean Up Your Act

Keep all of your papers neat, clean, and organized. Now is not the time to be sloppy. If you are not neat, now is an excellent time to begin.

Communication With Your Family

Good communication with your family members is very important. Try to get their support, cooperation, and encouragement during this time. Let them know the benefits and plan some special time with them during this preparation period.

Eye Care

It is very important to have your eyes checked. Human beings were not designed to do constant seeing less than an arm's length away. Our eyes were designed for survival, spotting food and enemies at a distance. Your eyes will be under stress because of prolonged, near-vision reading, which can result in headaches, fatigue, nausea, squinting, or eyes that burn, ache, water, or tire easily.

Reducing Eye Strain

Be sure to look up occasionally, away from near tasks to distant objects. Your work area should be three times brighter than the rest of the room. Don't read under a single lamp in a dark room, and try to eliminate glare. Mixing of fluorescent and incandescent lighting can be helpful. Sit straight, chest up, shoulders back, and weight over the seat so both eyes are an equal distance from what is being read.

Getting Organized

Our lives are so busy that simply making time for study and homework is almost impossible. You can't waste time looking for a pencil or missing paper. Maintain a folder for notes, exams and answer keys. It is very important that you have a private study area available at all times.

Learn How to Read

Before you get started, take a few moments to review the book's table of contents, index, and graphics. This will help you develop a better sense of the material and what you can expect.

Set Priorities

Once you begin your study, stop all phone calls, TV shows, radio, and other interruptions.

Speak Up in Class

If you are in a classroom setting, the most important part of the learning process is class participation.

If you don't understand the instructor's point, ask for clarification. Don't try to get attention by asking questions you already know the answer to.

Study Anywhere/Anytime

To make the most of your limited time, always keep a copy of the book with you. Any time you get a free minute, study and continue to study every chance you get. You can study at the supply house when waiting for your material, you can study during your coffee break, or even while you are at the doctor's office. Be creative.

You need to find your best study time, for some it could be late at night when the house is quiet. For others, it's the first thing in the morning before things get going.

Study With A Friend

Studying with a friend can make learning more enjoyable, you can push and encourage each other. You are more likely to study if someone else is depending on you. Those who study alone spend most of their time reading and re-reading the text and trying the same approach time after time even though it might be unsuccessful.

Support

You need encouragement in your studies and you need support from your loved ones.

Time Management

Time management and planning are very important. There simply are not enough hours in the day to get everything done. Make a schedule that allows time for work, rest, study, meals, family, and recreation. Establish a schedule that is consistent from day to day.

Have a calendar and immediately plan your homework. Try not to procrastinate, that is putting it off because you are afraid. Learn to pace yourself to accomplish as much as you can.

Training

Get plenty of rest and avoid intoxicating drugs, including alcohol. Stretch or exercise each day for at least 10 minutes. Eat light meals such as pasta, chicken, fish, vegetables, fruit, etc. Try to avoid heavy foods such as red meats, butter, and other high-fat foods. They slow you down and make you tired and sleepy.

ACKNOWLEDGMENTS

Mom, thank you for showing me the importance of God, the value of hard work and the importance of education.

I would like to thank Mike Culbreath, a master electrician, for helping me transform my words into graphics and Paul Bunchuk, master electrician, of PNB Graphics for design, layout, and typesetting. Thanks to the both of you for not sacrificing quality, and for the extra effort to make sure that this book is the best that it can be.

To my family, thank you for your patience and understanding. I would like to give special thanks to my beautiful wife Linda, and my children Belynda, Melissa, Autumn, Steven, Michael, Meghan, and Brittney (seven kids).

And thanks to those who helped me in the electrical industry, Ray Mullin for my big break in becoming an author with Delmar Publishers, Electrical Construction and Maintenance magazine, and Joe McPartland, my mentor, who was always there to help and encourage me. I would like thank the following for contributing to my success: John Calloggero, Dick Loyd, Mark Ode, D.J. Clements, Joe Ross, Tony Selvestri, James Stallcup, and Marvin Weiss.

The final personal thank you goes to Sarina, my friend, office manager, and Educational Director of Mike Holt Enterprises. Thank you for covering the office for me the past few years while I spent so much time writing books. Your love and concern for me has helped me through many difficult times.

The author would also like to thank those individuals who reviewed the manuscript and offered invaluable suggestions and feedback.

Paul Bunchuk
PNB Graphics
Ft. Lauderdale, Florida

James Cole
Coastal Carolina Community College
Jacksonville, North Carolina

Mike Culbreath
MC Gralx
Kewadin, Michigan

Greg Fletcher
Kennebec Valley Technical College
Fairfield, Maine

Richard Kurtz
Electrical Consultant
Boynton Beach, Florida

John Marcelli
Accubid Systems, Inc.
Toronto, Canada

Ray C. Mullin
Author - *Electrical Wiring, - Residential*
Author - *Electrical Wiring - Commercial*
Northbrook, Illinois

Jack McCormick
McCormick Systems, Inc.
Mesa, Arizona

Ron McMurtry
Kentucky Technical College
Paducab, Kentucky

Sarina Snow
Educational Director
Mike Holt Enterprises, Inc.

Robert L. Smith, P. E.
Author - Electrical Wiring - Commercial
Champaign, Illinois

Brooke Stauffer
National Electrical Contractors Association
Author - *SMART HOUSE Wiring*
Bethesda, Maryland

Chapter 1

Introduction

OBJECTIVES

After reading this chapter, you should understand

- the purpose of electrical contracting.
- why many electrical contractors are unsuccessful.
- why an estimate is required to create a project budget.
- how a contractor's success is impacted by competition, labor productivity, management skills, overhead, and risk.
- that the electrical industry is in constant change and which factors impact the electrical market.

INTRODUCTION

A great percentage of electrical work is acquired through the estimating process, and most jobs are awarded to the contractor who has the best perceived price, but not necessarily the lowest. Because of the demands to have the best perceived price, profit margins are limited. This permits you to have only a small margin for error in the estimate. A proper estimate must accurately determine your cost in completing the job according to the customer's needs. This price must be acceptable to your customers at a value that includes sufficient profit for you to stay in business. In addition to helping you determine the selling price for a job, the estimate is used as the foundation for project management.

1.01 ESTIMATING VERSUS BIDDING

Determining the selling price for a job is actually two separate components. The first component is called the estimate, which determines the cost of the job. The second component is the bid, which determines the job's selling price. It is critical that you understand the difference between an estimated cost and a bid price. Estimating is determining your cost and bidding is determining the selling price.

Estimating

The purpose of estimating is to determine the cost of a project before you actually do the work. Estimating must take into consideration variable job conditions, the cost of materials, labor cost, direct job expenses, and management costs (overhead).

Bid Process

Once you know the estimated cost of a project, you can determine the selling price of the job. Determining the selling price of a job is called bidding.

1.02 A GOOD ESTIMATING SYSTEM

A good estimating system should help you quickly and accurately determine the cost of a project, and includes all anticipated costs. The system must be efficient, accurate and attempt to prevent mistakes. It should have a method to verify that the estimate is accurate.

It is not uncommon, when estimating, for omissions, mistakes, or inaccuracies to occur. These can combine to exceed the job's profit margin, resulting in a net loss for the job.

1.03 OBJECTIVES AND PURPOSE OF AN ELECTRICAL CONTRACTOR

The purpose of an electrical contractor is to make a profit. To be a successful electrical contractor, you must provide a quality service to your customers at a competitive price that is higher than your cost. Successful electrical contracting is really just that simple, sell the job at a competitive price that exceeds your cost. Therefore, the first step to becoming a successful electrical contractor is knowing your cost.

1.04 WHY ARE SO MANY ELECTRICAL CONTRACTORS UNSUCCESSFUL?

Many electrical contractors have worked their way up through the trade from apprentice electrician. They have plenty of hands-on experience in installing electrical systems, but lack management skills.

Most electrical contractors wear many hats: secretary, warehouse person, truck driver, supervisor, electrician, salesperson, etc. There aren't enough hours in the day or night to keep up with all of the demands of running a successful contracting business. If an electrical contractor is not careful, he or she might forget the most important hat, managing the job to insure that a profit is made.

If you don't have an accurate estimate, you won't know what it will cost you to do the job, and you're likely to feel uncomfortable with your bid price. This lack of critical information has caused many electrical contractors to take contracts below their cost.

1.05 THE ESTIMATE MUST PROVIDE INFORMATION FOR PROJECT MANAGEMENT

With over 30,000 electrical contractors in the United States, competition is fierce. Successful electrical contractors must know how to manage their projects so they can compete profitably in today's highly competitive market place. The actual cost of any project is significantly impacted by how well the job is managed. To properly manage a job, the project manager must have a budget (estimate).

The job budget must have information as to what material is required, when it's required, and the labor required to complete each phase of the job. With proper information, the project manager will be better prepared to complete the project as planned.

1.06 JOB MANAGEMENT ULTIMATELY CONTROLS THE COST OF A PROJECT

Proper project management is often the difference between profit or loss so you must realize that effective job planning, labor scheduling, and material purchasing are all factors in determining the ultimate cost of a project.

You must be sure that the job foreman is informed on how the job was estimated. There must be continuous communication between the electricians performing the work, the project manager, and the estimator.

Without a proper estimate you won't know what materials or tools you need. You might not get the material on the job in a timely manner, which can result in your workers standing around wasting time and your money. If you improperly manage the material, you might have too much material on the job which increases the likelihood of it being wasted or stolen.

1.07 CAN I BE COMPETITIVE?

Most electrical contractors are concerned about their ability to be competitive, make money, and stay in business. In order to be competitive, the contractor must offer the customer a quality service at a reasonable price. To accomplish this, the electrical contractor must control the job costs so they are within the estimated budget. It is also critical that the contractor control overhead expenses to keep them to a minimum. Factors that affect a contractor's competitiveness include:

1. Competition
2. Cost of material (buying power)
3. Experience
4. Labor cost and productivity
5. Management skills
6. Overhead
7. Selling the job at your price

1. Competition

Make it a point to know who your competitors are and consider the number of contractors bidding the job. When possible, try not to bid jobs that have more than four contractors. On the other hand, if there are less than three contractors you can probably raise your profit margin.

Small contractors are often less competitive because of inexperience, inefficiency, and poor management. They are also less competitive because they have a higher overhead cost per job as compared to the larger contractor. But, many small contractors offer excellent service to their customers and are in great demand because of their performance.

2. Cost Of Material

Suppliers won't acknowledge it, but they offer different prices to different contractors for the same material. What can you do to get the best price? Start by becoming a good customer. Make it in their best interest to give you the best price. A few simple rules of getting the best price are; shop around and check your prices, pay your bills on time and take advantage of the 2% discount.

Did you know that it costs more to buy or pick up your material at the supply house counter than it does to have it delivered?

Author's Comment. Low price is not everything. You'll want to develop a relationship with suppliers who will help you solve your problems and who will be there when you need them.

Quotes

Lighting fixtures and switchgear prices are set by the factory sales representative at a percentage above cost. This price before the bid has been awarded to the electrical contractor is called the "street price." Once the job has been awarded, some suppliers will cut their price to the "buy price" which can be as much as 7 to 10% lower than the "street price."

Note. Be sure your supplier includes the cost for all accessories, freight, and delivery.

3. Experience

The more experience you and your employees have, the fewer mistakes you're likely to make. You should be able to complete the project in a more efficient and productive manner.

If you want to be competitive in a market you're not familiar with, find how you can gain the needed experience. One way to gain experience is through the school of hard knocks. But, maybe you can become more creative and find a way to educate yourself before you take that kind of risk. Talk to other contractors, attend seminars, read trade magazines, and watch training videos. Do whatever you can to reduce any kind of loss you might incur as a result of inexperience.

Another factor that must be considered when determining the profit margin of a job is the job risk, particularly labor. The greater the perceived job risk, the higher the profit margin needs to be to accommodate possible losses. If you bid work in which you are experienced, your profit margin can be lower and the bid will be more competitive.

4. Labor Cost And Productivity

The impact of competitiveness between contractors because of the difference in pay scale can be significant. Some contractors pay rock bottom salaries and other contractors pay above union scale. You should pay your electricians a competitive salary so as to discourage their desire to become an electrical contractor. The salary should compensate for their abilities and contribution to your company's bottom line. Pay top dollar and get highly skilled, motivated, educated and trained electricians. The result will be an increase in labor productivity. Pay a low salary and expect code violations, increased supervision requirements, and an increase in lower productivity.

To expand your work force, make an effort to hire quality people, pay them accordingly, and provide them with proper training. Many of today's successful contractors have an apprenticeship and continuing training program for their employees.

5. Management Skills

If you don't know how to manage the job according to the estimate, you are not likely to make the profit that was anticipated. If you don't have the necessary management skills to properly manage your jobs, attend management seminars, watch training videos, and get involved with a local contractors' organization.

Many electrical contractors do not realize that they're not alone in their experiences. By joining a contractors' organization you will gain the experience of those who have been there before you. Learning from another contractor's misfortune is always a better experience than making a mistake. In many parts of the country there are local electrical contractors' organizations; in addition there are two well respected national organizations for electrical contractors. They are:

Independent Electrical Contractors, Inc.
(IEC was established in 1958)
507 Wythe Street
Alexandria, Virginia 22314
1-703-549-7351, Fax 1-703-549-7448

National Electrical Contractor's Association
(NECA was established in 1901)
3 Bethesda Metro Center, Suite 1100
Bethesda, Maryland 20814-5372
1-301-657-3110, Fax 1-301-215-4500

6. Overhead

Overhead cost represents between 20% and 40% of a contractor's total sales. To become competitive you must keep your overhead cost as low as possible.

7. Selling The Job At Your Price

Confidence and professionalism are very important ingredients in selling the job at your price. Do you come across as a professional by your appearance and the appearance of your workers and vehicles?

If you offer excellent service at a reasonable price, then you won't be hesitant when justifying the prices to the customer. Is it a quality installation that you're selling, or is it low price? Low price, without quality service, often results in an unhappy customer. With proper management, it only costs a little more to provide a quality installation, and in the long run it's more cost effective for your customer.

1.08 THE MARKET

Many contractors develope a niche in the market such as tract housing, custom homes, remodeling work, stores, industrial maintenance, etc. The electrical industry is in constant change; new markets are developing and some are dying. At times the local market will be expanding or contracting, depending on technology, the economy, and customer needs.

New and expanding markets can be temporary or permanent and offer greater opportunities for developing new customers. These markets will likely have fewer competitors and greater profit margins, but the risk will also be greater. Shrinking or dying markets offer your customers an opportunity to reduce their cost because of increased competition.

Consider every estimate request as an opportunity to monitor the market's direction.

SUMMARY

Introduction

A great percentage of electrical work is acquired through the estimating process, and most jobs are awarded to the contractor who has the best perceived price, but not necessarily the lowest.

1.01 Estimating Versus Bidding

Determining the selling price for a job is actually two separate components. The first component is called the estimate, which determines the cost of the job. The second component is the bid, which determines the job's selling price.

1.02 A Good Estimating System

A good estimating system should help you quickly and accurately determine the cost of a job, and includes all anticipated costs. The system must be efficient, accurate and attempt to prevent common mistakes. It should have a method to verify that the estimate is accurate.

1.03 Objectives And Purpose Of An Electrical Contractor

The purpose of an electrical contractor is to make a profit. To be successful as an electrical contractor, you must provide a quality service to your customers at a competitive price that is greater than your cost.

1.04 Why Are So Many Electrical Contractors Unsuccessful?

Many electrical contractors have worked their way up through the trade from apprentice electrician. They have plenty of hands-on experience in installing electrical systems, but lack management skills.

1.05 The Estimate Must Provide Information For Project Management

The actual cost of any project is significantly impacted by how well the job is managed. To properly manage a job, the project manager must have a job budget (estimate). The job budget must have information as to what material is required, when it's required, and the labor required to complete each phase of the job.

1.06 Job Management Ultimately Controls The Cost Of A Project

Proper project management is often the difference between profit or loss. You must realize that effective job planning, labor scheduling, and material purchasing are all factors in determining the ultimate cost of a project.

1.07 Can I Be Competitive?

In order to be competitive, the contractor must offer the customer a quality service at a reasonable price. To accomplish this, the electrical contractor must control the job costs so they are within the estimated budget. It is also critical that the contractor control overhead expenses to keep them to a minimum. Other factors that affect a contractor's competitiveness include:

1. Competition
2. Cost of material (buying power)
3. Experience
4. Labor cost and productivity
5. Management skills
6. Overhead
7. Selling the job at your price

1.08 The Market

The electrical industry is in constant change, new markets are developing and some are dying. At times the local market will be expanding or contracting depending on technology, the economy, and customer needs. Consider every estimate request as an opportunity to monitor the market's direction.

Chapter 1

Review Questions

Essay Questions

1. What is the difference between estimating and bidding?

2. Give an example of three important qualities of a good estimating system.

3. What is the purpose of an electrical contractor?

4. Why are many electrical contractors unsuccessful?

5. How can not having a job budget (estimate) cause you to lose money on a job?

6. What does it take to be competitive?

7. Why should the electrical contractor have confidence and act in a professional manner?

8. What factors impact the electrical contractor's market?

Multiple Choice Questions

1. A proper estimate must provide the lowest possible price to the customer at a value that includes sufficient _____ for the electrical contractor to continue in business.
   (a) jobs   (b) employees   (c) profit   (d) none of these

2. The first step to becoming a successful and profitable electrical contractor is knowing your _____.
   (a) competition   (b) cost   (c) profit   (d) code book

3. The ultimate cost of a project is significantly impacted by how well the _____.
   (a) trades work together   (b) material is delivered
   (c) workers know the NEC   (d) project is managed

4. Most electrical contractors are concerned about their ability to _____.
   (a) be competitive (b) make money   (c) stay in business (d) all of these

5. When possible, try to avoid bidding jobs that have more than _____ contractors.
   (a) three   (b) four   (c) five   (d) six

13

6. It often costs _____ to buy electrical material at the supply house counter than it does to have the material delivered to the job site.
   (a) the same　　　　(b) significantly less　　　　(c) more　　　　(d) slightly less

7. Generally speaking, the greater the perceived job risk, the _____ is necessary to accommodate the possible losses.
   (a) greater the bid profit margin
   (b) lower the profit margin
   (c) lower the contingency
   (d) none of these

8. Overhead often can represent between _____ % of the total sales.
   (a) 20 and 40　　　　(b) 25 and 30　　　　(c) 25 and 45　　　　(d) 20 and 30

9. Consider all estimates as an opportunity to _____.
   (a) analyze the market
   (b) practice your estimating skills
   (c) improve your management abilities
   (d) keep your labor force busy

Chapter 2

About Estimating

OBJECTIVES

After reading this chapter, you should understand
- the qualities of a good estimator.
- the duties and responsibilities of the estimator.
- how to properly set up the estimating work space and what tools are required.
- the different types of estimating methods.
- what an accurate estimate must include.
- how accurate an estimate can be.
- the advantages and disadvantages of manual estimating, computer assisted estimating, and an estimating service.

2.01 QUALITIES OF A GOOD ESTIMATOR

The following is a list of qualities that identify a good estimator:
- A willingness to learn.
- A good knowledge of construction and the ability to visualize the electrical requirements.
- An orderly mind and a tendency to be careful, accurate, and neat.
- An open mind willing to change and take advantage of new products and new technology.
- Decisiveness and the ability to make decisions and not be intimidated by details.
- Fairness, honesty and integrity.
- Knowledge of electrical codes and the ability to read blueprints.
- Patience: you must be able to finish the estimate without losing your cool.
- Procedures, the ability to follow them.

2.02 DUTIES AND RESPONSIBILITIES OF THE ESTIMATOR

While the duties of estimators may vary from contractor to contractor, the basic principles remain the same. Generally, the duties of the estimator include but are not limited to:
1. Determining the cost of the job (estimate).
2. Purchasing material.
3. Insuring bid accuracy.
4. Project management/tracking.

1. Determining The Cost Of The Job (Estimate)

Some companies only have their estimators perform the take-off, develop a bill of material, and determine labor-hours and material costs. Management then recaps the estimate and applies overhead and profit adjustments. Other companies have their estimators include all of the costs, including profit.

2. Purchasing Material

The estimator is familiar with the job; he or she is often expected to order the material required to complete the job. The estimator is also expected to have good negotiating skills with suppliers so as to arrive at a competitive price.

Computer Assisted – A computer assisted estimate can quickly provide the material items and quantities for this purpose.

3. Bid Accuracy

The estimator must develop a system to insure that the bid is accurate, and verify that errors in the estimate have not been made.

4. Project Management/Tracking

The estimator should provide the project information required to manage the material and labor efficiently and effectively. This job management information is important for job costing and tracking as well.

2.03 THE ESTIMATING WORK SPACE AND TOOLS

Before making an estimate, you need to have the proper work space and tools.

The Work Space

The work space must be efficient, well illuminated, and located so that you will not be disturbed. The bigger the work area the better; there is never enough work space. The work space should be designed to reach everything from a sitting position. This includes books, paper work, telephone, fax machine, computer, etc.

Estimating Tools

When you estimate a job with improper tools it will take longer to complete the estimate and may have poorer results. Good estimating tools cut down on human errors, increases efficiency, and pay for themselves very quickly. The estimator should have the following items:

Adding Machine

Get a large adding machine with paper and large keys. Estimated cost $50.

Aspirin

Just in case.

Bookcases

Bookcases are required to hold catalogs, literature, and manuals from manufacturers and electrical distributors. You can build these yourself for about $50.

Calculator

Get a solar powered calculator so you don't have to worry about batteries. Make sure it has a large display, large keys with a memory and percent (%) features. Estimated cost $20.

Chair

Many estimators prefer to work with an adjustable height swivel chair between two large tables, or between a table and a desk. This permits the most efficient use of the work space. If you get a chair with arm rests that rolls, it shouldn't cost more than $200.

Colored Pencils or Pens

You will need colored pencils or pens to mark symbols on the blueprint that have been taken-off. You will want to get a good set of color markers with both large and small tips. Estimated cost $20.

Computer

Being competitive in today's market is very difficult without a computer. Estimated cost $2,000.

You'll want the following software to be competitive:

Estimating Software – A computer with good software will take about a month to get comfortable with and three months to master. Estimated cost $2,000 – $6,000. See Chapter 8 on how to select estimating software.

Internet and E-Mail Access – Having access to the Internet affords the opportunity to keep informed on material prices, the National Electrical Code, trends in the electrical industry, and answers to almost any question you might have.

Author's Comment. You can locate my Internet site at: http://www.mikeholt.com or reach me via e-mail at mike@mikeholt.com

Electronic mail (e-mail) is a significant form of communication. It offers the opportunity to communicate with your suppliers and customers in a more effective manner. You must keep on top of the changing nature of the information age. Estimated cost $25 per month.

Spreadsheets – Spreadsheet programs are very useful for keeping track of cash flow, job scheduling, financial information, budgets, and other uses. (They are often free with the purchase of a computer.)

Street Map – SelectStreet from ProCD produces an excellent CD-ROM that contains every street in the United States. Estimated cost $100.

Time and Material Pricing – Computer billing software will help you get your bills out faster, and improve your cash flow. You'll have information at your finger tips on who's been billed, who's paid, and how much your customers owe to you. Estimated cost $1,000 to $2,000.

Word Processing – Word processing software is important for letters, contracts, proposals, and many other purposes. Often free with computer.

Copy Machine

Costly, but very handy for making copies of panel schedules and fixture legends. Until you get a copy machine, use your fax machine to make some copies, but many Fax copies are not permanent. Estimated cost $700.

Counter

Counters are used to take-off lighting fixtures, switches, and receptacles. There are two types of counters, the mechanical type and the electronic counter. The mechanical type is designed to be used with your left hand. You put your finger through the hole, wrap your hand around it, and press the counter with your left thumb. Estimated cost $15.

With the electronic type you can use either hand, but it's not as convenient as the mechanical type. The electronic counter can interface with a computer for instant input.

Desk

You need a large desk to hold all of the paperwork, pencils, pens, computer, and everything else that seems to accumulate. A small work area will force you to continuously struggle to keep things organized, wasting your valuable time. Estimated cost $200.

Digital Plan Wheel And Counter

A digital plan wheel is used to measure circuit run lengths and have an electronic counter. A built-in digital plan wheel has the ability to change the scale quickly and is very easy to read and use. You can get a digital plan wheel and counter for about $60. The type that connects the digital device to a computer for instant input costs about $200.

Drafting Table

You should have an incline drafting table with enough space to lay the blueprints out flat. The table should have a lip on the bottom to keep the prints from sliding off. Build it yourself for about $40.

Erasers

Pick up several inexpensive erasers for about one dollar each. There are electric erasers available if you think one is necessary. These cost about $35.

Filing Cabinets

Filing cabinets are necessary so you can keep everything organized. They are handy for filing estimating records, completed estimates, quotations, forms, etc. Estimated cost $250.

Forms, Worksheets, And Proposals

You need to have your estimating forms readily available. If you have a computer you can create your own custom forms to accommodate your needs. See the Appendix.

Lighting

Get plenty of light for the room and for the work surface area. Use indirect halogen lighting fixtures, which can be purchased for less than $30 each.

Magnifying Glass

Some details on the blueprints might be difficult to decipher without a magnifying glass. Get a good one for about $10.

Pencil Sharpener

One that is battery powered will be sufficient, but if you're doing a lot of estimating you may want a 120 volt electric pencil sharpener. Estimated cost $5 to $15.

Personal Information Manager (PIM)

A personal organizer is a must and there are two types; the paper notebook type, or the electronic type. Electronic organizers can contain telephone numbers, memos, scheduling, a calendar, calculators, and much more. Estimated cost $100.

Personal Stereo With Head Phones

Head phones with music can be used to help you focus and prevent distractions. Estimate cost about $50.

Phone And Answer Machine

Two lines are a must with computer modems, faxes, e-mail and Internet access. Be sure you have call waiting, conference calling, and one telephone set for each person in the office. Each phone should have speed dialing; be sure to use it. The second line should not have call waiting since you're going to use it for electronic communication. Estimated cost $35 to $75 per month.

Plan Racks

Plan racks are used to store the blueprints and help keep the work area organized. You can make blueprint racks yourself. Estimated cost $50.

Scale Ruler

Make sure you have an architectural scale ruler to the inches, not an engineering ruler, which is based on the decimal. Careful, it's easy to make a mistake and use the wrong scale. Estimated cost $5.

Stapler

A stapler and a staple remover are necessary in any office, and don't forget to get staples. Estimated cost $5.

Wall Space

A large clear wall in front of your drafting table is very convenient for posting important estimate information.

White Board Or Chalk Board

A white board or chalk board can be used for many functions such as scheduling, keeping track of jobs, etc. Estimated cost $100.

2.04 TYPES OF BIDS

There are several types of bid requirements you might experience. They include:
1. Competitive bid
2. Design build
3. Negotiated work
4. Time and material (fixed fee)
5. Unit pricing

1. Competitive Bid

This type of bid can be for private, public, or government projects. These projects can be found in local newspapers in the classified section under Public Notices, as well as trade publications, such as *The Dodge Report*. These publications list projects by category such as residential, industrial, commercial, and by total expected project price range.

Competitive Bid Selected

Invitations to bid are limited to selected contractors meeting specific criteria. You can get on government lists to receive invitations to bid on upcoming projects. Often you must respond to the invitation, even if it is just to indicate your decision to decline the opportunity. If you don't respond, you might be removed from the bid list.

If your company is classified as a minority business, get information about set-aside projects for minority contractors.

Competitive Bid Pre-Qualified

This type of bid requires that contractors demonstrate their ability to complete the project. The qualifications might include: bonding capacity, previous experience, or financial capacity.

2. Design Build

Sometimes an electrical contractor is given a general floor layout without much detail and requested to design and construct the electrical wiring according to written specifications. To be successful with design build, you really need to know your customers' needs and the electrical code. For the sharp electrical contractor, this is an excellent way of doing business. Profit margins can often be quite high.

3. Negotiated Work

Negotiated work is generally not advertised and there are a limited number of contractors requested to negotiate the bid price. The electrical contractor and the customer negotiate a price that satisfies both parties. Negotiated jobs are often required to be completed within a tight construction schedule and within a flexible budget.

4. Time And Material Or Fixed Fee Proposal

Time and material pricing, sometimes called *fixed fee*, is required when existing conditions make it difficult to provide a fixed dollar bid. This type of bid is based on a given rate per hour for labor (including benefits, overhead and profit) with the material billed separately at an agreed markup, such as 20% above cost.

5. Unit Pricing

Some jobs are awarded on a unit price basis, where the unit price includes both material and labor cost. This is often the case when the customer is not quite sure of the quantities of the specific items. Unit price examples:

Single pole switch	$46.11
Duplex receptacle	$45.50
Lay-in fluorescent	$106.98
Temporary service	$330.63

Unit pricing is used for almost all types of construction such as renovations, office build-outs, change orders, etc. See Chapter 5, Part C on how to develop unit prices.

2.05 WHAT AN ACCURATE ESTIMATE MUST INCLUDE

An accurate estimate must include labor cost including burden (fringes), material cost including lighting fixtures and gear quotes, sales tax, subcontract and rental expenses, direct job expenses, and overhead.

> **Note.** To determine the bid price, always include a margin for profit.

2.06 IMPROPER ESTIMATING METHODS

There are many improper estimating styles of determining the selling price for a job. They include:
1. Ignoring the specifications
2. Meeting the lowest bid
3. Shot-in-the-dark
4. Square foot method (ball-park price)

1. Ignoring The Specifications

With this method, you provide a price that does not include the requirements of the specification.

2. Meeting The Lowest Bid

"Hey if they can do it for that price, so can I." This attitude will get you into a lot of trouble. How do you know what the lowest price is anyhow?

3. Shot-In-The-Dark

"Let's see, this house has 5 blueprint pages so it should go for about $8,500." Contractors often give a shot-in-the-dark price because they feel forced to give a price. Often the price is way out of line, either very high or very low. Either way the electrical contractor loses credibility with the customer. If the price is low, the electrical contractor must try to explain that he or she made a mistake and hopes the contractor won't award the job to him or her.

4. Square Foot Method (Ball-Park Price)

With proper historical data on past jobs, the square foot method can be used to get an approximate budget price, but not a bid price.

> **Example.** A contractor wants to build a 10,000 square foot warehouse and wants a ball-park price to determine the feasibility of the project. Historical data indicates that it costs between $3.50 and $4.00 per square foot without fixtures. Your ball-park price should be between $35,000 and $40,000.

Be careful, don't get careless and give a customer a price based on the square foot method because you didn't have time to estimate the job properly. When you give a ball-park price, be sure that you put it in writing and be clear that it's not a bid price.

> **Case Study.** Contractor bids a job with a ball-park-price of $75,000. He's awarded the job and finds out the fixtures and electronic dimmers cost $25,000, which he did not figure on.

Note. The square foot method may be helpful to check an estimate for bid accuracy

2.07 PROPER ESTIMATING METHODS

The proper methods of estimating include time and material pricing, unit price, and the detailed method.

Time And Material

Time and material pricing is often used to price change orders, or when conditions make it difficult to provide a fixed bid. A T&M bid is based on a given rate per labor-hour, with material billed separately at an agreed markup.

Common conflicts that develop when you bill customers based on time and material are:

1. They don't think your electricians spent the amount of time on the job that you billed them for.
2. They don't feel that your workers were productive and they don't want to pay for nonproductive time.
3. They want a detailed list of materials billed and question the cost of each item. Many of these conflicts develop because the customer does not understand construction and the many factors that impact labor productivity and material cost.

Unit Price

Some jobs are awarded on a unit price basis. By having a unit price, the customer knows in advance what it will cost. This is often the case when the customer is not quite sure of the quantities of the specific items required for the project. The unit price must include the labor cost, labor burden, material cost, sales tax, direct cost, overhead, and profit. The value of each unit must be updated regularly, which is often not the case. Profits on unit price estimates tend to decrease on each succeeding bid because material and labor costs may increase before the unit-prices have been re-priced.

The greatest disadvantage to a manually generated unit price estimate is that you are not able to generate a bill of material for project management, nor are you able to track job costs to be sure the job is operating within the projected budget.

Computer Assisted – Unit price estimating, when done with a computer, does not have any of the above disadvantages.

Detailed Method

The detail method estimate provides you with accurate material quantities. With this information you can account for variable job conditions to determine the total anticipated labor requirement.

Computer Assisted – The detail estimate will also provide you with all of the necessary details to manage a project profitably. In addition, this method will permit you to develop historical information for future jobs.

2.08 THE DETAILED ESTIMATING METHOD

Estimating And Bid Process

1. Understanding the scope of work.
2. The Take-off.
3. Determining the bill of material.
4. Pricing and laboring material.
5. Extending and totaling.
6. The estimate summary.
7. Applying profit and other costs.
8. Bid accuracy and bid analysis.
9. Bid proposal.

To properly estimate a job, you must have the ability to mentally visualize the mechanical requirements and materials required to complete the job. For the person who has the electrical experience required for the job, this shouldn't be much of a problem. However, if you have never done the type of work required to be estimated, this will be next to impossible. If you don't know what is required to complete the installation, then how can you estimate the job?

Let's assume that you do light commercial and residential work such as office remodels, strip stores, residential homes, etc. If you have never done a fire alarm system, or gas station, or an industrial job, it will be difficult for you to determine the selling price.

Not only do you need to know what is mechanically required to complete the job, you must have a good mathematical mind. This means that you must be comfortable with numbers, such as adding, subtracting, multiplying, dividing, and percentages.

Note. If you have difficulty with numbers, you should use a computer to help you with estimating and bidding.

How Difficult Is It To Bid A Job?

Some of these steps are straightforward while others require you to rethink your natural feelings.

Step 1. Understanding The Scope Of Work

Before you can estimate a job, you must understand the work to be completed. This means that you must have the ability to read and understand the symbols of the blueprints and conditions of the specification.

Step 2. The Take-Off

When you perform the take-off, you are mentally visualizing the installation of the proposed electrical system and you are counting and measuring blueprint symbols. For the person who has the specific electrical experience, performing the take-off is a very easy step.

Step 3. Determining The Bill Of Material

Determining the bill of material is very easy for the person who understands the material items required to complete the job.

Computer Assisted – If you are using a computer, this step is completed by the computer.

Step 4. Pricing And Laboring Material

Looking up prices and labor-units in a catalog can be rather time consuming. This step is easier for the experienced electrician because they are familiar with the material. With some assistance most people can learn how to perform this process. See the *Electrical Estimating Workbook*, Chapter 1 – Price And Labor Unit Catalog.

Computer Assisted – If you are using a computer, this step is completed by the computer.

Step 5. Extending And Totaling

Extending and totaling is a no brainer and can be performed by anyone who has reasonable math skills.

Computer Assisted – If you are using a computer, this step is completed by the computer.

Step 6. The Estimate Summary

The individual who performs the estimate summary must have an understanding of construction business practices, financial statements, historical labor productivity, and overhead calculations.

Step 7. Applying Profit And Other Costs

Once you have determined the estimated cost of the job, you must add profit and other final costs to determine the selling price. This requires that you have good business skills and that you understand the market.

Step 8. Bid Accuracy And Bid Analysis

When the estimate and bid is complete, you must verify that you did not make any common estimating errors. In addition, you must verify that your price is valid and accurate.

Computer Assisted – For all practical purposes, the bid analysis requires the use of a computer to validate the bid price.

Step 9. Bid Proposal

When you have completed the bid, you must submit a written proposal. The proposal must clarify what your bid price includes and what is not included.

Author's Comment. At the conclusion of my seminars, I often hear, "I still don't understand how to estimate fire alarms, hazardous locations, etc." It's not that students don't understand the estimating process, it's that they don't feel comfortable estimating a job that they are not familiar with. This feeling is quickly resolved as soon as they are familiar with the requirements of the job.

2.09 HOW ACCURATE CAN AN ESTIMATE BE?

Not all expenses can be anticipated in advance, but experienced estimators accept a satisfactory margin of error in the accuracy of the bid. As with anything in life, the more experience you gain, the more accurate and confident you become. With increased experience and practice, you will also increase your speed in completing an estimate accurately.

If you break the job down into its smallest possible parts, then the magnitude of each mistake will be reduced and, hopefully, the mistakes will cancel each other out.

Computer Assisted – Using a computer to determine the bill of material, perform pricing, laboring, extending, and totaling eliminates most estimating errors.

Accuracy Of Estimating Materials

Material is the most predictable part of the estimate. If you use a computer, you can generally calculate your material cost to within a few percentage points of the actual cost for new work and within 10% of the actual cost for remodel work. If you are estimating manually, errors in determining the bill of material, pricing, extending and totaling can be significant.

The estimator cannot anticipate changes in the wiring by the electrician, or unusual waste by workmen. With proper project management, however, material overruns can be reduced.

Accuracy Of Estimating Labor

Labor is more difficult to predict than material, but with experience, labor can be calculated to within 10% for new work and 20% for remodeling.

2.10 MANUAL ESTIMATE, COMPUTER ASSISTED, OR AN ESTIMATING SERVICE?

There are three primary methods of estimating a job.
1. The manual method.
2. Computer assisted method.
3. Estimating service.

Each method has its own set of advantages and disadvantages. You need to be honest and select the method that works best for you. Once you understand manual estimating you can determine which method is the most cost effective for you.

1. Manual Estimates

The manual method has been used by electrical contractors for decades. The advantage of this system is that it does not require you to gain any computer skills, nor does it require you to purchase a computer.

Because it takes so much time to estimate a job manually, you must learn to utilize your time and resources effectively. Quite often with this method, you only have the time to get the bottom line price and not much more. Also, manual estimating requires so much time that estimates often become backlogged and project management suffers.

Disadvantages

There are several disadvantages to estimating manually. They include:
1. Cost to complete each estimate.
2. Errors in math.
3. Lack of project management information.
4. Inability to respond to changes.
5. Time to complete the estimate.

1. Cost To Complete Each Estimate

It costs about $800 and takes about 32 hours for every job that you are awarded. This is assuming it takes an average 8 hours for each bid at $25 per hour and you're awarded one out of every four bids you submit.

2. Errors In Math

When estimates are backlogged, you begin to feel the pressure to get the estimate done and get on to the next estimate. This pressure can result in increased errors as you work with many numbers, especially with last minute changes.

3. Lack Of Project Management Information

Because of the time it takes to manually extract project management information, most contractors just don't do it. The result is that the job cannot be managed as estimated. In addition you will not be able to develop historical information for future jobs.

4. Inability To Respond To Changes

When you're manually estimating a job it can be very frustrating to receive last minute changes to the blueprints or specifications. Sometimes the change is so great that you don't have enough time to respond in time. This can result in your attempting an educated guess or giving up and not submitting the bid at all. Either way, this is not a good business practice.

5. Time To Complete The Estimate

If it takes an average of eight hours to estimate one job, you'll average thirty two hours of estimating time for every job you actually get.

Today, more than at any time in history, we operate in an age of instant information and expected response. We have cellular phones, e-mail, the Internet, and fax machines. Because of the instant information age, customers are demanding more and more information, and expecting it faster and faster. Long gone is the day when we could say, "I'll get that for you tomorrow." Because of the ready availability of computers, people expect answers and responses in a matter of minutes, not hours or days. Manual estimating often will not permit you to respond quickly enough for your customers' needs.

2. Computer Assisted Method

The computer assisted method of estimating is actually the same as estimating manually, except that a computer is used to perform the thousands of mathematical calculations millions of times faster and more accurately than a human. There are many advantages to using a computer for crunching numbers.

Advantages

The competitive advantages to a computer assisted estimate include:
1. Changes are easily accommodated.
2. Improved project management.
3. Improved bid accuracy.
4. Reduced estimating time.
5. Reduced management cost (overhead).

1. Changes Are Easily Accommodated

With a computer, last minute changes are easily accommodated and, as material costs change, the bid can be updated immediately. Last minute supplier lighting fixture and switchgear quotes are easily accommodated.

2. Improved Project Management

A computer assisted estimate with proper software should produce project management reports. Quality computer estimating software programs provide you with information for job management, job tracking, and bid analysis. This permits a reduction in errors before the bid is submitted and as the job progresses.

3. Improved Bid Accuracy

The margin for error when using a computer is reduced significantly. No transposing of numbers, no mistakes on the totals or when transferring numbers to the estimate summary. Your bids will become clearer, ledgible, and professional in appearance. Your estimates will not become backlogged and you won't feel the pressure to rush the estimate, especially with last minute fixture or switchgear quotes or changes to the bid.

4. Reduced Estimating Time

You can reduce your estimating time by as much as 75%, depending on the type of job, the software, and your experience. A computer permits you to produce up to four times more estimates in the same amount of time required to produce one manually. What would take 8 hours manually should take less than 2 hours, and the odds of getting the job are increased.

5. Reduced Overhead

Overhead is reduced since you can complete an estimate in a quarter of the time at a quarter of the cost. What would have cost $200 and taken 8 hours will now take 2 hours and cost only $50. This results in your prices becoming more competitive and an opportunity for greater profit.

3. Estimating Service

An estimating service is a temporary agency that you only pay when you need it. When an estimating service produces an estimate, you have the opportunity to review the bid to ensure that the estimate is accurate and complete.

You might use an estimating service to double check a bid, or if you don't have the time to do the bid yourself. An estimating service is an excellent tool to help you gain estimating experience at a reduced risk. Estimating services offer:
1. Low up-front costs.
2. Knowledge of bid cost.

1. Low Up-Front Costs

When you use an estimating service, your up-front cost is comparatively low. You can enjoy the benefits of computer generated estimates without investing in your own computer estimating system.

2. Knowledge Of Bid Lost

With an estimating service, you'll know in advance what it costs to estimate a job. Their fees are generally based on the total electrical bid dollar amount. The following table is an example of a typical estimating service fee schedule:

From	To	Fee
$0	$45,000	$160
$45,001	$100,000	$310
$100,001	$150,000	$400
$150,001	$200,000	$450
$200,001	$300,000	$625
$300,001	$400,000	$700
$400,001	$500,000	$925

SUMMARY

2.01 Qualities Of A Good Estimator

The following is a list of qualities that identify a good estimator:
- A willingness to learn.
- A good knowledge of construction and the ability to visualize the electrical requirements.
- An orderly mind and a tendency to be careful, accurate, and neat.
- An open mind willing to change and take advantage of new products and new technology.
- Decisiveness and the ability to make decisions and not be intimidated by details.
- Fairness, honesty and integrity.
- Knowledge of electrical codes and the ability to read blueprints.
- Patience: you must be able to finish the estimate without losing your cool.
- Procedures, the ability to follow them.

2.02 Duties And Responsibilities Of The Estimator

While the duties of estimators may vary from contractor to contractor, the basic principles remain the same. Generally, the duties of the estimator include but are not limited to:
1. Determining the cost of the job (estimate).
2. Purchasing material.
3. Insuring bid accuracy.
4. Project management/tracking.

2.03 The Estimating Work Space And Tools

Before you estimate your first job, you need to have the proper work space and tools. Your work space must be efficient, well illuminated, and located so that you will not be disturbed. When you estimate a job with improper tools it will take you longer to complete the estimate and you will often have poorer results. Good estimating tools cut down on human errors, increase efficiency, and pay for themselves very quickly.

2.04 Types Of Bids

There are several types of bid requirements you might experience. They include:
1. Competitive bid
2. Design build
3. Negotiated work
4. Time and material (fixed fee)
5. Unit pricing

2.05 What An Accurate Estimate Must Include

An accurate estimate must include labor cost including burden (fringes), material cost including fixture and switchgear quotes, sales tax, subcontract and rental expenses, direct job expenses, and overhead.

Note. To determine the bid price, always include a margin for profit.

2.06 Improper Estimating Methods

There are many improper estimating styles of determining the selling price for a job. They include:
1. Ignoring the specifications
2. Meeting the lowest bid
3. Shot-in-the-dark
4. Square foot method (ball-park price)

Note. The square foot method is helpful to check an estimate for bid accuracy.

2.07 Proper Estimating Methods

The proper methods of estimating include time and material pricing, unit price, and the detailed method.

2.08 The Detailed Estimating Method

There are six steps of an estimate and three steps of the bid process.

Estimating And Bid Process

1. Understanding the scope of work.
2. The take-off.
3. Determining the bill of material.
4. Pricing and laboring material.
5. Extending and totaling.
6. The estimate summary.
7. Applying profit and other costs.
8. Bid accuracy and bid analysis.
9. Bid proposal.

2.09 How Accurate Can An Estimate Be?

Not all expenses can be anticipated in advance, but experienced estimators accept a satisfactory margin of error in the accuracy of the bid. If you break the job down into its smallest possible parts, then the magnitude of each mistake will be reduced and, hopefully, the mistakes will cancel each other out.

Computer Assisted – Using a computer to determine the bill of material, perform pricing, laboring, extending, and totaling eliminates most estimating errors.

Accuracy Of Estimating Materials

Material is the most predictable part of the estimate and can generally be calculated to within a few percentage points of the actual cost for new work.

Accuracy Of Estimating Labor

Labor is more difficult to predict than material, but with experience, labor can be calculated to within 10% for new work and 20% for remodeling.

2.10 Manual Estimate, Computer Assisted, Or An Estimating Service?

Today there are three primary methods of estimating a job. They are:

1. the manual method,
2. computer assisted,
3. estimating service.

Each method has its own set of advantages and disadvantages.

Chapter 2

REVIEW QUESTIONS

Essay Questions

1. What are the general responsibilities of the electrical estimator?

2. Why is it important to have a proper work space and the proper estimating tools?

3. What costs must the estimate include? What must the bid include?

4. When should you use the ball-park price method for estimating and bidding?

5. What are the three most common proper methods of estimating?

6. What are the disadvantages of time and material estimates?

7. What are disadvantages of unit pricing if done manually?

8. How difficult is it to estimate a job properly?

9. How accurate can the detailed estimate be?

10. What are the three primary methods of estimating a job?

11. What are the disadvantages of estimating manually?

12. What are the advantages of estimating with a computer?

13. What are the advantages of an estimating service?

Multiple Choice Questions

1. A good estimator must have an understanding of _____, as well as electrical construction experience.
   (a) the National Electrical Code (b) building construction
   (c) local electrical codes (d) all of these

2. The estimator is expected to have good negotiating skills with _____ to arrive at a competitive price.
   (a) management (b) suppliers (c) customers (d) all of these

3. It is the responsibility of the estimator to develop a system to insure that the bid is _____.
   (a) accurate   (b) complete   (c) competitive   (d) none of these

4. Project management information, such as labor hours and material cost, is required for proper job _____.
   (a) tracking   (b) costing   (c) both a and b   (d) none of these

5. A competitive bid can be associated with _____ projects.
   (a) governmental   (b) public   (c) private   (d) all of these

6. Invitations to bid, called Competitive Bid Selected, are limited to selected contractors _____.
   (a) meeting specific criteria   (b) classified as a minority business
   (c) based on the size of the business   (d) none of these

7. The type of bid referred to as Competitive Pre-Qualified requires that contractors demonstrate their _____ prior to being requested to bid the project.
   (a) business expertise   (b) ability to complete the project
   (c) knowledge of the NEC   (d) none of these

8. To be successful with design build, it is essential you know your _____.
   (a) customers' needs   (b) electrical code   (c) inspector   (d) a and b

9. Unit price bids are often used for _____.
   (a) remodelling   (b) change orders   (c) office building build outs   (d) all of these

10. When you take-off a job, you are mentally visualizing the _____ of the proposed electrical system.
    (a) blueprints   (b) specifications   (c) installation   (d) none of these

11. When using a computer to produce estimates you can reduce your estimating time by as much as _____ % depending on the type of job and your experience.
    (a) 75   (b) 80   (c) 85   (d) 90

12. You might use an estimating service to _____.
    (a) double check a bid you've already put together
    (b) obtain a computer printout for job management information
    (c) get a bid done when you don't have time to do it yourself
    (d) all of these

Chapter 3

Understanding Labor-Units

OBJECTIVES

After reading this chapter, you should understand
- what a labor-unit is and why you use them.
- what's included and what's not included in a labor-unit.
- why standard labor-units must be raised or lowered according to specific job conditions.
- which labor-unit manual you should use.
- why you don't care about your competitors' labor-units.
- how to develop your own labor-units.
- how to adjust your labor-units to accommodate for variable job conditions.

INTRODUCTION

For many electrical contractors, determining the labor for a job is often very scary and intimidating. So before you actually estimate a job, you need to gain confidence in how to determine the expected labor. There are two methods of estimating the labor for a job: one is using your experience from previous jobs and the second is labor-units. Using your experience is fine as long as you have sufficient experience. That is, if you have wired houses or stores, you can get reasonably close on the estimated time to complete a similar job. With labor-units, you have a fixed-base to work from, unlike experience which is subject to mood or perception.

Estimating with labor-units is both a science and an art form. It is a science in the sense that the labor required to complete a task is a function of the materials to be installed and their quantities. If you know the quantity of each material required for a job, you can easily determine the labor required. It is an art form in the sense that you need to become creative in making some adjustments to the labor-units for the specific job conditions.

Many electricians feel like they have no control if they use labor-units to estimate labor. I hope that once you have completed this book you will gain a better understanding of how much control you really have by using labor-units for estimating electrical wiring.

3.01 WHAT EXACTLY IS A LABOR-UNIT?

A labor-unit represents the approximate time required to install an electrical product, component or equipment. Labor-units are based on the assumption that a skilled and motivated electrician is completing the task under standard installation conditions with the proper tools.

Labor-units serve as a standard value which must be raised or lowered according to specific job conditions. There are many factors that can cause an increase in the labor-units, and in some rare cases, favorable job conditions can cause a decrease.

Labor-units represent labor productivity under typical job site conditions with reasonable working conditions. They must be adjusted for management skill, as well as variable job site conditions. The methods of adjusting labor-units are discussed later in this unit.

> **Note.** According to the NECA Manual of Labor-Units, a typical electrical construction job includes a single building of no more than three stories not exceeding 100,000 square feet, located near a metropolitan area, with a job schedule of no more than 8 hours per day and 40 hours per week.

How Labor-Units Are Expressed

Labor-units are expressed in units of decimal hours: E – each, C – hundreds, or M – thousands. You don't use minutes when you estimate a job, you use decimal labor-hours.

> **Example.** Receptacle 18/C = .18 hour each receptacle or 60 minutes × .18 hour each = 11 minutes.
>
> Switch 3-Way 25/C = .25 hour per switch or 60 minutes × .25 hour each = 15 minutes.

You are probably thinking that 11 minutes to install a receptacle or 15 minutes to install a 3-way switch is too much time, and you'll never get a job if you use labor-units. I understand the feeling, but it's really not the case. I think you'll become convinced as you continue through this chapter.

Work Experience

Your electrical work experience is very important when you estimate electrical wiring, but you must know when and how to use that experience. Many electricians estimate a job based on how long they think it would take them to do the job. There are many factors that they might not have taken into consideration, such as supervision, job layout, tool management, handling of the material, nonproductive time and job conditions.

I'm not saying that work experience is not important—as a matter of fact it is critical—but you need to understand how to estimate labor in spite of your experience. If you could simply estimate the labor of any job with your experience, you probably wouldn't be reading this book.

Meeting Room Labor Hours

Take a few moments and review the Meeting Room Blueprint M–1 on the next page, Figure 3–1. How many labor-hours do you think it will take to complete the wiring?

Meeting Room Labor-Unit Hours

I have asked over one thousand master electricians in my classes to estimate the labor-hours required to rough and trim the meeting room. The answers range from a low of forty hours to over two hundred hours.

> *Author's Comment.* I now understand one of the reasons why there is such a vast difference in pricing between electrical contractors.

The estimated labor hours for the Meeting Room, according to the labor-units in Chapter 1 of the Electrical Estimating Workbook, is about 81 hours. Some electrical contractors will be more efficient and organized and possibly complete the job in 66 hours (225%), other contractors will struggle to complete it in 110 hours (125%).

3.02 HOW LABOR-UNITS ARE USED

To determine the labor required for a job, you must determine the material required and their quantities. With this information, you use the labor units to determine the total labor required. Table 3–1 demonstrates how labor-units are used to determine the Meeting Room's labor requirement.

3.03 WHAT'S INCLUDED IN THE LABOR-UNIT

A labor-unit is comprised of six major components. They include:

1. Installation	50%
2. Job layout	15%
3. Material handling and cleanup	10%
4. Nonproductive labor	5%
5. Supervision	10%
6. Tool handling	10%

Electrical Estimating — Chapter 3 – Understanding Labor-Units

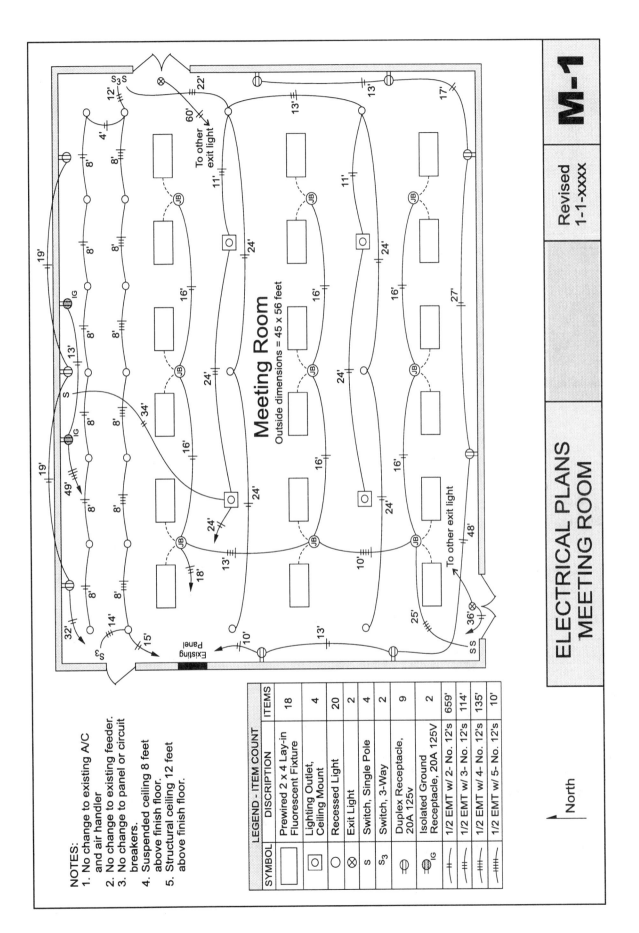

Figure 3–1
Meeting Room Blueprint M–1

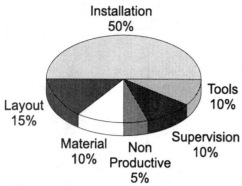

1. Installation 50%

The actual installation time represents approximately 50% of the labor-unit. This includes the installation of boxes, conduit, fittings, wiring devices, fixtures, switchgear, disconnects, panels, breakers, etc.

2. Job Layout 15%

The layout of the work to be installed represents about 15% of the labor-unit. This includes measuring and determining the type, size, and location of the conduit, wire, box, circuit, fixture, etc.

3. Material Handling And Clean-Up 10%

Material handling and clean-up is approximately 10% of the labor-unit. Material must be purchased, unpacked, received, counted, unloaded, and stored. Don't forget you must pick up and dispose of the packing containers that the items were shipped in.

4. Nonproductive Labor 5%

All jobs contain nonproductive labor and should be managed so as not to exceed 5% of the labor-unit.

Author's Comment. Be careful of early quitting time which is often linked with late starting time. You know how it goes, two guys look at their watches; one watch indicates 8:00 A.M., and the other indicates 7:55 A.M. Of course they wait five minutes to start the job. The opposite is true at the end of the day. This game results in a loss of 10 minutes a day per electrician. This calculates out to almost $600 per year per electrician at a labor rate of $10 per man hour with labor burden of 38%.

Calculation 10 minutes × 5 day per week × 50 weeks per year = 2,500 minutes/60 minutes per hour = 42 hours × $10.00 per labor-hour × 1.38 (benefits) = $580.

Table 3–1 Meeting Room Labor-Unit Sample

Boxes	Qnty	Labor	U	Total Hours
Box 4" × 4" Metal	30	18.00	C	5.40
Rings				
Square Round	6	4.50	C	0.27
1 Gang Device	13	4.50	C	0.59
2 Gang Device	2	5.00	C	0.10
Switches				
Switch 1-Pole 277 Volts	5	20.00	C	1.00
Switch 3-Way 277 Volts	2	25.00	C	0.50
Receptacles				
Recpt. 20Amp 125 Volt	9	19.00	C	1.71
Recpt. IG 20 Amp 125 Volt	2	0.25	E	0.50
Plates				
Switch 1 Gang	2	2.50	C	0.05
Switch 2 Gang	2	4.00	C	0.08
Receptacle Duplex	11	2.50	C	0.28
Blank 4" × 4" Metal	9	6.00	C	0.54
Raceway				
1/2" EMT	891	2.25	C	20.05
1/2" EMT Connector	100	2.00	C	2.00
1/2" EMT Coupling	89	2.00	C	1.78
Wire				
No. 12 THHN	2,393	4.25	M	10.17
Fixture Labor				
Flourescent Lay-in	18	0.75	E	13.50
Ceiling Mounted	4	0.50	E	2.00
Recessed	20	1.00	E	20.00
Exit	2	0.25	E	0.50
Total				81.02

5. Supervision 10%

Supervision represents about 10% of the labor-unit. It includes the review of the blueprints and specifications, ordering material, working out installation problems, and coordinating with other trades and subcontractors to avoid delays in the work. Supervision in the labor-unit includes inspection tours, crew makeup decisions, record keeping of material, job progress reports, and time cards.

6. Tool Handling 10%

The management, handling and layout of tools represent about 10% of the labor-unit. You must purchase, receive and store the tools, move the tools to the work area, set up, and remove and clean the tools. Can you imagine performing electrical work today without portable drills, sawzalls, or hole-saws? Be sure you always provide the proper tool so that the installation can be completed in the most efficient manner.

3.04 LABOR-UNITS DO NOT INCLUDE

Labor-units do not include the assembly of fixtures, switchboards, panels, or other equipment. Be careful when the owner supplies the fixtures or when you get great prices from your supplier; the equipment might come in a bag with assembly instructions. Other factors that are not included in a labor-unit are:

Cutting holes or openings.

Excavation, drilling, or blasting.

Heavy equipment (operator's time).

Hoisting above three floors.

Maintenance of temporary equipment.

Painting of conduits.

Testing or welding.

Pre-Assembled Equipment

Whenever possible, purchase pre-assembled equipment or equipment that is installer friendly. Equipment that is not preassembled will cost you more by the time you put it together. Order fluorescent 2 × 4 lay-in fixtures pre-lamped and pre-whipped.

For larger installations, order the wire in triplex or quadplex on the reel. This is a real time saver because the wire is already together and you're only pulling from one reel. The cost of triplex or quadplex might be slightly greater, but the labor savings can exceed the increased material cost.

Unusual Tasks

There will be times when you must determine the labor for a task that you have absolutely no idea how to do. Break the task down into as many small individual labor segments as possible. Try to establish a labor-unit for each segment, by comparing the individual segments to a similar tasks that you have a labor-unit for. If all else fails, make a reasonable guess, such as two hours per hundred pounds and .2 hours per termination.

Note. If your employees are not familiar with the wiring of a system, the labor required to manage, install, and troubleshoot will be significantly increased.

3.05 LABOR-UNIT MANUALS, WHICH SHOULD YOU USE?

Chapter 1 of the *Electrical Estimating Workbook* contains labor-units for residential, commercial, as well as some industrial installations. These labor-units have been developed by Mike Holt and have been used successfully by thousands of electrical contractors. These labor-units must be factored to the job conditions and your management experience.

You should consider purchasing a comprehensive labor-unit manual that contains at least 10,000 labor-units. They generally cost less than $100. Excellent labor-unit manuals are available in both book form and electronically from:

Electrical Resources, Inc.
7169 University Boulevard
Winter Park, Florida 32792
1-407-657-7001, Fax 1-407-657-0559

Minnesota Electrical Association
3100 Humboldt Avenue South,
Minneapolis, MN 55408
1-800-829-6117, Fax 1-612-827-0920

National Electrical Contractors Association
3 Bethesda Metro Center, Suite 1100
Bethesda, Maryland 20814-5372
1-301-657-3110, Fax 1-301-215-4500

There is no perfect labor-unit that can be applied to all jobs. With experience and historical data, you will develop techniques to help you adjust the labor-units to represent your productivity with specific job conditions.

3.06 YOUR LABOR-UNITS AS COMPARED TO YOUR COMPETITORS'

Productivity in the installation of electrical equipment is impacted by many factors. However, management determines the labor budget, level of supervision, if the labor force is skilled and motivated, and insures that the proper tools and material are on the job. How you manage your business impacts the labor productivity and your profitability.

Other factors that impact labor productivity are:
1. Computers
2. Labor
3. Material and tools
4. Supervision

The result of these factors is that no two electrical contractors will have the same labor-unit productivity.

1. Computers

A computer assisted estimate can provide you with an estimated labor budget so that you know how many hours each phase of the job should take. This permits you the opportunity to achieve that goal of completing the job within budget.

If you don't know how long it takes to complete a project, you can't track the actual job labor as the job progresses. You'll never know if your labor is productive, and you'll not be able to profitably move the project to completion.

2. Labor

Labor-units are based on the assumption that highly trained, skilled, and motivated electricians are completing the task. Some electrical contractors utilize an unskilled and unmotivated labor force, while other contractors cultivate a highly skilled and motivated team. The skill and motivation of your labor force will impact your labor productivity.

Training – A happy, properly skilled, and continuously trained labor force will be most efficient and productive. Money spent on training is offset by increased labor productivity and reduces down time due to accidents or injuries.

Do you have an ongoing training program? Do you send your employees to school to review the National Electrical Code and Code changes? Are your workers taught and shown ways to become more productive and customer responsive? If you fail an inspection, do you review the Code violations with the electricians so they won't make the same mistakes again?

Experience – Labor productivity is significantly affected by the experience of the installers. Be sure your employees are properly trained in all aspects of their job such as safety, the proper use of tools, the National Electrical Code, and efficient work practices.

With experience, workers can complete a project quicker. If you estimate jobs that you have experience with, you should have a more competitive labor-unit, as opposed to jobs with which you have little or no experience.

Example. Tract housing electrical contractors reduce labor-units in Chapter 1 of the *Electrical Estimating Workbook* by 40% and still make money. Commercial electrical contractors increase the labor-units by 25% and they're lucky to break even when they wire a house.

3. Material and Tools

To complete a project efficiently, you must have the proper material and tools on the job when they are needed. If not, your labor force will be wasting your money trying to find something to do.

Labor-units are based on a skilled electrician utilizing the most current labor saving tools. If you do not provide proper tools, and training on the safe use of the tools, your productivity will decrease. Make sure the tools are always properly maintained and are in a safe working condition.

4. Supervision

There's no predicting how long it will take to complete a job that is not properly managed. Your jobs must be managed to insure that they are installed as estimated and according to the job labor budget.

3.07 KNOWING YOUR COMPETITORS' LABOR-UNITS

It doesn't help you to know the labor-units of your competitors, they're not yours. To be competitive you must always strive to be an efficient and effective electrical contractor. Don't let crises dictate how to manage your jobs. Establish a job budget and be sure your field supervision is organized and aware of the job's labor budget. Make sure your electricians are skilled, properly trained, and have the material and tools on the job when needed. It's really not that complicated.

3.08 HOW TO DEVELOP YOUR OWN LABOR-UNITS

To develop your own labor-units, you must have information on past labor performance of similar jobs. Track job hours and compare them against the job's budgeted hours. After a while you will gain the knowledge necessary to adjust your labor for the next job. In addition, past job performance is useful for bid analysis.

Author's Comment. If you do not manage your job properly, your labor-units will need to be increased upward. If you increase your labor-units to reflect your inefficiency (instead of correcting the problem), you will be become less competitive.

3.09 VARIABLES THAT IMPACT LABOR-UNITS

There is no set of labor-units that can be applied to all jobs; they must be increased or decreased to accommodate varying job conditions. Some of these variables can be controlled and others must be accommodated. It generally makes more sense to adjust the total labor hours or sub-total labor hours rather than each individual labor-unit for the varying conditions.

The following factors need to be considered when adjusting labor-units for job conditions. The adjustments included are only a suggested guide so you must keep historical information on past jobs and develop your own adjustments. The following is not an all inclusive list, but it does cover the most important job variables.

1. Building conditions
2. Change orders
3. Concealed and exposed wiring
4. Construction schedule
5. Job factors
6. Labor skill (experience)
7. Ladder and scaffold
8. Management
9. Material
10. Off hours and occupied premises
11. Overtime
12. Remodel (old work)
13. Repetitive factor
14. Restrictive working conditions
15. Shift work
16. Teamwork
17. Temperature
18. Weather and humidity

You might be wondering, with all of these variable factors affecting the labor-unit, how will you ever be able to estimate a job. Keep in mind you'll only have a few of these variables on any one job.

1. Building Conditions

Complexity

It takes much longer to run 100 feet of $3/4$" EMT in a research laboratory than it does to run it down the wall of an unfinished office. Job complexity can cause confusion and often requires greater supervision.

Elevator

In some locations elevators can only be used to transport material during certain hours of the day.

Floor Conditions

The floor is covered with 4 × 4's, 2 × 4's, sheets of plywood, garbage, and water. Who pays for the labor required to clean up the work area before you get started? Be sure your contract is clear on this subject.

Height

The height of a building can have a significant impact on labor productivity. As the number of floors increase, there are added labor-hours due to the extra time required to move equipment, material, tools, and people to the work area. The decrease in productivity is primarily because of the time required to wait for the lifts at the start of the day, end of the day, breaks, lunch, etc.

Single Floor Adjustment

If you are working above the third floor, add 1% to the total labor for each floor.

> **Example.** If you're remodeling an office building on the ninth floor, increase your total labor by 9%.

Wiring The Entire Building

The following adjustment factors apply to the total building labor-hours. The reason these percentages appear low is because each floor in the building is similar and you gain labor productivity due to repetition.

1 to 2 Floors	+ 0%
3 to 6 Floors	+ 1%
7 to 8 Floors	+ 2%
9 to 14 Floors	+ 5%
15 to 19 Floors	+ 7%
20 to 30 Floors	+ 13%

Example. If you are wiring a 7-story building, increase the total labor by 2%.

Note. NECA has available a 24-page report entitled "The Effect of Multi-Story Buildings On Productivity."

2. Change Orders

Jobs are rarely completed exactly as originally planned. Changes to the electrical system are required because of changes in the building design or the owner's needs. Change orders can result in a delay in the job schedule, which can decrease labor productivity. Often, change orders do not affect the planning or scheduling of work, but they can create unfavorable worker attitudes. There are no specific adjustments for change orders, therefor you should track your jobs to see if this impacts your labor productivity.

Other factors you should consider about change orders are greater management time, interruption of job flow, compressed job schedule, overtime, revisions to as-builts, and possible restocking charges from your suppliers.

3. Concealed And Exposed Wiring

Concealed Wiring

When installing boxes and raceways concealed in concrete or masonry make the following adjustments:

Boxes/raceways in walls + 50%

Boxes/raceways in columns + 100%

Exposed Wiring

Installing exposed boxes and raceways takes more time to complete than concealed wiring. Running exposed raceways means that you must take the time to do nice offsets, insure that the pipe is installed parallel, level along the building lines. Multiple pipes all coming out of one panel must be carefully bent so they turn within each other.

+ 10% for the boxes

+ 20% for the raceways

4. Construction Schedule

The construction schedule must be taken into consideration when you prepare the estimate. If the job is expected to take a long time, labor productivity often decreases.

Accelerated Schedule

An accelerated job can require a need for new and less motivated employees who can significantly decrease labor efficiency. If you have too many electricians on the job, productivity can and often will decrease. There is no specific adjustment factor for this condition, but you should consider its impact on your labor productivity.

Example. The crew size for a 6-week job which is estimated to take 1200 hours would require five electricians: 1200 hours/6 weeks = 200 hours per week/40 hours × 5 electricians. If you accelerate the schedule to 3 weeks, you will need ten electricians instead of five; and your productivity will probably decrease.

Note. NECA has available a 32-page report entitled "Project Peak Workforce Report."

Overtime

If the general contractor is unable to properly manage the overall project, you may have to make emergency changes to your work schedule. Schedule changes from one job often impact other projects as well. Be sure you have a clear contract on who pays for overtime if the construction schedule is changed.

5. Job Factors

Duration

Normal job duration exists when building construction progresses rapidly in a manner that doesn't require overtime. When a job is accelerated and the job requires longer work days, or over-manning of the job, labor productivity will begin to decrease. Extended duration jobs resulting from delays due to weather, poor job management, or the electricians simply getting bored, can cause a reduction in labor productivity and an increase in careless mistakes.

Note. NECA has available a 32-page report entitled "Normal Project Duration."

Location

If the job is in a remote location, you will find it more of a challenge to manage the job properly. This can result in a job being ignored with disastrous results. In addition, the job location can affect the quality of the available work force. If the job is located in an area that requires extensive travel time, it will be more difficult to get the material to the job when needed. Worse yet, the job could be located in the middle of a big city where the roads are under construction and the traffic is horrendous.

Size

Job size must not be overlooked. Labor-units must be increased for smaller jobs because they often have frequent interruptions. Medium to larger jobs can result in savings of labor because of repetition and economies of scale.

6. Labor Skill (Experience)

Is qualified help available? Remember, the lower the pay scale, the lower the skill level, the longer it will take for the job to be completed.

Attitude

What happens when you have a job and it appears to the workers that there's no other work coming in? Do your workers start slowing down to stretch this one out? Think about it.

What is your attitude toward your employees? Are your employees company people? What is the attitude of your employees toward completing a quality task on time? Do you have employees who don't like the jobs they are assigned, and who complain continuously? How does this affect morale and productivity for the other workers?

Experience

Naturally, if you're experienced, the time it takes to complete a task is less than if you're not very experienced. This adjustment at best is an educated guess.

Skill

If you do not have motivated, skilled, and experienced electricians, you will need increased supervision. Poorly skilled labor results in an increase in costs to correct violations and to fix mistakes. Are you forced to hire the first warm body that shows up at the job? Your hiring policy will determine the quality of your employees and their attitude. Labor productivity begins at the hiring stage and ends with training and proper supervision.

Training

A happy, properly skilled and continuously trained labor force will be most efficient and productive. Money spent on training is offset by increased labor productivity, and reduced down time due to accidents or injuries. Are your workers taught and shown ways to become more productive and customer responsive? If you fail an inspection, do you review the Code violations with the electricians so they won't make the same mistakes again?

7. Ladder And Scaffold

Ladder

Work that requires you to work on a ladder will require an increase in the labor-unit. The following adjustments, based on working height, should be used until you develop your own units.

12 Feet	+ 3%
13 Feet	+ 5%
14 Feet	+ 8%
15 Feet	+10%
16 Feet	+ 13%
17 Feet	+ 16%
18 Feet	+ 19%
19 Feet	+ 22%
20 Feet	+ 25%

Scaffold

If you are required to work from a scaffold, the scaffold must be set up, moved, and dismantled. Before the scaffold can be moved, all tools, equipment and material must be brought down. If a large obstruction is in the way, the scaffold must be disassembled to move it around the obstruction. For safety consideration, one person should be at the bottom of the scaffold at all times. If you have a motorized scaffold, you won't have to increase the labor as much.

Add 40% to the labor-units plus the time for pickup, setup, moving, take-down, and return.

8. Management

Proper job management insures that the labor force is informed on what is to be accomplished and how long it is expected to take. Make sure that your labor force has the proper tools, material, and skills to complete the task efficiently. Your workforce will be contented and proud to work for your company, which means they'll give their best. Emergency ordering of material and tools, or changes of labor scheduling, often result in lost time.

Labor

The most efficient method of improving labor productivity is through proper training and efficient management. Is your foreman or supervisor trained for efficient and effective job management? Does the foreman or supervisor know the job budget for material and labor? Did you or your estimator take the time to explain to the foreman or supervisor in the field how the job was figured and planned? Do you have regular meetings between the office and the field to review job progress and to offer assistance? Or do you just let them hang to dry in the wind?

Tools

Labor-units assume that you have the correct tools for the job to be done. Use labor saving tools whenever possible to increase labor efficiency.

9. Material

Delivery

Planning your material orders is critical to insure the electricians are not held up. There are hundreds of minor items that they can run out of, it's the nickel and dime items that really can stop a job. Improper material management can contribute to reduced labor productivity. Most of us have worked jobs where we stood around waiting for the shop or the supplier to get the material to the job. Do you order and receive your material on a timely basis? Planning is important, material won't be delivered to remote locations as often as to easily accessible locations.

Storage/Disposal

Where can the job's material be stored? Is a secure storage space available at the job site? Will there be nonproductive time allowed for the electricians to walk to the job trailer for material and tools?

10. Off Hours and Occupied Premises

Off Hours

Does the job require you to work during off hours, or only during certain hours of the day? An example would be a department store where you can only work from 10 P.M. at night until 9 A.M. the next morning.

Occupied Premises

Does the job require you to work in an existing occupied premise around people, furniture, and equipment?

11. Overtime

Overtime interrupts established life patterns and causes fatigue, boredom, and lower productivity. Workers have a tendency to forget safety procedures which can result in serious accidents and possibly fatalities. Consider the following adjustments:

<u>Six 8-Hour Days</u>
One Week: +15% for the sixth day
Every Week: +15% all labor for week

<u>Saturday And Sundays After Five Days</u>
One Weekend: + 30% for weekend hours
Every Weekend: + 30% all labor for week

<u>Extended Regular Hours</u>
Six 10-hour days: + 18% all labor
Seven 10-hour days: + 30% all labor

> **Note.** NECA has available a 20-page report entitled "Overtime and Productivity in Electrical Construction."

12. Remodel (Old Work)

If you're working in an existing building and trying to fish-in wire or cut in boxes, the labor adjustment might be as much as 200%. The only way that you can know with any degree of accuracy is to do a job and track the labor for future reference. Once you have some experience and job history, you will be better prepared to adjust the labor for the next job.

13. Repetitive Factor

When you perform the same function over and over again, you can complete each task a little bit faster than the previous time. In a short period of time you can significantly improve your labor productivity. Consider the following labor adjustments:

1 to 2 Repeats	0%
3 to 5 Repeats	− 15%
6 to 10 Repeats	− 25%
11 to 15 Repeats	− 35%
16 + Repeats	− 45%

14. Restrictive Working Conditions

Sometimes many trades must work in a tight space all at once. This results in a significant reduction in labor efficiency.

15. Shift Work

When working in shifts, you need to take into consideration the lifestyles of your employees. Generally, single employees prefer to work the second shift and married employees prefer the third shift. Ask them what they prefer and try to accommodate them when possible.

	Overall	Single	Married
First Shift	0%	0%	0%
Second Shift	+20 to 25%	+15 to 20%	+25 to 30%
Third Shift	+15 to 20%	+20 to 25%	+15 to 20%

16. Teamwork

The general contractor's lack of coordination of the other trades can decrease your labor productivity. If the general contractor doesn't coordinate effectively, it's going to cause everyone problems and money.

When there is stress between the various trades, you can expect problems and reduced labor productivity. Jobs where all the trades interact like one big happy family have a great attitude and labor is more efficient. When possible, try to stimulate and encourage positive teamwork habits between the trades.

Productivity is often increased simply by giving the job foreman a bonus if the job is completed in less than the budgeted hours. Successful electrical contractors have incentive plans or reward systems so that foremen or project managers will attempt to complete the project within the labor budget. The incentive plan could be paid time off, or perhaps a bonus based on hours saved for the foreman, project manager, and even the electricians.

17. Temperature

Optimum labor efficiency can be achieved when the temperature is between 40°F and 70°F and the relative humidity is below 80%. Studies have shown that extreme temperatures cause workers to concentrate on themselves rather than on the job to be performed. This can lead to an increase in accidents, a deterioration in workmanship, and lower labor productivity.

Cold

Extremely cold conditions can cause a significant reduction in labor productivity because of warm-up breaks. Many electricians hate working under these conditions and have a tendency to get head colds and feel drained, run down, and tired.

Heat

Elevated temperatures cause a decrease in labor productivity due to the time required to wipe body perspiration from face and hands and perspiration getting on work surfaces making it difficult to handle material, equipment and tools. When the temperature is above 101°F, electricians become more fatigued, belligerent, and irritable.

If you're used to working in a cold ambient temperature, you will need to increase the labor-units when the temperature increases. If you're used to working in a warm or hot environment, you'll need to increase labor-units when the ambient temperature decreases.

Note. NECA has available a 32-page report entitled "The Effect of Temperature On Productivity."

18. Weather And Humidity

Humidity has a small impact on labor productivity, but rain can ruin an otherwise productive day. Water runoff can destroy all of your work in a ground floor slab. Leaky interiors can slow down interior production to a standstill. Are you going into the rainy season?

3.10 LABOR-UNIT SUMMARY

At this point you should have a better understanding of what comprises a labor-unit, and what's not included. You might even be thinking that 11 minutes to install a receptacle might not be enough time after all; at least not when you consider supervision, material handling, lay-out, installation, tooling, material lay-out, clean up, breaks, nonproductive factors and job conditions.

As you can see, it requires a lot of thought to estimate the labor for a given job, but with labor-units and experience you should begin to pull ahead of your competitors.

SUMMARY

Introduction
Estimating with labor-units is both a science and an art form. Science in the sense that the labor required to complete a task is a function of the materials to be installed and their quantities. If you know the quantity of each material required for a job, you can easily determine the labor required. An art form in the sense that you need to become creative in making some adjustments to the labor-units for the specific job conditions.

3.01 What Exactly Is A Labor-Unit?
A labor-unit represents the approximate time required to install an electrical product, component or equipment. Labor-units are based on the assumption that a highly trained, skilled, and motivated electrician is completing the task under standard installation conditions with the proper tools.

Labor-Units Are Expressed
Labor-units are expressed in units of decimal hours: E - each, C - hundreds, or M - thousands. You don't use minutes when you estimate a job, you use decimal labor-hours.

Work Experience
Your electrical work experience is very important when you estimate electrical wiring, but you must know when and how to use that experience. Many electricians estimate a job based on how long they think it would take them to do the job. There are many factors that they might not have taken into considered such as supervision, job lay-out, tool management, handling of the material, nonproductive time and job conditions.

3.02 How Labor-Units Are Used
To determine the labor required for a job, you must determine the material required and their quantities. With this information you use the labor-units to determine the total labor required.

3.03 What's Included In The Labor-Unit

A labor-unit is comprised of six major components. They include:
1. Installation   50%
2. Job layout   15%
3. Material handling and cleanup 10%
4. Nonproductive labor   5%
5. Supervision   10%
6. Tool handling   10%

3.04 Labor-Units Do Not Include

Labor-units do not include the assembly of fixtures, switchboards, panels, or other equipment. Other factors that are not included are:

Cutting holes or openings.
Excavation, drilling, or blasting.
Heavy equipment (operator's time).
Hoisting above three floors.
Maintenance of temporary equipment.
Painting of conduits.
Testing or welding.

Unusual Tasks

There will be times when you must determine the labor for a task that you have absolutely no idea how to do. Break the task down into as many small individual labor segments as possible. Try to establish a labor-unit for each segment, by comparing the individual segments to a similar tasks that you have a labor-unit for. If all else fails, make a reasonable guess.

3.05 Labor-Unit Manuals, Which Should You Use?

You should consider purchasing a comprehensive labor-unit manual that contains at least 10,000 labor-units. They generally cost less than $100. There is no perfect labor-unit that can be applied to all jobs. With experience and historical data, you will develop techniques to help you adjust the labor-units to represent your productivity with specific job conditions.

3.06 Your Labor-Units As Compared To Your Competitors'

How you manage your business impacts the labor productivity and your profitability. Other factors that impact labor productivity include:
1. Computers
2. Labor
3. Material and tools
4. Supervision

3.07 Knowing Your Competitors' Labor-Units

It doesn't help you to know the labor-units of your competitors, they're not yours.

3.08 How To Develop Your Own Labor-Units

To develop your own labor-units, you must have information on past labor performance of similar jobs. Track job hours and compare them against the job's budgeted hours. After a while you will gain the knowledge necessary to adjust your labor for the next job. In addition, past job performance is useful for bid analysis.

3.09 Variables That Impact Labor-Units

There is no set of labor-units that can be applied to all jobs; they must be increased or decreased to accommodate varying job conditions. Some of these variables can be controlled and others must be accommodated. It generally makes more sense to adjust the total labor hours or sub-total labor hours rather than each individual labor-unit for the varying conditions.

3.10 Labor-Unit Summary

At this point you should have a better understanding of what comprises a labor-unit, and what's not included. As you can see, it requires a lot of thought to estimate the labor for a given job, but with labor-units and experience you should begin to pull ahead of your competitors.

Chapter 3

Review Questions

Essay Questions

1. In what manner is estimating labor a science and an art form?

2. In what way is electrical experience a hindrance when estimating electrical work?

3. Explain what's included in the labor-unit.

4. What is not included in a labor-unit?

5. How do your determine the labor required for a task that you have absolutely no idea how to do?

6. What factors contribute to the difference in labor-units between electrical contractors?

7. Why doesn't it help for you to know your competitors' labor-units?

8. How can you develop your own labor-units?

Multiple Choice Questions

1. A labor-unit represents the approximate time required to install an electrical product, component or equipment. Labor-units are based on the assumption that a _____ electrician is completing the task under standard installation conditions with the proper tools.
   (a) highly trained   (b) skilled   (c) motivated   (d) all of these

2. Labor-units must be adjusted for variables of labor _____ to account for labor skill, management organization, and job site conditions.
   (a) overhead   (b) productivity   (c) training   (d) none of these

3. Labor-units are expressed in units of _____ hours.
   (a) fraction   (b) quarter   (c) decimal   (d) all of these

4. The actual installation time represents slightly more than _____% of the labor-unit and includes the installation of boxes, conduit, fittings, wiring devices, fixtures, switchgear, disconnects, panels, breakers, etc.
   (a) 10   (b) 30   (c) 50   (d) 70

5. The layout of the work to be installed represents about _____% of the labor-unit.
   (a) 15 (b) 20 (c) 30 (d) 40

6. Material handling represents about _____% of the labor-unit.
   (a) 10 (b) 20 (c) 35 d) 50

7. Non-productive labor should be managed so as not to exceed _____% of the labor-unit.
   (a) 2 (b) 3 (c) 4 (d) 5

8. Supervision represents approximately _____% of the labor-unit and includes the review of the plan, ordering material, working out installation problems, and coordinating with other trades.
   (a) 10 (b) 20 (c) 35 (d) 50

9. Tool handling and tool management represent approximately _____% of the labor-unit.
   (a) 5 (b) 10 (c) 15 (d) 20

10. Among the variables that affect labor productivity are _____.
    (a) the workers' attitude, the job complexity, the labor rates, and floor conditions
    (b) weather, temperature extremes, split shifts, site congestion, and overtime
    (c) the crew size, the construction schedule, the job location, and security restrictions
    (d) all of these

11. If you are wiring a four-story building, the labor should be increased by _____%.
    (a) 1 (b) 2 (c) 3 (d) 4

12. If you are installing conduit and boxes concealed in concrete walls, the labor-unit needs to be increased by _____%.
    (a) 10 (b) 20 (c) 40 (d) none of these

13. If you are installing conduit exposed, the labor-unit needs to be increased by _____%.
    (a) 10 (b) 20 (c) 30 (d) 40

14. Work located up twelve feet requires an increase of _____% in the labor-unit.
    (a) 1 (b) 3 (c) 5 (d) none of these

15. If your job requires that electricians work six weeks, six days per week, you must increase the labor by _____%
    (a) 15% for the sixth day
    (b) 15% all labor
    (c) 30% for weekend hours
    (d) none of these

16. When you perform the same function over and over again, you can complete each task a little bit faster than the previous time you completed that task. If a task requires 15 repeats, the labor should be decreased _____%.
    (a) 35 (b) 45 (c) 55 (d) none of these

17. When working in shifts, you need to take into consideration the lifestyles of your employees. It's more efficient to have _____ employees work the second shift.
    (a) married (b) single (c) either a or b (d) none of these

Chapter 4

The Estimating Process

OBJECTIVES

After reading this chapter, you should understand
- the difference between estimating and bidding.
- why you must be selective on the jobs you bid.
- how to prepare the estimate.
- what to do if you don't understand the blueprints or specifications.
- about estimating forms and worksheets and how to use them.
- what a take-off is all about.
- how to determine the bill-of-material for the job.
- how to price material, determine labor, perform extensions and determine totals.

INTRODUCTION

Determining the selling price of a job consists of three parts, the estimate, the estimate summary, and the bid process.

The Estimate And Bid Process

Part 1 – The Estimate

The estimate consists of determining the approximate material cost and labor man-hours for the job. To properly estimate a job, you must have the ability to mentally visualize the mechanical requirements and material required to complete the job. The estimating process, in its most basic form, consists of:
1. Understanding what is required to be completed within the contract price.
2. Counting and measuring symbols.
3. Determining a material required for pricing and laboring.
4. Pricing and laboring the material.
5. Extending and totaling the cost of material and labor-hours.

Part 2 – The Estimate Summary – Chapter 5, Part A

The estimate summary is used to determine the job's break even cost. Take into consideration the cost of labor, materials, sales tax, direct job costs, and overhead.

Part 3 – The Bid Process – Chapter 5, Part B

The bid process consists of applying profit and other cost to determine the selling price. Once you have determined the selling price, you need to verify that it is valid and you must create a proposal.

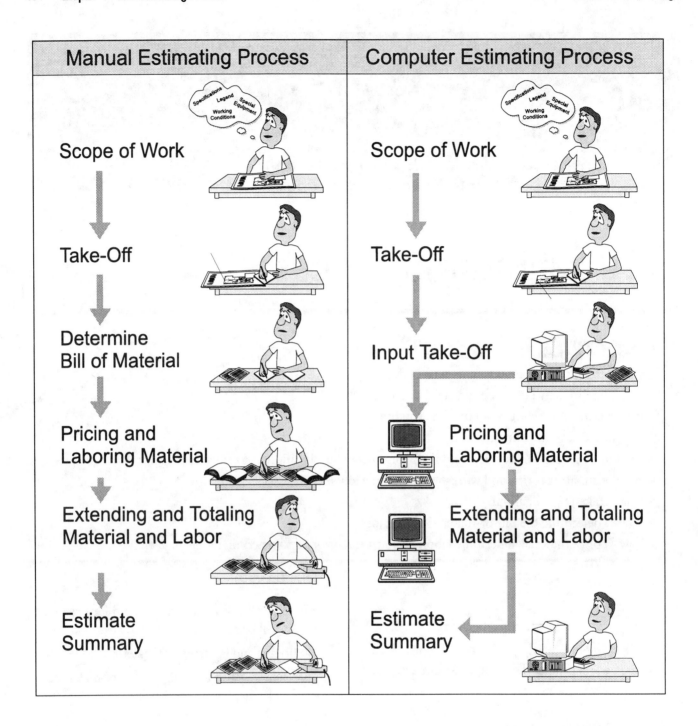

4.01 JOB SELECTION

Before you estimate any job, you should always ask yourself, "Is this the most efficient use of my time, talents, and skills, at this time?" Because of limited resources such as energy, time, and money, you can't estimate all jobs. You must become selective about which jobs you feel will better meet your financial goals. Only consider jobs in which you have a reasonable chance of making the profit margin you want. To be profitable, you need to be an efficient electrical contractor, not the largest. Treat your employees and customers fairly and honestly. Don't worry about getting every job, simply manage your resources as carefully as possible at all times.

Before you bid the job, what does your gut tell you? What are the chances of getting the job? Do you think the job will be profitable? Do you have the money to finance the job?

Listen To Your Gut Feelings

Remember, the only reason you are in business is to make money, and you don't need to waste your time with someone who is going to give you a hard time. Before estimating any job, determine if you want to do work for this particular contractor. Run a simple credit check, get references, talk with some current and past subcontractors. Does the contractor have a good reputation for paying their subcontractors on time?

If you are not comfortable with a potential contractor and your gut feeling is telling you "there's something wrong," don't even think of bidding the job.

What Are The Chances Of You Getting The Job?

Have you ever bid on job you knew you weren't going to get? You knew you weren't going to get the job, but you bid it anyway? This is not the way you should run your business! If you don't think the chance of you getting the job is very good, don't waste your time.

Can This Job Be Profitable?

How many times have you bid on a job you didn't think you were going to make money on? Many electricians feel an obligation to bid a job, even when they know that the job won't be profitable. If you don't think you're going to make money on the job, don't bid it!

Financial Resources

Careful, don't bid a job that is greater than you can financially handle. Some experts suggest that you never bid on job that is twice as large as your typical size job. Others suggest that you never bid on a job that is greater than 25% of your total annual sales.

When you bid on a large job, you need to be sure you have all the resources to complete the job profitably. Remember, this is a commitment of your limited time and resources. Consider the financial impact of any job before you bid it.

Should You Bid This Job?

Review the following questions to help you decide if you want to bid the job. Remember, this is only a guide and you should create a set of questions that take into consideration factors that are important for your decision.

Questions To Consider

1. How many electrical contractors are bidding this job?
2. Will this job work well with my present work load?
3. Do we excel in this kind of work and can I be competitive?
4. Have I handled projects like this before?
5. Do I have the financial resources?
6. Do I have the proper work force?
7. Do I have enough time to properly estimate this job, and am I qualified to estimate this kind of work?
8. Can I manage this job profitably?
9. Do I have the proper tools and equipment?
10. What are the chances of getting the job?

Do not take the job if you answer yes to either of the following questions:

11. Do I think this person is going to hurt me (gut feeling)?
12. Will my bid be used for price shopping?

Just Say No

If the job doesn't meet your needs, call the contractor or owner and let that person know you are not interested in the job. No matter how much whining you hear, don't bid the job. In this business you must learn to say no and mean it. When you decline to bid a job, consider giving the contractor your competitor's phone number. You kill two birds with one stone; you don't waste your time, and you keep your competitors busy.

I realize that when you're getting started in business you'll feel you must bid every job that comes along. This is natural and it's okay. This gives you plenty of opportunity to practice your estimating skills. However, if you provide your customers with a quality job at a fair price, you will become busy and will need to be selective of the jobs you bid.

Note. Keep information on past job performances so you know which jobs you made the most profit on.

4.02 UNDERSTANDING THE SCOPE OF WORK (STEP 1)

Understanding The Scope Of Work

Before you begin the estimate, you must understand the scope of work to be completed according to the contract. To accomplish this, be sure you have all of the information you can about the job. You must have a complete and current set of blueprints and specifications.

Plans (Working Drawings)

Be sure you have every page of the blueprints, including the architectural elevations, mechanical, and structural. The first thing you should do when you receive the blueprints is to be sure they are readable and clear. Give them a quick scan to help you get an overview of the project's requirements.

Specifications

Specifications are additional requirements that govern the material to be used and the work to be performed. Specifications are designed to simplify the task of interpreting the blueprints and should clear up many issues. A typical table of contents of the specifications would include:

Division 0	Bidding Requirements
Division 1	General Requirements
Division 2	Sitework
Division 3	Concrete
Division 4	Masonry
Division 5	Metals
Division 6	Wood and Plastics
Division 7	Thermal/Moisture Protection
Division 8	Doors And Windows
Division 9	Finishes
Division 10	Specialties
Division 11	Equipment
Division 12	Furnishings
Division 13	Special Construction
Division 14	Conveying Systems
Division 15	Mechanical and Plumbing
Division 16	Electrical

Author's Comment. For the past few years, the quality of blueprints and specifications have significantly deteriorated. Poor blueprints and specifications often result in confusion and an increase in the time it takes to complete the estimate and job. Keep track of all bid inconsistencies and be sure your contract is clear.

Visit The Job Site

Be sure to visit the job site before you bid the job. Some contractors use a video camera and/or a Polaroid camera to document the job site conditions. That way they don't make any mistakes with the estimate.

Preparing For The Estimate

Job Folder

You need to create a Job Folder to contain your bid notes, job information sheet, take-off worksheets, bill-of-material worksheets, quote sheets, summary sheet, and other papers associated with the estimate.

Estimate Record Worksheet

Create and complete an Estimate Record Worksheet. This form contains pertinent job information such as job name, job location and address, phone numbers, names of the owner, contractor, architect, engineer, and to whom the bid is to be submitted.

Other information that should be included is the telephone company and the electric utility contact names and phone numbers.

Once you have completed the Estimate Record Worksheet, hang it up on the wall over your take-off bench for handy reference. This information will be useful when creating a bid proposal. See the next page for a sample of the Estimate Record Worksheet.

Note. A blank sample copy is also contained in the Appendix.

Plan And Specification Review

Before you get any deeper, you need to take some time in a quiet place to carefully read the specifications and all notes on the blueprints. Give both the blueprints and specifications a good review to help you get a clear understanding of the scope of the project and the bid requirements before you begin the actual take-off.

ESTIMATE RECORD WORKSHEET

Job Information Detail   Estimate Job Number:

Job Name: _____	Bid Due Date: _____
Job Location: _____	City: _____
Job Owner: _____	Phone: _____

Plans And Specifications Detail

Date Of Plans: _____   Blueprint Page Numbers: _____

Comments: _____
_____

Contractor Information Detail

Contractor: _____	Contact: _____
Address: _____	City: _____
	E-mail: _____
Phone: _____	Mobil: _____
Beeper: _____	FAX: _____

Important Contacts

Architect: _____	Phone: _____
Engineer: _____	Phone: _____
Telephone Utility: _____	Phone: _____
Electric Utility: _____	Phone: _____
Electrical Inspector: _____	Phone: _____

Notes: _____
_____
_____
_____
_____

Plans

Take a close look at all blueprint legends, details, notations, and symbols. Watch for the electrical requirements for control wiring, underground wiring, area lighting, signs, and outdoor equipment. Become familiar with the entire installation and check for any special or unusual features such as unusual ceiling heights. Look to see if proper working space is provided for the electrical equipment and watch for the location of the utilities, etc.

Case Study No. 1 – A friend of mine, who was just getting started in business, did not read the note on the blueprint that required him to replace 180 feet of No. 4/0 service conductors with 500 kcmil.

Result – He under-bid the job by $2,100 on a $28,000 dollar job and he got it.

Case Study No. 2 – The blueprints indicated that the electrical contractor was required to install three of the owners fixtures. The contractor figured three hours for each fixture.

Result – The actual fixtures weighed over 500 pounds each and required three men three days to install.

Specifications

Underline or highlight important and/or unusual items that can impact the estimate. Determine who is to be responsible for such things as painting exposed conduits, trenching, back-fill, concrete work, patching, clean up, temporary power, etc. Determine if the use of special equipment or overtime is required. Don't take anything for granted.

Case Study No. 3 – I know of an electrical contractor who misinterpreted the specifications about gross receipts taxes for a job on an Indian reservation in Arizona. He did not figure the 3.5% tax on a $12,000,000 dollar job = $420,000.

Result – He got the job.

Case Study No. 4 – Another contractor was required to provide a video projector, according to the specifications. He guessed that it would cost $5,000.

Result – It actually cost $18,000.

Estimate And Bid Notes

If you don't understand something in the blueprints or specifications, you might have the tendency to put the estimate aside until the last minute. This will result in you not submitting the bid in a timely manner, or worse yet, estimating the job at the last moment without confidence.

Keep a note pad handy so you can keep track of any questions you have about the estimate. If there is anything you don't understand, get the answer as soon as possible, don't wait until the last moment. Contact the architect or engineer, but be sure you get your questions answered as soon as possible. If not, don't bid the job. See the Estimate and Bid Notes worksheet on the next page.

Note. A blank sample copy is also contained in the Appendix.

Specification Check List

In an effort to keep track of the plan and specification details, you will want to complete the Specification Check List worksheet and hang it up over the take-off bench for quick reference while you do the estimate. See page 52 for a sample of a Specification Check List worksheet.

Note. A blank sample copy is also contained in the Appendix.

4.03 THE TAKE-OFF (STEP 2)

Estimating Forms And Worksheets

To quickly and accurately estimate a job, you must use proper estimating forms and worksheets. Proper forms and worksheets will save you time, create consistency, and assist in the reduction of errors. The forms and worksheets should also help serve as a reminder of items that are easily omitted or forgotten.

Different types of construction styles, such as residential, commercial, or industrial lend themselves to different types of estimating forms or worksheets. In addition, different types of forms and worksheets are required for different parts of the estimate. For example, a worksheet used to determine the lighting requirements would be different from the worksheets needed to record the feeders and service equipment requirements.

If you have a computer you would greatly benefit from designing your own custom worksheets, or they can be ordered from:

Estimate And Bid Notes Page ____ of ____

Blueprint Questions or Comments:

Specification Questions or Comments:

Specification Check List

Labor-Unit Adjustment (See Section 3.09)

1. Building Conditions _____
2. Change Orders _____
3. Concealed And Exposed Wiring _____
4. Construction Schedule _____
5. Job Factors _____
6. Labor Skill (Efficiency) _____
7. Ladder and Scaffold _____
8. Management _____
9. Material _____
10. Off Hours And Occupied _____
11. Overtime _____
12. Remodel (Old Work) _____
13. Repetitive Factor _____
14. Restrictive Working Conditions _____
15. Shift Work _____
16. Teamwork _____
17. Temperature _____
18. Weather And Humidity _____

Labor Adjustment

Additional Labor (See Section 5.01)

1. As-Built Plans _____
2. Demolition _____
3. Energized Parts _____
4. Environmental Hazardous _____
5. Excavation, Trenching And Backfill _____
6. Fire Seals _____
7. Job Location _____
8. Match-Up Of Existing Equipment _____
9. Miscellaneous Material Items _____
10. Mobilization (Startup) _____
11. Nonproductive Labor _____
12. OSHA Compliance _____
13. Plans and Specifications _____
14. Public Safety _____
15. Security _____
16. Shop Time _____
17. Site Conditions _____
18. Sub-Contract Supervision _____
19. Temporary, Stand-By Power _____

Hour Adder

Direct Job Expenses (See Section 5.05)

1. As-Built Plans $_____
2. Bus. & Occupational Fees $_____
3. Engineering Drawings $_____
4. Equipment Rental $_____
5. Field Office Expenses $_____
6. Fire Seals $_____
7. Guarantee $_____
8. Insurance – Special $_____
9. Miscellaneous, Labels $_____
10. Mobilization $_____
11. OSHA Compliance $_____
12. Out of Town Expenses $_____
13. Parking Fees $_____
14. Permits/Inspection Fees $_____
15. Public Safety $_____
16. Recycle Fees $_____
17. Storage & Handling $_____
18. Sub Contract:
    _____ $_____
19. Supervision Cost $_____
20. Temporary Wiring:
    Lighting $_____
    Power $_____
    Maintenance $_____
21. Testing $_____
22. Trash $_____
23. Utility Cost $_____

Total Direct Cost

Other Final Costs (See Section 5.12)

1. Allowances/Contingency $_____
2. Back-Charges $_____
3. Bond $_____
4. Completion Penalty $_____
5. Finance Cost $_____
6. Gross Receipts Or Net Profit Tax $_____
7. Inspector Problems $_____
8. Retainage $_____

Total Other Cost

Other Considerations

1. Conductor Size - Minimum? _____
2. Raceway Size - Minimum? _____
3. Control Wiring Responsibility? _____
4. Cutting Responsibility? _____
5. Demolition Responsibility? _____
6. Excavation/Back Fill Responsibility? _____
7. Painting Responsibility? _____
8. Patching Responsibility? _____
9. Special Equipment? _____
10. Specification Grade Devices Or Fittings? _____

Minnesota Electrical Association
3100 Humboldt Avenue South
Minneapolis, MN 55408
1-800-829-6117, Fax 1-612-827-0920

National Electrical Contractors
 Association (NECA)
3 Bethesda Metro Center, Suite 1100
Bethesda, Maryland 20814-5372
1-301-657-3110, Fax 1-301-215-4500

Note. Sample estimating forms are contained in the Appendix.

Take-Off

A take-off is the action of counting and measuring the symbols. A proper take-off should insure that there will be little need to refer to the plans or specifications to determine the bill-of-material required to complete the project. To accomplish this, you must follow an orderly, organized, methodical routine that is complete and consistent for each and every job.

Use colored pencils or pens to identify each item that you have taken-off. When you have finished the take-off, your blueprints should be a colored representation of the electrical work required to be installed. The following is a sample of a sequence and color code to identify those items taken-off.

Sequence	Color
1. Fixtures	Yellow
2. Switches	Blue
3. Receptacles	Light Green
4. Miscellaneous	Purple
5. Circuit Conductors	Orange
6. Separate Circuits	Pink
7. Special Systems	Red
8. Feeders	Brown
9. Transformers	Pick a color
10. Service	Pick a color
11. Other	Pick a color

If you're not permitted to mark up the blueprints, you have three choices. One, go to a copy center and make a copy for yourself at 100% scale; two, get a large plastic film, place it over the blueprints and mark up the plastic film; or three, don't bid the job. If you attempt to estimate a job without color marking the blueprints, you're more likely to make a mistake and you won't have the confidence that your price is accurate.

Counting Symbols

The purpose of counting symbols is to identify their quantity so that you can determine the jobs bill-of-material. When counting, you must be capable of reading and interpreting all blueprint symbols and notes. You must also understand the language on the drawings and specifications and your counting accuracy is expected to be 100%.

When you count symbols, it is best to use a hand held counter to keep track of the count. After you count a symbol, mark it with a colored pencil or pen. This insures that no drawing symbol is counted more than once, and that no symbol is missed.

Counting Sample

The following table demonstrates the symbol count for the Meeting Room as shown in Blueprint M–1, Figure 4–1.

Fixtures	Quantity
Lay-In Flourescent Fixtures	18
Ceiling Mounted Fixtures	4
Recessed Fixtures	20
Exit Fixtures	2
Switches	
Switch - Single Pole (1 Gang)	1
Switch - 3 Way (1 Gang)	1
Switch & Switch (2 Gang)	1
Switch & Switch 3-Way (2 Gang)	1
Receptacle	
Convenience Receptacles	9
IG Receptacles	2
Total Outlets	**59**

Measuring Circuits

Measuring consists of determining the circuit length for branch circuits, feeders, and service raceways. You need to be sure you accurately measure all circuits, because if you're wrong, you will either lose money or not get the job at all.

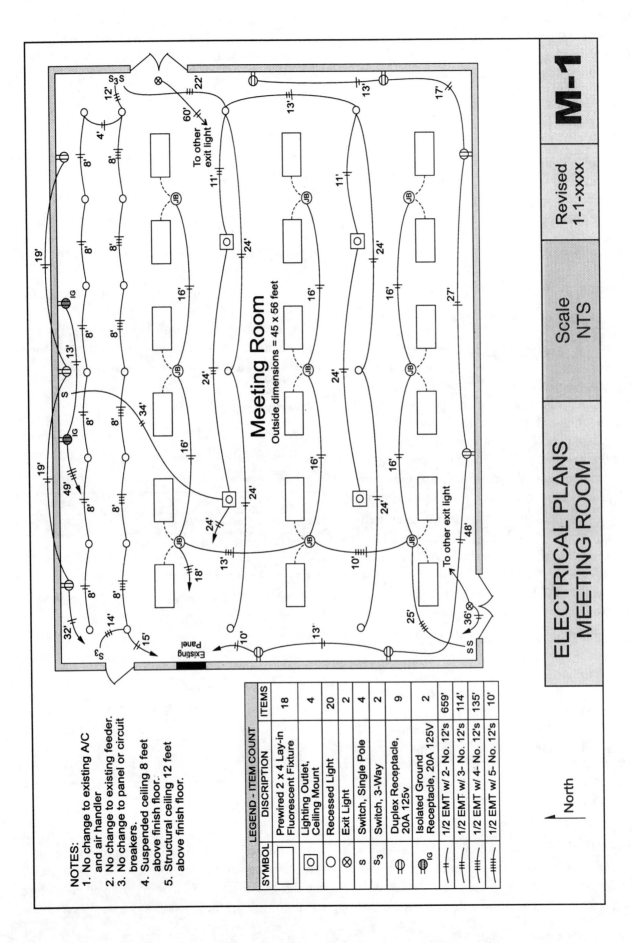

Figure 4–1
Meeting Room Blueprint M–1

Verify the architectural scale listed in the blueprints before you measure any circuit. Test the scale to insure its accuracy, since sometimes a blueprint is duplicated at a copy center at a reduced size.

Most scale dimensions are noted in the title block of drawings; however, the scale might be different on different pages, or different on the same page. If you're not careful, your measurement can be off by as much as 100%; that is, if you thought you were working a 1/4 inch per foot scale, and it was actually 1/8 inch per foot scale.

Tools used to measure the circuits include an architectural ruler, a scaled measuring tape, a mechanical measuring device, or an electronic measuring wheel. The architectural rule is fine for a few quick measurements. I really like the measuring tape for 1/8 inch and 1/4 inch scale drawings. However, the electronic scale wheel is the most convenient for multiple scales.

Measuring Sample

Once you have counted the fixtures, switches, and convenience receptacles, you can measure the branch circuit wiring for these outlets. When you measure each run, don't forget to consider the drops for the switches, receptacles and panel. Many estimators will place a scaled line on the blueprint to represent the distance of the drops. This way when they come across a drop for a switch, they simply scale the distance from the pre-scaled line.

> **Note.** Electronic scale wheel devices permit you to press one key to add a constant distance to the totals. But if your drops are of different lengths, the constant will have to be reset.

When you measure branch circuits, be sure to take-off the two-wire circuits first, the three-wire circuits next, and then the four-wire circuits, etc. When you measure the circuit wiring, trace the line that represents the wiring with a color pen or pencil. It doesn't matter what color you use, just be consistent. The following is just a suggestion:

- Two-wire circuits – Yellow
- Three-wire circuits – Blue
- Four-wire circuits – Green
- Five-wire circuits – Orange

The following table demonstrates the measured circuit lengths for the Meeting Room as shown in blueprint M–1. Using graphic symbols rather than writing out the description can save time.

Branch Circuits	Pipe	Wire (+ 10%)
1/2" EMT w/ 2 No. 12	645 Feet	1419 Feet
1/2" EMT w/ 3 No. 12	108 Feet	356 Feet
1/2" EMT w/ 4 No. 12	128 Feet	563 Feet
1/2" EMT w/ 5 No. 12	10 Feet	55 Feet
Totals	**891 Feet**	**2,393 Feet**

If the working drawings do not indicate the circuit layout or wiring configuration (most do not), you must perform this task first and then measure the circuit wiring.

Take-Off Sequence

There is no set sequence for performing the take-off, but you will develop a system that fits your personal style and needs. Your style will need to be adjusted for new types of construction or jobs.

Whatever system you develop, be sure to use the same procedures every time you estimate a job. Consistency will help you reduce the time it takes to estimate a job, as well as help you eliminate some errors.

> **Note.** Some estimating sequences are more efficient if you use a computer rather than estimating by hand.

The following are three typical take-off sequences.

1. One Section/Page At A Time

Taking-off one page of the blueprint at a time, or all of the wiring of the first floor, second floor, etc. This method is excellent if you use a computer, but impractical if you estimate manually.

2. Start At Service And End With Lighting

With this method you start the take-off at the utility service location, continue taking off the feeders, the branch circuits, and complete the take-off with counting lighting fixtures. This method is time consuming and requires many movements between many pages of drawings and for most, it is unnatural. This style is generally not recommended, but if you have a computer estimating system, it can work okay.

3. Start With Lighting And End With The Service

With this method, you start the take-off with the lighting fixtures and finish the take-off with the service equipment.

This take-off sequence generally follows:
A. Count lighting fixtures and develop quantities for fixture quotes.
B. Count switches.
C. Count receptacles.
D. Measure branch circuit wiring.
E. Count and measure individual branch circuits and home runs. Develop quantities for switchgear quotes.
F. Count and measure special systems such as television, phone, CATV, alarm, security, sound, etc.
G. Count panels and measure feeders, service runs, service equipment. Develop quantities for switchgear quotes.

This method permits a quick overview of the job as you count the fixtures, switches, and then the convenience receptacles, which permits you to ease into the estimate. After taking off the homeruns and special circuits, you should have a good idea of the scope of the project and you will be better prepared to deal with the more complex portions of the take-off. Using graphic symbols rather than writing out the description can save time.

4.04 DETERMINING THE BILL-OF-MATERIAL (STEP 3)

Once you have completed the take-off, you need to determine the bill-of-material. The manual take-off should provide enough information to determine the type, size, and quantity of all material items required to complete the installation.

Determining the bill-of-material requires that you have the ability to visualize all of the material items required for each symbol. If you don't understand the electrical wiring requirements or you don't have the ability to visualize the electrical components required, you cannot determine the bill-of-material.

Computer Assisted – If you are estimating the job with computer assemblies, the bill-of-material can be automatically generated and this step can be mostly completed by the computer.

Material Spread Sheet

Determine Bill of Material

Input Take-Off

If estimating the job without a computer, you must determine the bill-of-material in the most efficient manner. Some parts of the take-off such as fixtures, switches, and receptacles lend themselves to the use of a spreadsheet to determine the most common items.

Take a few moments at this time to review the Meeting Room Bill-of-Materials Spreadsheet on the next page.

Note. The following spreadsheet is based on the Meeting Room Blueprint Figure 4–1, and the symbol graphic details as shown in Figures 4–2 and 4–3 (pages 58 and 59).

4.05 PRICING AND LABORING (STEP 4)

Pricing and Laboring Material

Once you have determined the estimated bill-of-material, transfer the items and their quantities to the price/labor worksheet.

Computer Assisted – Pricing and laboring is automatically completed by the computer.

Pricing

Pricing consists of locating the material cost for each item in the manufacturers price catalog and penciling this cost to the price/labor sheet. To save time, locate the cost for all of the material items before you perform the cost extensions. You must be sure you indicate the unit of measure that the cost reflects, such as *E* for each, *C* for one hundred, *M* for one thousand. If there is no cost required for a particular material item, just draw a line through the cost field. See the Pricing And Laboring Worksheet 4–1 on page 60 for a sample of pricing.

Meeting Room Bill-Of-Material Spreadsheet 1 of 1														
		Boxes	EMT	Rings		Devices			Plates			
Description	Qnty.	4" x 4"	Conn.	Sq. Rnd.	1 Gang	2 Gang	Switch	3-Way	Recpt.	IG Recpt.	Switch	2G Switch	Recpt.	Blank
Fixtures														
Lay-in Flourescent Fixture	18	9	18	*	*	*	*	*	*	*	*	*	*	9
Ceiling Mounted Fixture	4	4	8	4	*	*	*	*	*	*	*	*	*	*
Recessed Fixture	20	*	40	*	*	*	*	*	*	*	*	*	*	*
Exit Fixture	2	2	4	2	*	*	*	*	*	*	*	*	*	*
Switches														
Switch - 1 Pole	1	1	2	*	1	*	1	*	*	*	1	*	*	*
Switch - 3 Way	1	1	2	*	1	*	1	1	*	*	1	*	*	*
Switch & Switch	1	1	2	*	*	1	2	*	*	*	*	1	*	*
Switch & 3 Way	1	1	2	*	*	1	1	1	*	*	*	1	*	*
Receptacles														
Convenience Receptacles	9	9	18	*	9	*	*	*	9	*	*	*	9	*
IG Receptacles	2	2	4	*	2	*	*	*	*	2	*	*	2	*
Totals	59	30	100	6	13	2	5	2	9	2	2	2	11	9

Laboring

Laboring consists of locating labor-units that are associated with material items and penciling these values to the price/labor worksheet. You can save time by locating the labor-units for all of the material items before you perform the labor-hour extensions. If there is no labor required for a particular material item, just draw a line through the labor-hour field. Be sure you are aware of the unit of measure that the labor-hour reflects such as *E*, *C*, or *M*. See the Pricing And Laboring Worksheet 4–1 for a sample of laboring.

4.06 EXTENSIONS AND TOTALS (STEP 5)

Once you have penciled in the material cost and the labor-hour value for each item, you can determine the material cost and labor-hour extension for each item. Then you can determine the totals for the worksheet.

Computer Assisted – Extending and totaling is automatically completed by the computer.

Material Cost Extension

The material cost extension for each item is determined by dividing the material cost by the unit, then multiplying this value by the quantity of material items.

Example. The total cost of thirty metal boxes that cost $59.00 per hundred boxes is calculated according to the following:

Cost per hundred boxes = $59.00

Cost for each box = $59.00/100 = $0.59

Cost of 30 boxes = $0.59 × 30 boxes

Cost of 30 boxes = $17.70

See the Pricing And Laboring Worksheet 4–2 on page 61 for a sample of material extensions.

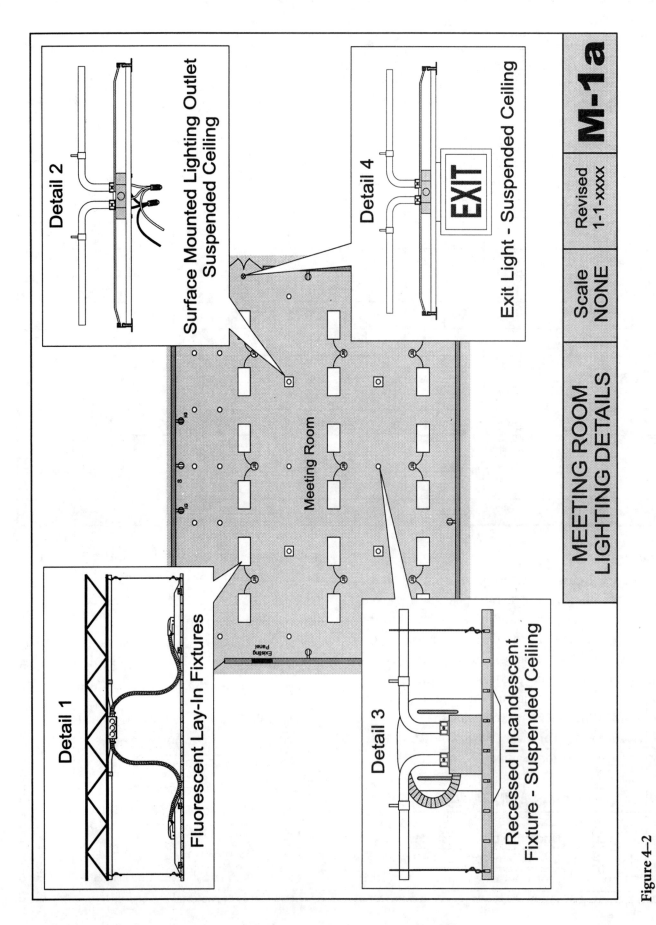

Figure 4–2
Meeting Room Blueprint Lighting Details M–1a

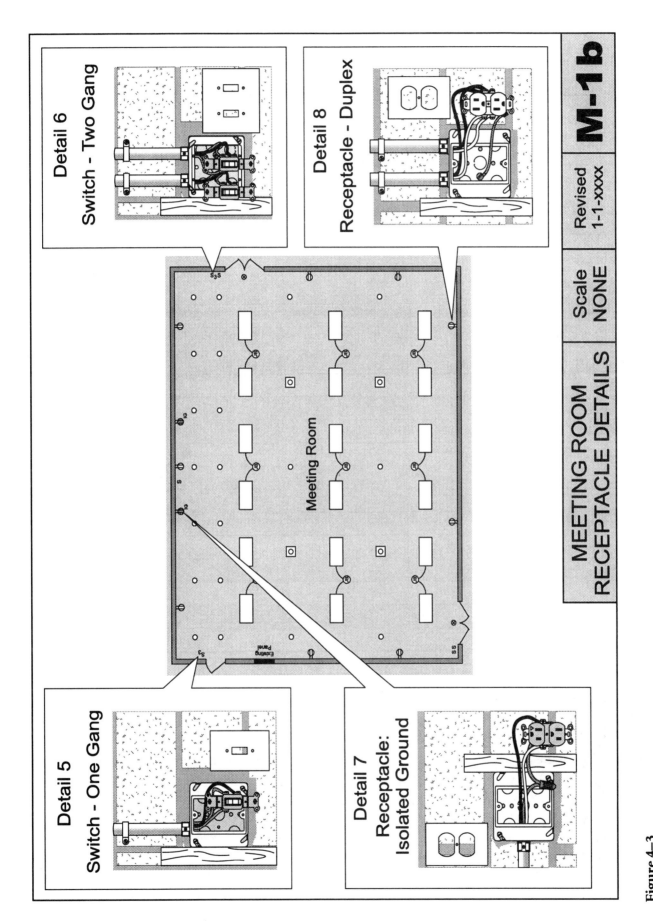

Figure 4–3
Meeting Room Blueprint Receptacle Details M–1b

PRICING AND LABORING WORKSHEET 4-1							
Meeting Room Pricing And Laboring Sheet Page 1 of 1							
Description	Quantity	Cost	U	Total Cost	Labor	U	Total Hours
Boxes							
Boxes 4" X 4" Metal	30	$59.00	C		18.00	C	
Rings							
Square Round	6	$64.00	C		4.50	C	
1 Gang Device	13	$39.00	C		4.50	C	
2 Gang Device	2	$52.00	C		5.00	C	
Switches							
Switch 1 Pole 277 Volt	5	$258.00	C		20.00	C	
Switch 3 Way 277 Volt	2	$375.00	C		25.00	C	
Receptacles							
Receptacle 20 Amp 125 Volt	9	$180.00	C		19.00	C	
Receptacle IG 20 Amp 125 Volt	2	$11.00	E		0.25	E	
Plates							
Switch 1 Gang	2	$47.00	C		2.50	C	
Switch 2 Gang	2	$81.00	C		4.00	C	
Receptacle Duplex	11	$47.00	C		2.50	C	
Blank 4" X 4" Metal	9	$99.00	C		6.00	C	
Raceways							
1/2" EMT	891	$13.00	C		2.25	C	
1/2" EMT Connector	100	$21.00	C		2.00	C	
1/2" EMT Coupling	89	$24.00	C		2.00	C	
Wire							
No. 12 THHN	2,393	$48.00	M		4.25	M	
Fixture							
Flourescent Lay-in	18	$50.00	E		0.75	E	
Ceiling Mount (By Owner)	4	*	*		0.50	E	
Recessed	20	$25.00	E		1.00	E	
Exit	2	$25.00	E		0.25	E	

Labor-Hour Extension

The labor-hour extension for each item is calculated by dividing the labor-hours by the unit, then multiplying this value by the quantity of the material items.

Example. The total labor of thirty metal boxes, labor-unit 18.00 per hundred boxes is equal to:

Labor per hundred boxes = 18 hours

Labor each box = 18/100 boxes, = .18 hour

Labor for 30 boxes = .18 hours × 30

Labor for 30 boxes = 5.40 hours

See the Pricing And Laboring Worksheet 4–2 for a sample of labor-hour extensions.

Pricing And Laboring Worksheet 4-2							
Meeting Room Pricing And Laboring Sheet Page 1 of 1							
Description	Quantity	Cost	U	Total Cost	Labor	U	Total Hours
Boxes							
Boxes 4" X 4" Metal	30	$59.00	C	$17.70	18.00	C	5.40
Rings							
Square Round	6	$64.00	C	$3.84	4.50	C	0.27
1 Gang Device	13	$39.00	C	$5.07	4.50	C	0.59
2 Gang Device	2	$52.00	C	$1.04	5.00	C	0.10
Switches							
Switch 1 Pole 277 Volt	5	$258.00	C	$12.90	20.00	C	1.00
Switch 3 Way 277 Volt	2	$375.00	C	$7.50	25.00	C	0.50
Receptacles							
Receptacle 20 Amp 125 Volt	9	$180.00	C	$16.20	19.00	C	1.71
Receptacle IG 20 Amp 125 Volt	2	$11.00	E	$22.00	0.25	E	0.50
Plates							
Switch 1 Gang	2	$47.00	C	$0.94	2.50	C	0.05
Switch 2 Gang	2	$81.00	C	$1.62	4.00	C	0.08
Receptacle Duplex	11	$47.00	C	$5.17	2.50	C	0.28
Blank 4" X 4" Metal	9	$99.00	C	$8.91	6.00	C	0.54
Raceways							
1/2" EMT	891	$13.00	C	$115.83	2.25	C	20.05
1/2" EMT Connector	100	$21.00	C	$21.00	2.00	C	2.00
1/2" EMT Coupling	89	$24.00	C	$21.36	2.00	C	1.78
Wire							
No. 12 THHN	2,393	$48.00	M	$114.86	4.25	M	10.17
Fixture							
Flourescent Lay-in	18	$50.00	E	$900.00	0.75	E	13.50
Ceiling Mount (By Owner)	4	*	*	*	0.50	E	2.00
Recessed	20	$25.00	E	$500.00	1.00	E	20.00
Exit	2	$25.00	E	$50.00	0.25	E	0.50
Totals				$1,825.94			81.02

Totals

Once you have extended the material cost and the labor-hours for all items, you can proceed to determine the total material cost and the total labor-hours for each worksheet.

Important. You must have someone double check your extensions and totals for each page. By having the cost and labor-hours checked twice, you gain confidence that your material cost and your labor-hours are probably correct. See Worksheet 4-2 for a sample of worksheet totals.

SUMMARY

Introduction

The estimate consists of determining the estimated material cost and labor man-hours for the job. To properly estimate a job, you must have the ability to mentally visualize the mechanical requirements and material required to complete the job. The estimating process, in its most basic form, consists of:

Step 1. Understanding what is required to be completed within the contract price.

Step 2. Counting and measuring symbols.

Step 3. Determining a material required for pricing and laboring.

Step 4. Pricing and laboring the material.

Step 5. Extending and totaling the cost of material and labor-hours.

4.01 Job Selection

Before you estimate any job, ask yourself, "Is this the most efficient use of my time, talents, and skills, at this time?" You must become selective about which jobs you feel will better meet your financial goals. Only consider jobs in which you have a reasonable chance of making the profit margin you want. Before you bid the job, what does your gut tell you? What are the chances of getting the job? Do you think the job will be profitable? Do you have the money to finance the job?

4.02 Understanding The Scope Of Work (Step 1)

Before you begin the estimate, you must be sure you have as much of the information about the job as possible. You must have a complete and current set of blueprints and specifications. Be sure to visit the job site before you bid the job. You need to create a Job Folder to contain your bid's paperwork. Create and complete an Estimate Record Worksheet and hang it up on the wall over your take-off bench for useful reference.

Before you get any deeper, you need to take some time in a quiet place to carefully read the specifications and all notes on the blueprints. Keep a note pad handy so you can keep track of any questions you have about the estimate. If there is anything you don't understand, get the answer as soon as possible, don't wait until the last moment.

In an effort to keep track of the plan and specification details, you will want to complete the Specification Check List located in the Appendix, and hang it up over the take-off bench for quick reference while you do the estimate.

4.03 The Take-Off (Step 2)

To quickly and accurately estimate a job, you must use proper estimating forms and worksheets. Proper forms and worksheets will save you time, create consistency, and assist in the reduction of errors. The forms and worksheets should also help serve as a reminder of items that are easily omitted or forgotten.

A take-off is the action of counting and measuring the symbols. A proper take-off should insure that once the take-off is finished, there will be very little need to refer to the drawings or specifications again. To accomplish this, you must follow an orderly, organized, methodical routine that is complete and consistent for each and every job. Use colored pencils or pens to identify each item that you have taken-off. When you have finished the take-off, your blueprints should be a colored representation of the electrical work required to be installed.

Counting Symbols

The purpose of counting symbols is to identify their quantity so that you can determine the job's bill-of-material.

Measuring Circuits

Measuring consists of determining the circuit length for branch circuits, feeders, and service raceways. Verify the architectural scale listed in the blueprints before you measure any circuit.

There is no set sequence for performing the take-off, but you will develop a system that fits your personal style and needs. Your style will need to be adjusted for new types of construction or jobs.

4.04 Determining The Bill-Of-Material (Step 3)

Once you have completed the take-off, you need to determine the bill-of-material. The manual take-off should provide enough information to determine the type, size, and quantity of all material items required to complete the installation.

Computer Assisted – Bill-of-material is automatically completed by the computer.

4.05 Pricing And Laboring (Step 4)

Once you have determined the estimated bill-of-material, transfer the items and their quantities to the price/labor worksheet. Pricing consists of locating the material cost for each item in the manufacturer's price catalog and penciling this cost to the price/labor sheet. Laboring consists of locating labor-units that are associated with material items and penciling these values to the price/labor worksheet.

Computer Assisted – Pricing and laboring is automatically completed by the computer.

4.06 Extensions And Totals (Step 5)

Once you have penciled in the material cost and the labor-hour value for each item, you can determine the material cost and labor-hour extension for each item. Once you have extended the material cost and the labor-hours for all items, you can proceed to determine the total material cost and the total labor-hours for each worksheet.

Computer Assisted – Extensions and totaling is automatically completed by the computer.

Important. Have someone double check your extensions and totals for each page.

Chapter 4

Review Questions

Essay Questions

1. In its most basic form, what does the estimating process consist of?

2. Because of limited resources such as energy, time, and money, you can't estimate all jobs. What are some of the factors that you should consider before you decide which jobs to bid?

3. In deciding whether to bid a job or not, what are some of the most important questions you should ask yourself before you begin the estimate?

4. What should you do if you don't want to bid the job?

5. What is the purpose of the estimate Job Folder?

6. What do you do with the Estimate Record Worksheet once it is completed?

7. What is the purpose of the Specification Check List and what's it used for?

8. What is the purpose of estimating forms and worksheets?

9. Explain what a take-off consists of.

10. What do you do if the blueprints can't be marked-up?

11. What is the purpose of counting symbols?

12. Why should you verify the architectural scale before you measure the circuit and feeders?

13. What is the proper Take-Off sequence?

14. If you are estimating manually, what do you do once you have determined the Bill-Of-Material?

Multiple Choice Questions

1. Bid only jobs that interest you and offer you the greatest _____.
   (a) challenge   (b) profit margin   (c) flexibility   (d) all of these

2. Be sure you understand the _____ of the bid and be careful to read the specifications and blueprint notes carefully.
   (a) language   (b) layout   (c) scope   (d) none of these

3. A note pad is useful to list _____ as you proceed through the estimate.
   (a) questions   (b) answers   (c) both a and b   (d) neither a nor b

4. The take-off must be quick and detailed enough so it can be used to determine the project's _____ for pricing and laboring.
   (a) bill-of-material   (b) estimated man-hours
   (c) overhead   (d) profit

5. Pricing consists of locating the material _____ for each item in the manufacturer's price catalog, and transferring this value to the pricing worksheet.
   (a) quantity   (b) cost   (c) markup   (d) discount

6. When completing the laboring portion of your estimate, be sure that you indicate the _____ the labor-hour reflects.
   (a) unit E, C or M   (b) adjustment   (c) material item   (d) all of these

7. When you know the material cost and labor-hour extension for each item, you can determine the totals for _____.
   (a) the owner   (b) each worksheet   (c) the quote sheets   (d) none of these

Chapter 5

The Estimate Summary And Bid Process

OBJECTIVES

After reading this chapter, you should understand

Part A – The Estimate Summary

- how to adjust labor for job site conditions.
- how to determine total labor cost, including labor burden (fringes).
- how to determine material cost, including miscellaneous items, waste, and theft.
- how to get the best fixtures and switchgear prices.
- how to account for direct job expenses such as small tools, equipment rental, etc.
- how to properly apply overhead to the estimate.
- how to determine the job's break even cost.

Part B – The Bid Process

- how to apply profit and other final cost to determine the bid price.
- how to verify that common estimating mistakes were not made.
- how to analyze the bid's accuracy.
- how to produce a bid proposal.

Part C – Unit Pricing

- how to determine unit pricing

INTRODUCTION

As explained in Chapter 4, determining the selling price of a job consists of three parts, the estimate, the estimate summary and the bid process. Chapter 4 contained the details of the estimate such as the take-off, determining the bill-of-material, pricing, laboring, extending and totaling. This chapter contains the details of the estimate summary in Part A, and the bid process in Part B and unit pricing in Part C. The estimate summary is used to determine the project's break even cost. The bid process consists of applying profit and other costs to determine the selling price. Once you have determined the selling price, you need to verify that the bid is valid and then you must create a job proposal. Unit pricing contains a few additional items (variables) which, when included in your bid, make it just that much more precise. With this extra information you are more likely to succeed at clearing a reasonable profit—if you get the job.

PART A – THE ESTIMATE SUMMARY

Summarizing the estimate to determine the break even cost can be overwhelming. One simple mistake at this point can be costly. This phase of the estimating process requires you to make judgments on intangibles such as job conditions, labor productivity, miscellaneous material requirements, waste, theft, small tools, direct job expenses, and overhead.

Estimate Summary

The first step is to transfer the totals for material cost and labor hours to the Summary Worksheet which is used to help organize the different cost so you don't make a mistake. The summary worksheet contains eight major sections, they include:

Step A – Labor-Hours
Step B – Labor Cost
Step C – Material Cost
Step D – Direct Job Cost
Step E – Estimated Prime Cost
Step F – Overhead
Step G – Estimated Cost

Review the Summary Worksheet below, for the Meeting Room Blueprint M–1. The details of summary worksheet, Steps A through G, are contained on the pages that follow the worksheet.

	SUMMARY WORKSHEET	
	Labor Summary	Hours
	Labor-Hours (Price/Labor Worksheet)	81.02
	Labor-Hour Adjustment - Miscellaneous, + 8%	6.48
	Total Adjusted Labor-Hours	87.50
	Additional Labor, + 9.50 Hours	9.50
Step A	Total Adjusted Labor	97.00
Step B	Labor Cost at $10.65 per labor-hour	$1,033.05
	Material Cost Summary	
	Material Cost (Pricing Worksheet)	$1,825.94
	Miscellaneous Material Items, + 10%	$182.59
	Waste And Theft, + 5%	$91.30
	Small Tools, +3%	$54.78
	Quote - Gear	$0.00
	Total Taxable Materials	$2,154.61
	Sales Tax, + 7%	$150.82
Step C	Total Material Cost	$2,305.43
Step D	Total Direct Job Cost	$100.00
Step E	Estimated Prime Cost ($1,033 + $2,305 + $100)	$3,438.00
Step F	Overhead at $13.00 Per Man-Hour	$1,261.00
Step G	Estimated Break Even Cost	$4,699.00
	Profit, + 15%	$704.85
	Bid Price	$5,403.85

5.01 LABOR-HOURS (STEP A)

Total Labor Cost – Meeting Room	
Labor-Hours (Price/Labor Worksheet)	81.02
Labor Hour Adjustment, + 8%	6.48
Total Adjusted Labor-Hours	87.50
Additional Labor	9.50
Total Adjusted Labor	**97.00**

Labor-Unit Adjustment

The first step of the estimate summary is to transfer the labor-hours and material cost from the price/labor worksheets to the estimate summary worksheet. The estimated labor-hours must be adjusted for working conditions based on a percentage of total man-hours or a fixed man-hour for specific conditions.

Determining the labor adjustment comes with experience and is part of what makes estimating an art form. With experience from past performances, the adjustment of labor man-hours for job conditions can be quite accurate; however, at times it is simply an educated guess. The following factors must be considered for every job that you estimate. For details on these factors see Chapter 3, Section 3.09.

1. Building conditions
2. Change orders
3. Concealed and exposed wiring
4. Construction schedule
5. Job factors
6. Labor skill (experience)
7. Ladder and scaffold
8. Management
9. Material
10. Off hours and occupied premises
11. Remodel (old work)
12. Overtime
13. Repetitive factor
14. Restrictive working conditions
15. Shift work
16. Teamwork
17. Temperature
18. Weather and humidity

Additional Labor

After you have adjusted the total labor-unit hours, you must consider if there are any additional labor requirements for the job. These conditions are not adjustments of the labor-unit, but additional labor required:

1. As-built plans
2. Demolition
3. Energized parts
4. Environmental hazardous material
5. Excavation, trenching, and backfill
6. Fire seals
7. Job location
8. Match-up of existing equipment
9. Miscellaneous material items
10. Mobilization (Startup)
11. Nonproductive labor
12. OSHA compliance
13. Plans and specifications
14. Public safety
15. Security
16. Shop time
17. Site conditions
18. Sub-contract supervision
19. Temporary, emergency, and stand-by power

1. As-Built Plans

As-built plans are intended to show in accurate detail the actual location of all feeders, branch circuits, and the size of equipment. Be sure to include the labor to create and maintain as-built plans.

> **Note.** Be sure you include the cost for revising as-builts when you invoice for change orders.

2. Demolition

Some jobs require you to remove the old electrical wiring before you begin adding anything new. The labor for demolition must be considered, and at times it's just an educated guess. However, with experience you'll get a feel as to what's reasonable to apply.

3. Energized Parts

Working on energized systems requires proper equipment and employees who have been trained and certified by OSHA to work "hot." This type of work requires special precautions to insure personal safety and will take more time to complete. How much more time? It's anybody's guess.

Author's Comment. The dangers of working on live energized electrical parts far exceed the potential profits.

4. Environmental Hazardous Material

Be sure your bid includes the labor required to handle environmental hazardous material. This includes preparation, packaging, shipping and proper disposal of ballasts, electric discharge lamps, radioactive exit signs, etc. Better yet, subcontract it to a company that specializes in this work.

5. Excavation, Trenching, And Backfill

There are places in the country where you can't plant a bush without a pick. Conversely, if you try to dig a trench in sugar sand, the more you dig the wider the trench gets, but it doesn't get any deeper. Poor soil conditions can make a simple job turn into a career. Don't forget about the requirements of core drilling, asphalt cutting, digging and back-filling. Often it's more cost effective to subcontract this type of work, rather than take the entire responsibility of cutting fiberoptic cable, telephone wires, underground high voltage utility lines, sprinkler, water, or a gas main.

6. Fire Seals

The National Electrical Code, in Section 300–21, requires a fire seal whenever you penetrate a fire rated wall, ceiling or floor. Be sure to include the labor for this condition.

7. Job Location

If the job is located in an area that requires travel time, it will be more difficult to manage the job and get the material to the job when needed. Some projects do not have water, electric power, toilets, or telephone service. This can make life more complicated and expensive, and these jobs will take longer to complete.

Travel Time

The following example should help you understand how to determine travel time.

> **Example.** What is the total travel time required for a 212 hour job that has three workers? The travel time for each day is 1/2 hour per worker.
>
> **Answer.** 212 hours/24 hours per day (3 workers) = 8.84 or 9 travel days. The travel time = 1/2 hr./day per worker. Nine travel days $\times$ 1/2 hr/day $\times$ 3 workers = 13.5 hours.

8. Match-Up Of Existing Equipment

Maybe you have a situation where you are required to match existing equipment, colors, or fixtures. This can become a time consuming adventure.

9. Miscellaneous Material Items

It is impractical to determine all labor required for the job. You must either calculate the expected miscellaneous labor required when you determine a bill-of-material, or add a percentage to the total labor hours for this factor.

Manual Estimate – When you estimate manually, increase the total labor hours by 8% for miscellaneous items not counted or measured.

Computer Assisted Estimate – There is no need to make any labor hour adjustment for miscellaneous material items.

10. Mobilization (Startup)

Don't forget to include the labor required to set up the job such as getting the job trailer prepared or installing temporary wiring.

11. Nonproductive Labor

Labor-units only include 5% for nonproductive labor. Does the job have the potential for excessive nonproductive time?

Breaks – When you're on a job for a while, your employees get to know the other trades and breaks tend to get longer and more frequent.

Distracting Job Conditions – Is this job going to be on a beach or other location with a high labor distraction factor?

Case Study – In Daytona Beach, electrical contractors increase the labor-hours by 10% for work performed during spring break.

Inspection Tours – Inspection tours are a fact of life and the larger the job, the more frequent and longer the tours. Sometimes projects have multiple inspectors for the different systems, often by different inspection agencies.

12. OSHA Compliance

Are you required to pay your electricians to attend a training and certification program? OSHA at times requires training and/or certification for emergency medical, working in confined spaces, working on energized parts, or tool and equipment handling.

13. Plans And Specifications

If adequate blueprints and specifications are not provided, you'll need to add a factor to account for anticipated nonproductive time to figure out what's required. Labor-units assume that you have clear and conflict-free blueprints and specifications. If this is not the case, you need to inform the owner that your estimate includes additional labor as a contingency.

14. Public Safety

Public safety is a factor, especially when doing work for city, county, state, or federal government agencies. Are you required to install safety cones, barricades, or security gates? Be sure to read the specifications closely.

15. Security

When working in some governmental and private facilities, you are required to jump through hoops to get clearance for your employees to enter the premises. Some facilities require security be notified well in advance of persons desiring entry in the premises. This requires proper supervision to insure that the workers are not waiting at the gate too long.

16. Shop Time

Sometimes work slows down and we keep our key employees busy. You'll have them clean out the shop, wash the trucks, etc. You'll have them do anything, just to keep them busy. Naturally you can't charge this nonproductive labor cost to a specific job, but you should consider the effects on net profit.

17. Site Conditions

Because of traffic conditions, projects in downtown areas of large cities can cause significant lost time. Traffic conditions and narrow streets make it difficult to unload material and equipment. Parking and a lack of storage space also is a problem. Add additional labor to cover this condition.

18. Sub-Contract Supervision

Don't forget to include the labor required for your electricians in the field to supervise and direct the sub-contractors.

19. Temporary, Stand-By, And Emergency Power

Labor required to insure that temporary, stand-by or emergency power is available and safe must be included in your bid.

5.02 LABOR COST (STEP B)

The estimated labor cost for a job is determined by multiplying the total adjusted labor man-hours by the labor rate per man-hour. The labor rate per man-hour can be determined by one of two methods: the shop average or the job average.

Total labor cost = 97.00 hours × $10.65
Total labor cost = $1,033.05

Author's Comment. The labor cost per man-hour is significantly different in different parts of the country. In some areas a journeyman electrician is paid less than $10 per hour and in other areas, the rate is over $35 per hour.

Note. For out of town and shift work many workers do not like to be separated from their families so you might pay a premium wage to overcome family pressure. Take this into consideration when you determine your labor rate.

Shop Average Labor Rate

The shop labor rate is determined by dividing the total field labor cost by the total number of field man-hours for the last six months. The average shop labor rate might be too low for jobs that require greater skill such as control wiring, or too high for simple jobs that require less skill such as residential wiring.

The following is an example of the shop average labor rate calculations.

Shop Average Labor Rate Calculation			
Month	Labor Cost	Hours	Rate/Hr
April	$14,200	1,166	$12.18
May	$17,750	1,666	$10.65
June	$21,300	2,167	$9.83
July	$14,200	1,250	$11.36
August	$17,750	1,584	$11.21
Sept.	$21,300	2,167	$9.83
Six Months	$106,500	10,000	$10.65

This example does not include fringes.

Job Average Labor Rate

Another approach is the job labor rate. This requires you to consider how you plan on manning the job. For an industrial job you will probably pay the electricians more than the electricians on a housing project. For prevailing wage jobs, be sure to use the rate as required in the specifications.

Note. The shop and job average labor rate for some contractors includes fringes and benefits and for other contractors it does not.

The following demonstrates an example of calculating the job average labor rate.

Job Average Labor Rate Calculation			
Crew	Rate	# Persons	Extension
Foreman	$14.00	1 Person	$14.00
Journeyman	$12.25	1 Person	$12.25
Apprentice	$9.00	3 Persons	$27.00
Total		5 Persons	$53.25

Weighted Rate-Per-Hour $53.25/5 persons = $10.65

This example does not include fringes.

5.03 LABOR BURDEN

You must include the cost of labor benefits, taxes, and labor insurance in the estimate. Labor burden cost typically represents 38% of total labor cost. In general, NECA contractors include these costs as part of their labor rate, and nonunion contractors recover labor burden costs when overhead is applied. This is explained in Sections 5.07 through 5.09 of this chapter.

5.04 TOTAL MATERIAL COST (STEP C)

The first step in determining the total material cost is to transfer the material cost from the pricing worksheets to the summary worksheet. Once this is accomplished, you need to make the necessary adjustments for:

1. Miscellaneous material
2. Waste and theft
3. Small tools
4. Lighting fixture and switchgear quotes
5. Sales tax

1. Miscellaneous Material

It is impractical to determine every material item required for the job. You must either calculate the expected miscellaneous items required when you determine a bill-of-material, or add a percentage to the total cost of material for this factor.

Manual Estimate – When you estimate manually, increase the total material cost by 10% for small miscellaneous items not counted or measured.

Computer Assisted Estimate – A 2% to 3% adjustment of material cost should be sufficient for miscellaneous material items not counted or measured.

2. Waste And Theft

In addition, you must be sensitive to the impact of the workers wasting material and theft on the job. This occurs when the electrician drops wire nuts, boxes or just leaves material in the work area overnight, or puts it in his own tool box. What's a reasonable factor to include in the estimate for waste or theft? The more organized and efficient you are at managing your material, the lower this factor will become, but a reasonable factor would be 5%.

3. Small Tools

Cost for ladders, cords, cordless drills, screw guns, and countless other small tools must be included in the estimate. A factor of 3% of the total material cost is considered an acceptable value.

4. Lighting Fixture And Switchgear Quotes

Equipment manufacturers, in an attempt to keep the greatest profit margins, will hold prices until the last moment before the bid, so their competitors won't have a chance to undercut them. You need to have the estimate summary worksheet completed ahead of time. This way, when the last minute price quotes arrive, you can determine the bid price.

Note. Be careful you don't make a mistake when you transfer the supplier prices to the summary worksheet. Insist on written quotes and don't accept verbal quotes.

Substitutions

If you plan on substituting equipment of equal value, be sure to submit cut sheets with your bid proposal.

Author's Comment. Material and quotes cost should not be marked-up. The cost for handling material is part of the labor-unit for field labor and part of overhead for management. These costs should not be marked up for profit, because profit has nothing to do with determining the true cost of the job.

5. Sales Tax

Sales tax varies from state to state, county to county, and also from city to city. Be sure you are familiar with the sales tax rules in the area of the bid and check with your accountant on how to handle tax exempt jobs. Careful, some states apply sales tax to labor cost.

The following example demonstrates the application of determining total material cost for the Meeting Room, Blueprint M–1.

Sales Tax, Meeting Room	
Material Cost (Pricing Worksheet)	$1,825.94
Miscellaneous Material Items, + 10%	$182.59
Waste And Theft, + 5%	$91.30
Small Tools, + 3%	$54.78
Quote - Gear	$0.00
Total Taxable Materials	$2,154.61
Sales Tax, + 7%	$150.82
Total Material Cost	**$2,305.43**

5.05 DIRECT JOB EXPENSES (STEP D)

Direct job expenses are expenses often not shown on the blueprints, but probably indicated in the specifications. The failure to consider direct job expenses result in these costs being absorbed out of profits. Consider the following expenses and be sure you have included their cost in the estimate.

1. As-built plans
2. Business and occupational fees (B&O)
3. Engineering/working drawings
4. Equipment rental
5. Field office
6. Fire seals
7. Guarantee
8. Insurance – special
9. Miscellaneous material items
10. Mobilization
11. OSHA compliance
12. Out of town expenses
13. Parking fees
14. Permits and inspection fees
15. Public safety
16. Recycling fees
17. Storage/storage handling
18. Sub-contract expenses
19. Supervision cost
20. Temporary wiring
21. Testing and certification fees
22. Trash disposal
23. Utility charges and fees

1. As-Built Plans

As-built plans are intended to show in accurate detail the actual location of all feeders, branch circuits, and the size of equipment.

> **Note.** Be sure you include the cost for revising as-builts when you invoice for change orders.

2. Business And Occupational Fees (B&O)

In some areas, you are required to purchase an additional business license or pay an occupational fee. Be sure your bid includes this often overlooked expense.

3. Engineering/Working Drawings

If you intend to make changes to the job's design, particularly for value engineered jobs, you will need to submit revised drawings. In addition, you might need to have those drawings sealed by an architect or engineer. Careful, this cost might also occur with change orders.

4. Equipment Rental

Many times it's more cost effective to rent equipment with an operator than it is to purchase the equipment outright. However, if you own the equipment, be sure you determine a fair value for its use and apply this value to the estimate.

> **Note.** Check with your accountant to determine the equipment value for estimates.

5. Field Office

All of the related expenses of the job office must be calculated and applied to the job. These expenses include trailer rental, office equipment, office personnel, trucks, gas and oil, phone, rent, etc.

6. Fire Seals

The National Electrical Code requires a fire seal whenever you penetrate a fire rated wall, ceiling or floor. Be sure to include this cost in the estimate.

7. Guarantee

All electrical installations require a warranty of the electrical system for a period of at least one year. At other times, there can be a political obligation for a longer period (special customer). Be careful, some states require guarantees of longer duration for condominiums.

8. Insurance—Special

Do the specifications indicate that the job requires fire insurance to cover the value of material in place? Be sure to include this expense.

9. Miscellaneous Material Items

The cost for common miscellaneous items such as phase tape, wire nuts, straps, spray paint, etc. has already been accounted for. However, some estimates have additional direct cost. Be sure to include these costs.

10. Mobilization

Some jobs require expenses to get the job started such as trailers, tools, etc.

11. OSHA Compliance

Can you think of any cost required to comply with OSHA regulations? What about safety rails, straps, or ladder tie downs? What about the cost of training and certification classes?

12. Out Of Town Expenses

When you are bidding a job that is out of your local area, you must consider travel expenses, toll, gas, lodging, meals, and long distance phone calls.

13. Parking Fees

If you're bidding a job in a large city, the cost of parking per day can add up.

14. Permits And Inspection Fees

Expenses such as permits and special inspections can range from a few dollars to many thousands of dollars. The cost of a permit for a 3,500 square foot house can range from $150 – $1,500. Remember to include these costs.

> Note. If possible, don't include permit costs in your bid but indicate this in your job proposal. If possible, have the owner/contractor responsible for permit fees.

15. Public Safety

Public safety is a factor, especially when doing work for city, county, state, or for federal government agencies. Are you required to purchase or rent safety cones, barricades, or security gates?

16. Recycling Fees

Federal and state laws often require the proper disposal of environmentally hazardous products. The following costs assume that you have prepared, packed and shipped these items to the recycling contractor.

Fluorescent Lamps – 4 Foot
1–999 . . . . $0.75 each
1,000 – 4,999 . . . . $0.55 each
5,000 – 29,999 . . . . $0.39 each
30,000 + ... $0.32 each

Ballasts
With PCB . . . . $0.65 each pound
Without PCB . . . . $0.25 each pound

17. Storage/Storage Handling

Storage space on the job site, or additional warehouse space, may be required for bulk material purchases such as fixtures, poles, switchgears, generators, etc.

> *Author's Comment.* If the job turns bad, you might not be permitted back on the job site to remove your material, tools or equipment. Be careful, if the job appears to be going sour, get your equipment off the job to a safe location.

> *Case Study* – I know of many cases where the owners did not pay the electrical contractor and kept the contractors' tools, equipment, fixtures, gear, and generator. By the time the contractors received a court order permitting them to get on the job, their equipment was gone. The owner claimed that he had no knowledge of the material and/or tools' whereabouts.

18. Sub-Contract Expenses

From time to time, it is more efficient to sub-contract a portion of the job such as fire alarm, control, security, rigging, setting of poles, digging trenches, core drilling, etc.

19. Supervision Cost

Some jobs by law or agreement require a non-working foreman or project manager. If the non-working labor is not productive, you must determine the cost required and include this cost in the estimate.

Example. Your union agreement requires a nonproductive project manager whenever there are more than twelve workers on the job site. The job is expected to require more than twelve workers for 60 days. Supervision cost is at $32.42 per man-hour, including all related expenses.

Supervision Cost – $32.42 × 8 hours × 60 days = $15,562

A project manager who oversees multiple jobs is generally considered overhead and this cost will be recovered when overhead is applied to the job.

20. Temporary Wiring

When including temporary wiring, be sure you are clear in your proposal as to what you intend to provide. Don't agree to vague terms such as "provide temporary wiring as needed." Be specific in your proposal: indicate the number of poles, lights, receptacles, switches, and any other item. If your bid does not include maintenance and repairs, or utility connection charges, be sure to say so in your proposal. If you can, just include a fixed dollar allowance in your bid for temporary wiring.

Case Study – The contract specified that he would supply all temporary power. The electrical contractor failed to visit the job site and assumed a basic temporary pole in the estimate.

Result – No electric utility power was available within two miles of the job and the electrical contractor had to supply four generators for one year including fuel.

21. Testing And Certification Fees

Does the job specifications require the electrical system be tested and/or certified, such as testing and certifiying fiber optic cables? If it does, be sure your bid includes these costs.

22. Trash Disposal

Must trash be removed from the site? Can you throw it in the dumpster? If you throw it in the dumpster, will you be charged by the general contractor? Make sure your contract is clear on this subject.

23. Utility Charges And Fees

Who is responsible for paying the electric deposit, monthly electric charge, or impact fees?

5.06 ESTIMATED PRIME COST (STEP E)

Now you must determine the total prime cost, which is simply adding up the cost of labor, material, plus direct job expenses.

The following example demonstrates the calculation of estimated prime cost for the Meeting Room.

Estimated Prime Cost for Meeting Room	
Labor Cost - 97 Hours @ $10.65	$1,033
Material Cost Including Quotes and Tax	$2,305
Direct Costs - Permit	$100
Estimated Prime Cost	**$3,438**

5.07 OVERHEAD

The cost of managing labor is called an overhead expense. Applying overhead expenses to the estimate is an art form. You need to consider the business volume and the job size. Most overhead expenses are fixed and do not vary much with changing business volume. As the business volume decreases, the ratio of overhead cost to labor will increase, and as the business volume increases, the ratio of overhead cost to labor will decrease.

Smaller jobs and/or smaller contractors have a higher ratio of overhead cost to labor as compared to larger jobs or larger contractors. On larger jobs, you may want to reduce the overhead rate, and on smaller jobs you may want to increase your overhead rate.

5.08 OVERHEAD CALCULATION

There are two methods of applying overhead expenses to an estimate:
1. Percentage method
2. Rate-per-hour method

1. Percentage Method

Most contractors apply overhead as a percentage of prime cost anywhere from 15% to 40%. The percentage method works well if you're doing similar jobs, like houses, office buildings, remodel work, etc.

The Wrong Percent

The overhead percent ratio listed in financial statements is in relationship to sales. The proper method of determining overhead is to use a markup value, not the percent listed in financial statements.

Example No.1. Let's say your prime cost is $7,000 and your overhead is 25% of sales. Many electrical contractors would make the following mistake in determining the job's overhead cost.

(1) Determining Overhead Cost Improperly
Prime cost	$7,000	80%
Overhead ($7,000 × 25%)	+ $1,750	+ 20%
Estimated cost =	$8,750	100%

The proper method of applying 25% overhead to a job is to use a markup of 33%.

(2) Determine Overhead Cost Properly
Prime cost	$7,000	75%
Overhead ($7,000 × 33%)	+ $2,310	+ 25%
Estimated cost	$9,310	100%

The following example lists the multiplier of prime cost required to cover a given overhead percent as it relates to sales.

> **Note.** The markup multiplier is the reciprocal of 100% less overhead percent to sales.

Large Quote Cost

There is a danger in using the percent method with a job that has large ticket items such as generators or lighting fixtures.

Overhead Multiplier			
% of Sales	Multiplier	% of Sales	Multiplier
15%	0.18	28%	0.39
16%	0.19	29%	0.41
17%	0.20	30%	0.43
18%	0.22	31%	0.45
19%	0.23	32%	0.47
20%	0.25	33%	0.49
21%	0.27	34%	0.52
22%	0.28	35%	0.54
23%	0.30	36%	0.56
24%	0.32	37%	0.59
25%	0.33	38%	0.61
26%	0.35	39%	0.64
27%	0.37	40%	0.67

Example No. 2. Let's say you are bidding a job to install a generator. You estimate that it will take 100 labor-hours at $15.00 per man-hour to install, the material is estimated at $2,000, and the generator at $50,000. Your overhead markup is 33%.

The owner wants two prices:
(1) A price if the owner supplies the generator.
(2) A turn-key price, where you supply the generator.

(1) Owner Supplies Generator
Labor (100 hours at $15.00)	$1,500
Material	+ $2,000
Estimated prime cost	$3,500
Overhead ($3,500 × 33%)	+ $1,155
Estimated cost =	$4,655

(2) Turn-Key Price
Labor (100 hours at $15)	$1,500
Material	$2,000
Generator	+ $50,000
Estimate prime cost	$53,500
Overhead ($53,500 × 33%)	+ $17,655
Estimated cost =	$71,155

Author's Comment. In the above example, the overhead would be $1,155 if the owner supplies the generator, and $17,655 if the contractor supplies the generator. It's not reasonable to apply almost $16,500 more overhead, just because the electrical contractor bought the generator.

2. Rate-Per-Hour Method

You're probably confused about how to apply overhead using the percentage method. That's okay because I really don't want you to use this method.

Since overhead is related primarily to the management of labor, overhead should be calculated as a rate per man-hour. The rate per man-hour method permits you to recover overhead regardless of the cost of material.

The overhead rate per man-hour is calculated by dividing the overhead dollars for at least the past six months by the field man-hours for the past six months.

The following example demonstrates the calculation of determining the overhead rate-per-hour method.

Overhead Cost Per Hour			
Month	Overhead Cost	Labor Hours	Labor Per Hr.
April	$13,333	1,166	$11.43
May	$16,666	1,666	$10.00
June	$20,000	2,167	$9.23
July	$13,333	1,250	$10.66
August	$16,666	1,584	$10.52
September	$20,000	2,167	$9.23
Six Months	$100,000	10,000	$10.00

5.09 APPLYING OVERHEAD (PART F)

If you are just getting started in business, or if you don't know your overhead cost, then use the lesser of either 40% of prime cost or $13 per man-hour. If you use these factors, do not apply a labor burden factor to labor cost.

The following example demonstrates the application of overhead cost to the Meeting Room Blueprint M–1.

Overhead – Meeting Room	
Labor Cost - 97 hours @ $10.65	$1,033
Material Cost Including Quotes And Tax	$2,305
Direct Job Cost: Permit	$100
Estimated Prime Cost	$3,438
Overhead (97 Hours At $13 Per Man-Hour)	$1,261

Overhead should not exceed:

$3,438 × 40% = $1,376.00, or

97 hours × $13 = $1,261.00

5.10 UNDERSTANDING BREAK EVEN COST (PART G)

Well, you're finally here, at the top, you've made it. You have estimated all of the costs associated to install the electrical wiring for the Meeting Room. All you need to do now is simply add up the costs and you can determine your estimated break even cost.

The following example demonstrates the estimated break even cost for the Meeting Room Blueprint M–1.

Break-Even Cost – Meeting Room	
Labor Cost - 97 hours @ $10.65	$1,033
Material Cost Including Quotes and Tax	$2,305
Direct Job Cost	$100
Estimated Prime Cost	$3,438
Overhead Dollars (@ $13.00 Per Man Hour)	$1,261
Estimated Cost (Break Even)	$4,699

Author's Comment. The argument I hear at this point in class is, "if I consider all of these factors, I won't get the job." Please remember you are only determining what you think the project will cost. If you don't think you can sell the job for what it costs you, you're left with only a couple of options. You either sell the job for less than it cost you, or you reduce your costs by becoming more efficient and effective. I think that the second option will produce better results.

Electrical contractors are not in the business of selling the job at or below cost. Once you determine the estimated break even cost of the job, you need to determine the selling price. This step is called bidding and is covered in the next part of this chapter.

PART B – THE BID PROCESS

This part completes the process of determining the job's selling price, and how to avoid common estimating errors to insure that your bid is valid. Also covered is how to perform proper bid analysis to validate that your bid price is accurate, and how to create a job proposal.

5.11 PROFIT

Net profit is the bottom line, it's your report card. A responsible estimate will include a reasonable amount for profit. You can lose money on a job that has a projected profit margin of 25%, if the estimate doesn't cover all expenses. If you have a poor estimate, a high profit margin will be meaningless. Errors in the estimate, or a poorly managed job, will result in unplanned expenses such as labor overruns, overtime, increased material use, or unanticipated direct costs.

How Much Profit Is Reasonable?

If your estimate is complete and you manage your company efficiently, you don't need a lot of profit to stay in business. A reasonable profit margin would be 15% to 20% of break even cost.

The following example demonstrates the application of profit for the Meeting Room Blueprint M–1.

Bid Price – Meeting Room	
Estimated Prime Cost	$3,438
Overhead at $13.00 per man-hour	$1,261
Estimated Break Even Cost	$4,699
Profit, + 15%	$705
Direct Job Cost	$0
Total Bid Price	$5,404

Factors to consider when applying profit include:

1. Competition and the economy
2. Management
3. Job size
4. Risks

1. Competition and the Economy

Consider the number of contractors bidding the job. When the construction market is in a recession or the market is shrinking, a highly competitive market will develop and competition will drive down the selling price. If the economy is strong or the market is expanding, profits will be permitted to be increased.

The law of supply-and-demand dictates that when you're busy raise your prices, and when you're slow, lower your prices.

2. Management

Profitable contracting demands that the job be sold for more than what it cost. In order to accomplish this goal, you must remember to manage your resources efficiently and effectively to keep your cost down.

Many inexperienced electrical contractors do not operate an efficient or organized company, and profit margins are often very low or nonexistent. Yes, the market does limit the selling price, but your efficiency determines to a great extent your profit margin.

3. Job Size

The generally accepted practice is that a small job will permit a greater profit margin as compared to a larger job. A job with 100 hours or less should have a profit margin of at least 25%, but the profit of a job of over 1,000 hours will seldom be much more than 15%. The reasons for the lower profit margin for larger jobs is competitive pressure.

> **Note.** If you supply expensive equipment, such as generators, UPS systems, etc., don't apply the same profit margin to these items. The profit margin on these items must be very low, such as 2% or 4%.

4. Risks

If you're going to do a job that's unfamiliar, you need to consider increasing your profit margin to cover the risk of estimate and field errors. However, the risk might not be as great for your competitors, causing them to provide a lower bid. Use caution when bidding high risk jobs, or better yet educate yourself to reduce the risk and become more competitive.

5.12 OTHER FINAL COSTS

Be sure you review the specifications and blueprints one final time before you submit your bid. Make sure you understand all the bid conditions. Are there any costs you didn't consider such as:

1. Allowances/Contingency
2. Back-charges
3. Bonds
4. Completion penalty
5. Finance cost
6. Gross receipts or net profit tax
7. Inspector problems
8. Retainage cost

1. Allowances/Contingency

If the specifications require that you provide an allowance, be sure to include this cost in your bid. Contingency might include copper wire price increases.

2. Back-Charges

Back-charges are deductions from your payments to cover an expense incurred by another contractor. Do you think that this job will have back-charges?

3. Bonds

There are many different types of bonds that might be required by the specification such as the bid bond. A bid bond guarantees the owner that the project will be completed on time, free of liens according to the contract's conditions. The cost of a bid bond is typically between 1.5% and 3% of the total selling price of the job.

4. Completion Penalty

Some jobs charge a penalty per-day, such as $100 to $500, for each day past the schedule completion date. Think about this cost when you determine your final bid price.

5. Finance Cost

If you have a job that will require special financing, calculate this cost and apply it at this time.

6. Gross Receipts Or Net Profit Tax

In some parts of the country, an annual gross receipts tax is charged between 2 to 4% of total sales. In addition, some areas have a net profit tax (this is not the same as the gross profit tax). Generally this tax must be paid before you pull the permit and it's value can range from 1% to 3% of the job cost.

7. Inspector Problems

Some inspectors enforce opinions on how they feel the installation should be installed. It's generally not cost effective to argue with the inspector for nickel and dime items, just make the change and include the expense as a cost of doing business.

8. Retainage Cost

Some jobs required that a portion of each payment (typically 10%) be held for period of time (typically 90 to 180 days) after final electrical inspection. The purpose of holding back the money (retainage) is to guarantee the owner that the electrical system has been installed correctly and according to the contract. You might want to add a few dollars to cover this factor, or just consider it as a cost of doing business which will be recovered by anticipated profits.

5.13 BID ACCURACY

To assure that your bid is accurate, you must verify that you have not made any of the following errors in the estimate:

- Assuming standard grade devices, when specification grade is required.
- Errors in multiplication or addition.
- Failing to include outside or underground work.
- Failure to determine job site conditions, especially for retrofit jobs.
- Failure to comply with the specifications or blueprint notes.
- Forgetting a major item, such as a switchgear quote.
- Forgetting to include special equipment.
- Forgetting to include changes to the original specifications or blueprints.
- Leaving a page out of the total.
- Not double checking all figures.
- Not transferring totals to the summary worksheet properly.
- Omitting a section of the estimate.

- Thinking that typical floors 6–12 is six floors, when it's seven floors.
- Using improper estimating forms.
- Using supplier take-off quantities for quotes.
- Relying on verbal supplier quotes.
- Wrong extensions or totals.
- Wrong scale on reduced blueprints.
- Wrong unit for labor-unit.
- Wrong unit for material cost.

Case Study – One of my students relied on a verbal supplier fixture quote of $75,000, rather than insisting on a written one.

Result –When he ordered the fixtures, the price was $90,000.

5.14 BID ANALYSIS

To insure that the bid is valid you should perform a bid analysis before you submit your bid. You are limited in your analysis when you estimate manually, but the following calculations should be performed and compared against past jobs.

Cost Distribution
(Cost divided by total cost)
Labor . . . . . $1,033/$5,406 = . . .19.11%
Material . . . $2,307/$5,406 = . . .42.67%
Direct Cost . . . $100/$5,406 = ...1.85%
Overhead . . . $1,261$5,406 = . . .23.33%
Profit . . . . . . $705/$5,406 = . . .13.04%

Cost Per Square Foot = $2.15
(Total cost divided by square foot)
($5,406/2,520 sq. ft.)

Labor Analysis
Total Hours . . . . .97 hours
Total 8 hr. Days (2 men)
(97 hours/16 hours) . . . . .6.06 Days
Total 5 Day Weeks
(8 days/5 days) . . . . .1.21 Weeks

5.15 BID PROPOSAL

When you have completed the bid, you must submit a written proposal that clarifies what your bid includes and what is not included. Preprinted wiring contracts are available from:
Minnesota Electrical Association
3100 Humboldt Avenue South
Minneapolis, MN 55408
1-800-829-6117, Fax 1-612-827-0920

If you have a computer, you can create your own proposal. The following are typical components that you might consider including in your proposal.

Note. You need to sit down with an attorney and review the following conditions.

Acceptance Of Proposal
Be clear that the proposal may be withdrawn if not accepted within _____ days from date of submission. Indicate that when signed by both parties, the proposal, including the conditions on the front and reverse side, constitutes a legal and binding contract. (Make sure the line for the signature is the last item on the front page.)

Change Orders
Any deviation or alteration from the blueprints or specifications should be executed only on receipt of written orders, and will become an extra charge. Charges for extras will be based on a labor rate of _____ dollars per man-hour. This includes labor cost, labor taxes, labor insurance, labor benefits, supervision, overhead and profit. Material shall be charged at Electrical Contractor's list price as published by Trade Service Corporation.

The Electrical Contractor must receive written authorization by an individual listed below prior to commencement of the work.

NO WORK SHALL COMMENCE UNTIL WRITTEN AUTHORIZATION IS RECEIVED BY THIS ELECTRICAL CONTRACTOR.

Individuals authorized to sign written change orders shall be:

Name: _____

Name: _____

Date Of Proposal

Include the date you submit the bid.

Default

Owner/contractor will be in default if any payment called for under this agreement and all authorized change orders becomes past due, if any written agreement made by the owner/contractor is not promptly performed, if any conditions warranted by the owner/contractor prove to be untrue, or the failure of owner/contractor to comply with any of the conditions of this agreement. In the event of owner/contractor default, the Electrical Contractor may:

1. Suspend work and remove un-installed Electrical Contractor's material or equipment from the premises. The owner/contractor agrees that Electrical Contractor may enter upon owner/contractor property for the purpose of repossessing such material or equipment without liability to owner/contractor for trespassing or any other reason.
2. The Electrical Contractor may retain all money paid hereunder, regardless of the stage of completion of the work and bring any appropriate action in court to enforce its rights.
3. The owner/contractor agrees to pay all costs and expenses, attorney's fees, court costs, collection fees (including fees incurred in connection with appeals) incurred by Electrical Contractor in enforcing his rights under this proposal.

Electrical Contractor Shall Not Be Liable

Electrical Contractor shall not be liable for failure to perform if prevented by strikes or other labor disputes, accidents, acts of God, governmental or municipal regulation or interference, shortages of labor or materials, delays in transportation, non-availability of the same from manufacturer or supplier, or other causes beyond the Electrical Contractor's control. In no event shall the Electrical Contractor be liable for special or consequential damages whatsoever or however caused.

Exclusions

This proposal does not include cost of trash removal, concrete, forming, painting, patching, trenching, core drilling, venting and sealing of roof penetrations. All waste created by Electrical Contractor will be removed to a specific area on the construction site as instructed by the owner/contractor.

Fixtures And Equipment Supplied By Others

This agreement includes the installation of fixtures furnished by others, if fixtures are on the job at the time of electrical trim out. Fluorescent fixtures supplied by others shall be pre-assembled, pre-whipped, and pre-lamped with in-line fuses.

Electrical Contractor shall not be responsible for owner-supplied fixtures or equipment due to losses related to theft, damage, vandalism, warranty, or any associated storage expenses. This agreement does not include:

1. Warranty of fixtures and equipment supplied by others.
2. Assembly of fixtures and/or equipment supplied by others.
3. Fixtures weighing more than fifty (50) pounds.

Equipment supplied by other trades shall not be required to be installed by this Electrical Contractor.

Installation Practices

All material and equipment supplied by the Electrical Contractor shall be as warranted by the manufacturer and will be installed in a manner consistent with standard practices at this time.

Job Name

Your proposal should include the name of the project or the name of the job.

National And Local Codes

Electrical installation shall meet the National Electrical Code and local electrical codes. Errors in design by the architect and/or engineer in the blueprints or specifications are not the responsibility of the Electrical Contractor. Any additional electrical work required by others and not indicated on blueprints and specifications shall not be part of this agreement.

Non-Compete Clause

Owner/contractor and all authorized representatives of owner/contractor are not to contract or employ any employees of this Electrical Contractor for one year from the completion of any electrical work performed by this Electrical Contractor with said owner/agent within an area of fifty (50) miles radius from this job site.

Payment Schedule And Terms

The proposal must be clear on a payment schedule, such as 30% slab, 50% rough and 20% final.

Any payments not received within 30 days of invoice date should be considered past due and will accrue an additional interest charge at 1.5% per month of the unpaid balance until paid in full. No work shall be performed (including warranty) if any invoice is past due (including change orders). In addition, no release of lien shall be signed unless all payments are paid in full.

> *Caution.* Never sign a contract that contains a contingency payment clause. This clause specifies that you only get paid if the general contractor gets paid. You have no control of the general contractor or his subcontractors and your money should not be dependent on the performance of others.

Performance

Reasonable time shall be given to the Electrical Contractor to complete each phase of the electrical job. However, Electrical Contractor agrees that where a written construction schedule is provided with the signing of this agreement, the Electrical Contractor shall pay all overtime costs necessary to complete construction in a timely manner to comply with the written construction schedule.

If a written construction schedule is not provided with the signing of this proposal, the Electrical Contractor shall not pay for any overtime to complete project. Any overtime required shall be considered a change order and written authorization shall be required in advance.

Plans And Specifications

Make sure you're very clear that your proposal is based on specific blueprints. List the page numbers including the last revision date. If there should be a conflict between the blueprints and specification, be sure your proposal notes these discrepancies.

Price

Indicate that the price is to remain in effect until _____ (a specific date). Any work required after this date is not covered within the contract and will be considered an extra. You don't want to be responsible to complete a job three years after you bid it, and the job never got started or finished.

Returned Checks

Customer agrees to pay a service charge of $25.00, each time a check is returned to Electrical Contractor.

Scope Of Work

Indicate what your proposal includes, such as complete electrical installation including the installation of lighting fixtures. Get as detailed as possible.

Submitted By

The proposal must be clear as to who submitted the proposal. Include your company name, contact name, address, office phone number, beeper, cellular phone number, fax number, e-mail address, or any other details necessary for the customer to contact you.

Submitted To

Be sure that you know the person you're submitting the proposal to, and don't misspell their name.

Temporary Wiring

Be specific on what you intend to supply and what your bid does not include.

Temporary wiring shall include:

_____

_____

_____

This Proposal Does Not Include:

_____

_____

Termination Of Agreement

This agreement shall remain in effect for _____ days from the signing of this agreement. Any work required under this agreement after this date is not covered and will be considered a change order and charged and paid accordingly.

Warranty

Warranty shall only apply to the electrical installation of the material, fixtures, equipment, and other items supplied by the Electrical Contractor. Warranty shall not apply to material, fixtures, equipment, and other items supplied by others. Warranty shall not apply to extensions or additions to the original installation if made by others.

Warranty shall commence from the final electrical inspection date for a maximum period of one year.

No warranty work shall be performed if any invoice is past due, including change orders.

PART C – UNIT PRICING

Now that you understand what it takes to determine the selling price of a job, let's learn how to bid a job using unit pricing. Unit pricing is used for almost all types of construction such as renovations, office build-outs, change orders, etc.

Actually unit pricing is the same as bidding a job; that is you must understand the scope of work, determine the bill-of-material, price, labor, extend, total, etc. In addition you must account for varying job conditions, material adjustment, direct cost, overhead and profit. Therefore before you can proceed, you need to know the following:

1. Labor-unit adjustment
2. Labor rate per man-hour
3. Material adjustment
4. Overhead rate
5. Profit margin

5.16 UNIT PRICE EXAMPLE

It's probably easier to demonstrate an example rather than explain the details. See Figure 5–1.
Lets assume the following:
Labor-Unit Adjustment . . . . None
Labor Rate Per Man-Hour: $10.65
Material Adjustment: . . . . + 15%
Direct Cost, Permit $50.00
Overhead Rate: . . . . + 33%
Profit Margin: . . . . + 15%

Determine the unit price for a temporary power pole that contains two duplex receptacles protected by a GFCI circuit breaker and one 30-ampere 240-volt receptacle. We will assume that the power pole can be used for at least five different installations. See Figure 5–1.

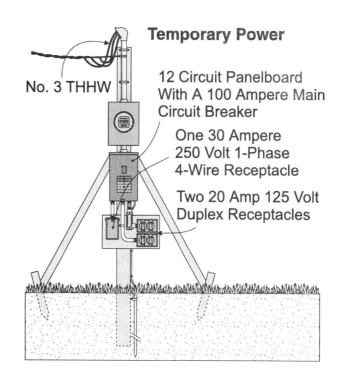

Figure 5–1
Temporary Power

Temporary Power Pole							
Construction Details	Qnty.	Cost	U	Extension	Labor	U	Extension
Weather Head 1 1/4"	1	$399.00	C	$3.99	35.00	C	0.35
Intermediate Metal Conduit 1 1/4"	10	$108.00	C	$10.80	5.25	C	0.53
Meter with Hub	1	$50.00	E	$50.00	2.00	E	2.00
Wire No. 3 THHN	50	$375.00	M	$18.75	7.50	M	0.38
Offset Nipple 1 1/4"	1	$317.00	C	$3.17	9.00	C	0.09
Panel 12 Circuit With 100 ampere Main	1	$56.00	E	$56.00	1.00	E	1.00
Breaker 100 Ampere 2-pole	1	$36.00	E	$36.00	0.20	E	0.20
Breaker 30 Ampere 2-pole	1	$10.00	E	$10.00	0.15	E	0.15
Breaker 20 Ampere GFCI	2	$12.00	E	$24.00	0.20	E	0.40
T Conduit Body 1/2"	1	$5.00	E	$5.00	0.35	E	0.35
T Conduit Body 1/2" Cover	1	$151.00	C	$1.51	4.00	C	0.04
Offset Nipple 1/2"	1	$127.00	C	$1.54	5.00	C	0.06
Weather Proof Boxes	3	$389.00	C	$11.67	21.25	C	0.64
Weather Proof Cover - Receptacle	2	$389.00	C	$7.78	4.00	C	0.08
Weather Proof Cover - 30 Ampere	1	$343.00	C	$3.43	8.00	C	0.08
Receptacle - 20 Ampere Duplex, 125 Volt	2	$180.00	C	$3.60	19.00	C	0.38
Receptacle - 30 Ampere 240 Volts	1	$7.00	E	$7.00	0.25	E	0.25
2" X 4" X 14' Wood Member	2	$3.00	E	$6.00	0.25	E	0.50
Initial Construction Total				$260.24			7.48
Installation Details							
No. 6 Bare	10	$203.00	M	$2.03	7.25	M	0.07
Ground Rod	1	$800.00	C	$8.00	50.00	C	0.50
Ground Rod Clamp	1	$217.00	C	$2.17	14.00	C	0.14
2" X 4" X 14' Wood Member	3	$3.00		$9.00	0.25		0.75
Travel Time (Round Trip)	1	*		*	1.00		1.00
Installation Labor	1	*		*	2.00		2.50
Installation Total				$21.20			4.96

Labor Cost Calculations

Labor-hours = 7.48 hours divided by five installations = 1.50 hours + 4.96 hours = 6.50 hours (rounded)

Labor Cost = 6.50 hours × $10.65 = $69.00

Material Cost Calculations

Material cost = $260 divided by five installations = $52.00 + $21.20, = $73.20

Adjusted material cost = $73.20 + 15% = $84.00 (rounded)

Unit Price Calculation Summary

Labor cost	$69.00
Material cost	$84.00
Direct cost, Permit	+ $50.00
Total prime cost	$203.00
Overhead, @ $13.00	+ $84.50
Break even cost	$287.50
Profit, + 15%	+ $43.13
Unit price	$330.63

SUMMARY

PART A – THE ESITMATE SUMMARY

Introduction

The estimating process requires you to make judgments on intangibles and the Summary Worksheet is used to help organize the different costs so you don't make a mistake. The summary worksheet contains eight major sections. They include:

Step A – Labor-Hours
Step B – Labor Cost
Step C – Material Cost
Step D – Direct Job Cost
Step E – Estimated Prime Cost
Step F – Overhead
Step G – Estimated Cost

5.01 Labor-Hours (Step A)

The estimated labor-hours must be adjusted for working conditions. Determining the labor adjustment comes with experience and is part of what makes estimating an art form. After you have adjusted the total labor-unit hours, you must consider if there are any additional labor requirements for the job such as:

1. As-built plans
2. Demolition
3. Energized parts
4. Environmental hazardous material
5. Excavation, trenching, and backfill
6. Fire seals
7. Job location
8. Match-up of existing equipment
9. Miscellaneous material items
10. Mobilization (startup)
11. Nonproductive labor
12. OSHA compliance
13. Plans and specifications
14. Public safety
15. Security
16. Shop time
17. Site conditions
18. Sub-contract supervision
19. Temporary, stand-by, and emergency power

5.02 Labor Cost (Step B)

The estimated labor cost for a job is determined by multiplying the total adjusted labor man-hours by the labor rate per man-hour. The labor rate per man-hour can be determined by one of two methods: the shop average or the job average.

5.03 Labor Burden

You must include the cost of labor benefits, taxes, and labor insurance in the estimate. In general, NECA contractors include these costs as part of their labor rate, and nonunion contractors recover labor burden costs when overhead is applied.

5.04 Total Material Cost (Step C)

The first step in determining the total material cost is to transfer the material cost from the pricing worksheets to the summary worksheet. Once this is accomplished, you make the necessary adjustments for items left out of the estimate, material waste, theft, and small tools. Then you include quote prices when they arrive and apply sales tax.

5.05 Direct Job Expenses (Step D)

Direct job expenses are expenses often not shown on the blueprints, but probably indicated in the specifications. Consider the following expenses and be sure you include their cost.

1. As-built plans
2. Business and occupational fees (b&o)
3. Engineering/working drawings
4. Equipment rental
5. Field office
6. Fire seals
7. Guarantee
8. Insurance – special
9. Miscellaneous material items
10. Mobilization
11. Osha compliance
12. Out of town expenses
13. Parking fees
14. Permits and inspection fees
15. Public safety
16. Recycling fees
17. Storage/storage handling
18. Sub-contract expenses
19. Supervision cost
20. Temporary wiring
21. Testing amd certification fees
22. Trash disposal
23. Utility charges and fees

5.06 Estimated Prime Cost (Step E)

The total prime cost, which is simply adding up the cost of labor, material, plus direct job expenses.

5.07, 5.08, And 5.09 Overhead

The cost of managing labor is called overhead and there are two methods of applying this cost to the estimate: the percentage method and the cost per hour method. Most contractors apply overhead as a percentage of prime cost and this method works well if you're doing similar jobs, like houses, office buildings, remodel work, etc. Since overhead is related primarily to the management of labor, overhead should be calculated as a rate per man-hour.

If you are just getting started in business, or if you don't know your overhead cost, then use the lesser of either 40% of prime cost or $13 per man-hour. If you use these factors, do not apply a labor burden factor to labor cost.

5.10 Understanding Break Even Cost (Part G)

All you need to do now is add up the costs and you can determine your break even cost.

PART B – THE BID PROCESS

5.11 Profit

A responsible estimate will include a reasonable amount for profit. If your estimate is complete and you manage your company efficiently, you don't need a lot of profit to stay in business. A reasonable profit margin would be 15% - 20% of break even cost.

5.12 Other Final Costs

Be sure you review the specifications and blueprints one final time before you submit your bid and make sure you considered all cost such as:

1. Allowances/Contingency
2. Back-charges
3. Bonds
4. Completion penalty
5. Finance cost
6. Gross receipts or net profit tax
7. Inspector problems
8. Retainage cost

5.13 Bid Accuracy

To insure that your bid is accurate, you must verify that you have not made any common errors in the estimate.

5.14 Bid Analysis

To insure that the bid is valid you should perform a bid analysis before you submit your bid. With a computer you can perform greater bid analysis.

5.15 Bid Proposal

When you have completed the bid, you must submit a written proposal that clarifies what your bid includes and what is not included.

PART C – UNIT PRICING

Unit pricing is the same as bidding a job: that is you must understand the scope of work, determine the bill-of-material, price, labor, extend, total, etc. In addition you must account for varying job conditions, material adjustment, direct cost, overhead and profit. Before you can provide a Unit Price bid, you need to know the following:

1. Labor-unit adjustment
2. Labor rate per man-hour
3. Material adjustment
4. Overhead rate
5. Profit margin

Chapter 5

Review Questions

Essay and Fill-in-the-Blank

Part A - The Estimate Summary

1. The estimated labor-hours must be adjusted for working conditions. List as many conditions that impact the labor-unit as you can think of.

2. After adjusting the labor-unit hours, you must consider if there are any additional labor requirements for the job. What things would this include?

3. What determines the estimated labor cost for a job?

4. Why are fixture and switchgear quotes often held until the last minute by the suppliers?

5. Why don't you mark up the material or quotes for profit?

6. Do all jobs require sales tax on material? What about labor?

7. Direct job expenses are expenses often not shown on the blueprints but which need to be included in the estimate. List as many of these expenses as you can think of.

8. The failure to consider direct cost expenses will impact profit in what way?

9. What is the cost of managing labor called?

10. Generally the greater the business volume or the larger the job, the _____ the overhead expense per man-hour?

11. What are two methods of applying overhead expenses to an estimate?

12. What are some of the dangers of the percentage method of applying overhead?

13. Overhead rate-per-hour can be determined by dividing the total overhead cost for at least the past six months by the _____.

14. Break even cost is the total cost of _____.

Electrical Estimating

Part B - The Bid Process

15. Profit is impacted by _____.

16. The purpose of a bid bond is to _____.

Multiple Choice Questions

Part A – The Estimate Summary

1. Summarizing the estimate requires you to make judgments on intangibles such as _____.
   (a) job conditions (b) direct cost (c) overhead (d) all of these

2. The first step in the estimate summary is to transfer the total labor-hours and material cost from the price/labor worksheet to the _____.
   (a) Switchgear Quote (b) Proposal
   (c) Take-Off Worksheet (d) Estimate Summary

3. If you are estimating manually, a/an _____% factor should be included in the estimate for miscellaneous material items.
   (a) 2.5 (b) 5 (c) 7.5 (d) 8

4. The value of sub-contract work must be included in the estimate at your cost and the bid must include the labor-hours required for _____.
   (a) completing the work (b) subcontractor supervision
   (c) obtaining the subcontract (d) all of these

5. The shop average labor rate per man-hour is determined by dividing the total shop labor cost by the total number of man-hours based on the last _____ months.
   (a) six (b) nine (c) twelve (d) eighteen

6. An alternative to the shop average labor rate of determining the labor rate is the job average labor rate, which requires you to consider how you intend on _____ the job.
   (a) financing (b) manning (c) supervising (d) all of these

7. You must include the cost of labor benefits, taxes, and labor insurance in the estimate. Generally _____ contractors recover labor burden costs when overhead is applied.
   (a) NECA (b) nonunion (c) all (d) none of these

8. If your are estimating manually, _____% of the total material cost should be added to the estimate to cover miscellaneous material.
   (a) 4 (b) 10 (c) 12 (d) 14

9. Typically _____% of the total material cost should be an acceptable value for waste and theft.
   (a) 2 (b) 3 (c) 4 (d) 5

10. Typically _____% of the total material cost should be an acceptable value for small tools.
    (a) 1 (b) 2 (c) 3 (d) 4

11. Be sure to become familiar with the sales tax rules when you bid a job in a new area because sales tax varies from _____.
    (a) state to state (b) county to county (c) city to city (d) all of these

12. Direct job expenses are expenses not shown on the blueprints, but are probably indicated in the _____.
    (a) proposal
    (b) meetings with the owners
    (c) specifications
    (d) meetings with the general contractor

13. Applying overhead (job management) expense for each job is _____ and you need to consider the business volume and the job size.
    (a) precise
    (b) an art form
    (c) consistent
    (d) luck

14. The _____ method for applying overhead works well if you're doing similar types of job such as houses, office buildings, remodel work, etc.
    (a) percentage
    (b) cost per hour
    (c) shot in the dark
    (d) conversion

15. Knowing what it costs you to do a job is called the estimated cost. This is also called the _____ value.
    (a) sound management
    (b) effective business control
    (c) profit
    (d) break even

Part B – The Bid Process

16. You can lose money on a job that has a 25% projected profit margin, if the bid doesn't cover all expenses.
    (a) True
    (b) False

17. Before you submit your bid you must _____
    (a) verify its accuracy and perform an analysis
    (b) check with other contractors to find out what they are submitting
    (c) create and sign the proposal
    (d) all of these

18. The _____ clarifies what your bid includes and what is not included.
    (a) plan
    (b) specifications
    (c) bid summary
    (d) written proposal

Part C - Unit Pricing

For the following three questions assume the following:

Labor-unit adjustment, None

Labor Rate, $10.65 per man-hour

Material Cost Adjustment, + 15%

Overhead, $13 per man-hour

Profit, + 15%

19. The unit price for the single pole switch is approximately _____. See Figure 5-2.
    (a) $28.09
    (b) $33.42
    (c) $36.79
    (d) $46.11

20. The unit price for the duplex receptacle is approximately _____. See Figure 5-3 (page 92).
    (a) $45.50
    (b) $30.07
    (c) $33.94
    (d) $35.42

21. The unit price for each lay-in fluorescent fixture is approximately _____. See Figure 5-4 (page 93).
    (a) $102.91
    (b) $106.43
    (c) $112.22
    (d) $119.19

Unit Cost - Single Pole Switch						
	Qnty.	Cost	Extension Cost	Labor Unit	U	Extension Hours
Box 4" × 4"	1		$			
Ring 1 Gang fi"	1		$			
1/2" EMT Connector	2		$			
1/2" EMT Pipe	20		$			
1/2" EMT Coupling	3		$			
Straps	3		$			
Switch – 15 Ampere	1		$			
Plate 1 Gang Plate	1		$			
No. 12 Wire	50		$			
Total			$8.34			1.29
Unit Price Summary						
Labor Cost at $10.65			$			
Material Cost + 15%			$			
Total Prime Cost			$			
Overhead at $13 Per Hour			$			
Break Even Cost			$			
Profit, +15%			$			
Unit Price			$46.11			

One Gang Switch

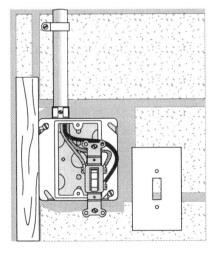

Figure 5–2

Chapter 5 – The Estimate Summary And Bid Process Questions

Unit Cost - Duplex Receptacle 15 Amperes 125 Volt						
	Qnty.	Cost	Extension Cost	Labor Unit	U	Extension Hours
Box 4" X 4"	1		$			
Ring 1 Gang 1/2"	1		$			
1/2" EMT Connector	2		$			
1/2" EMT Pipe	20		$			
1/2" EMT Coupling	3		$			
Straps	3		$			
Duplex Receptacle	1		$			
Plate 1 Gang Plate	1		$			
No. 12 Wire	50		$			
Total			$8.29			1.27
Unit Price Summary						
Labor Cost at $10.65			$			
Material Cost + 15%			$			
Total Prime Cost			$			
Overhead at $13 Per Hour			$			
Break Even Cost			$			
Profit, +15%			$			
Unit Price			$45.50			

Receptacle Duplex

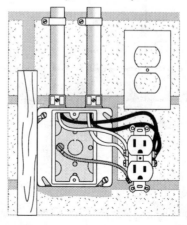

Figure 5–3

Unit Cost - Lay-In Fixture						
	Qnty.	Cost	Extension Cost	Labor Unit	U	Extension Hours
Box 4" × 4"	1		$			
Ring 1 Gang ½"	1		$			
½" EMT Connector	2		$			
½" EMT Pipe	20		$			
½" EMT Coupling	3		$			
Straps	3		$			
Lay-In Fixture w/Whip	1		$			
No. 12 Wire	50		$			
Total			$107.90			2.58
Unit Price Summary						
Labor Cost at $10.65			$			
Material Cost + 15%			$			
Total Prime Cost			$			
Overhead at $13 Per Hour			$			
Break Even Cost			$			
Profit, +15%			$			
Total Cost			$212.87			
Unit Price Per Fixture			$			

Fluorescent Lay-In Fixtures

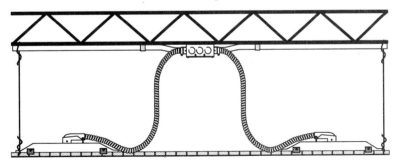

Figure 5–4

Chapter 6

Estimating Residential Wiring

OBJECTIVE

After reading this chapter, you should
- be able to manually estimate residential wiring and understand how residential estimating is accomplished by the use of a computer.

INTRODUCTION

This chapter explains in detail the manual and computer assisted estimating process for residential wiring. If you understand the wiring and material requirements for the electrical symbols shown on blueprints E–1 and E–2, Figures 6–1 and 6–2 that follow, you should be able to estimate this house in about 8 or 10 hours. It shouldn't take much more than 1 hour if you estimate with a computer. If you don't understand the blueprint details, be sure to order the residential wiring book published by Delmar Publishers, titled *Electrical Wiring – Residential*, authored by Ray Mullin.

Note. The estimate in this chapter is based on blueprints extracted from *Electrical Wiring – Residential* (slightly modified). A complete set of blueprints at/inch scale are contained in *Electrical Wiring – Residential*.

PART A – PLANS AND SPECIFICATIONS

This part contains the blueprints and blueprint symbol wiring details required to understand both the manual and computer residential estimate contained in Parts B and C of this chapter. Take a few moments at this time to review the blueprints and then review the blueprint symbol wiring graphics that follow, Figures 6–3 through 6–12. This chapter does not contain any specifications. However, Chapter 7 – Estimating Commercial Wiring does contain specifications.

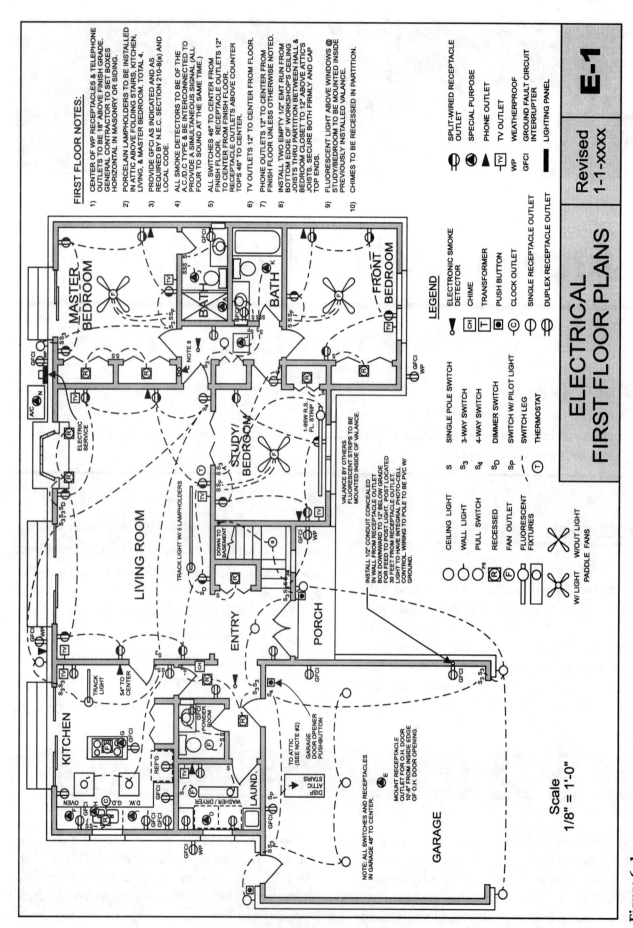

Figure 6–1
Residential Blueprint First Floor E–1

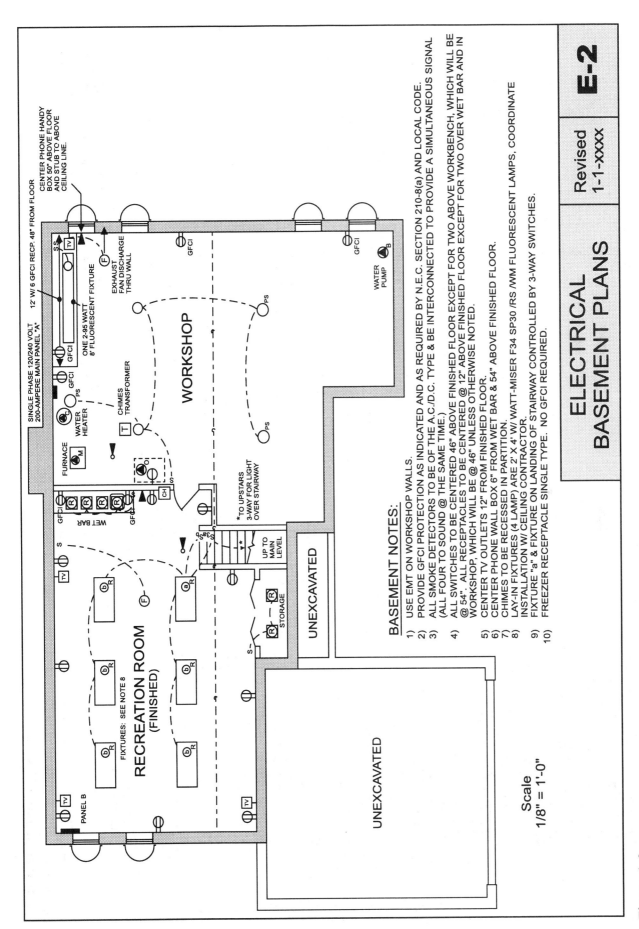

Figure 6–2
Residential Blueprint Basement E-2

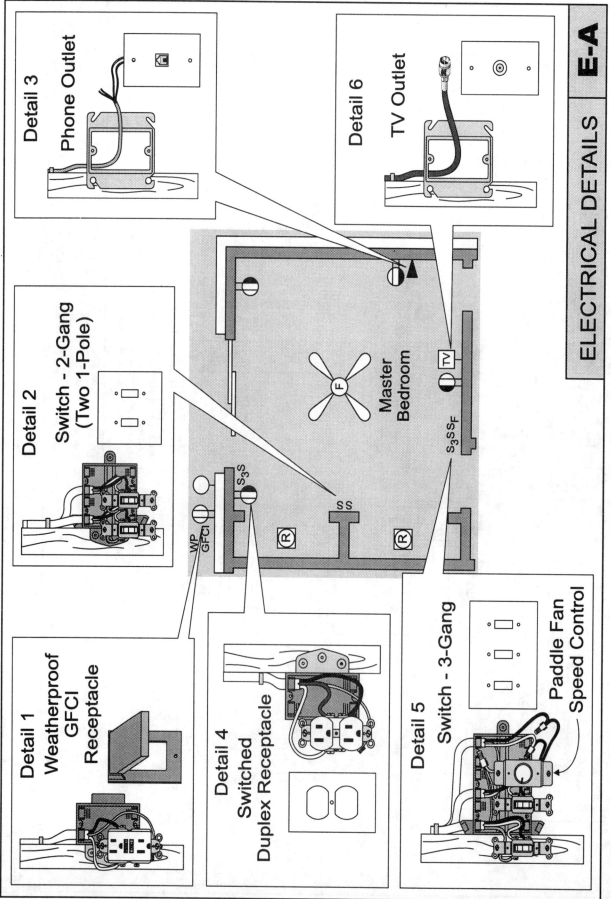

Figure 6–3
Blueprint Symbol Graphics

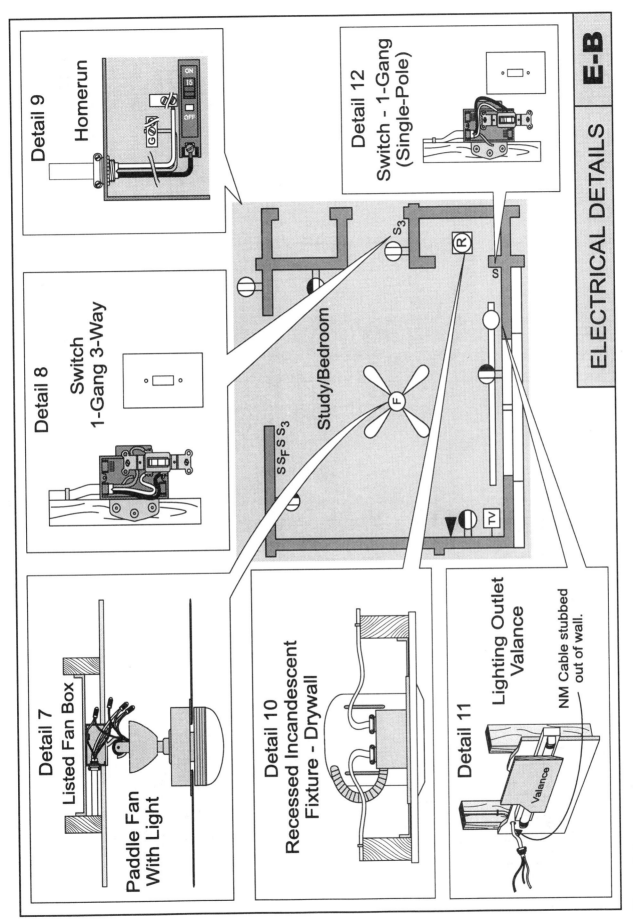

Figure 6–4
Residential Blueprint Basement E-2

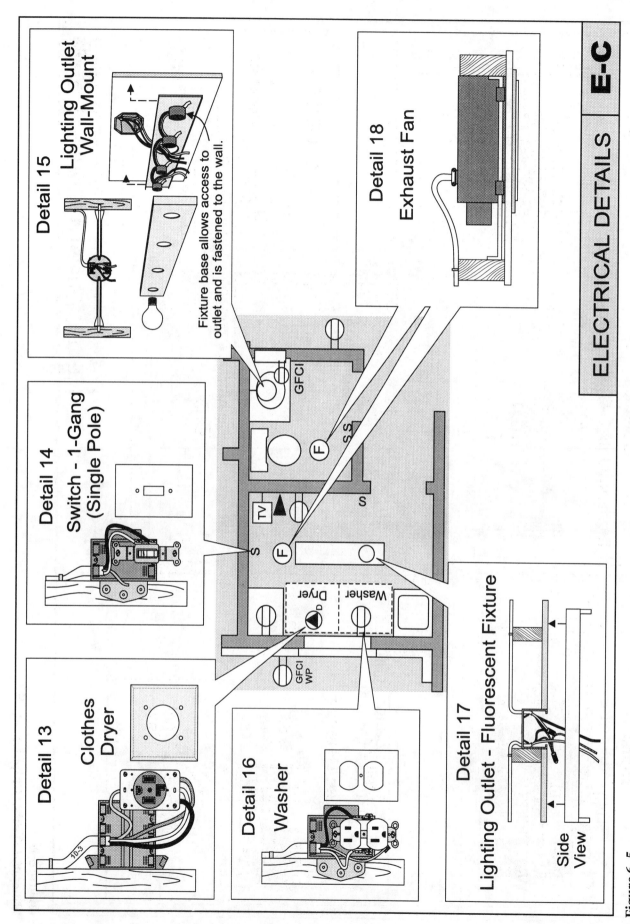

Figure 6–5
Blueprint Symbol Graphics

Electrical Estimating　　　　　　　　　　　Chapter 6 – Estimating Residential Wiring　　101

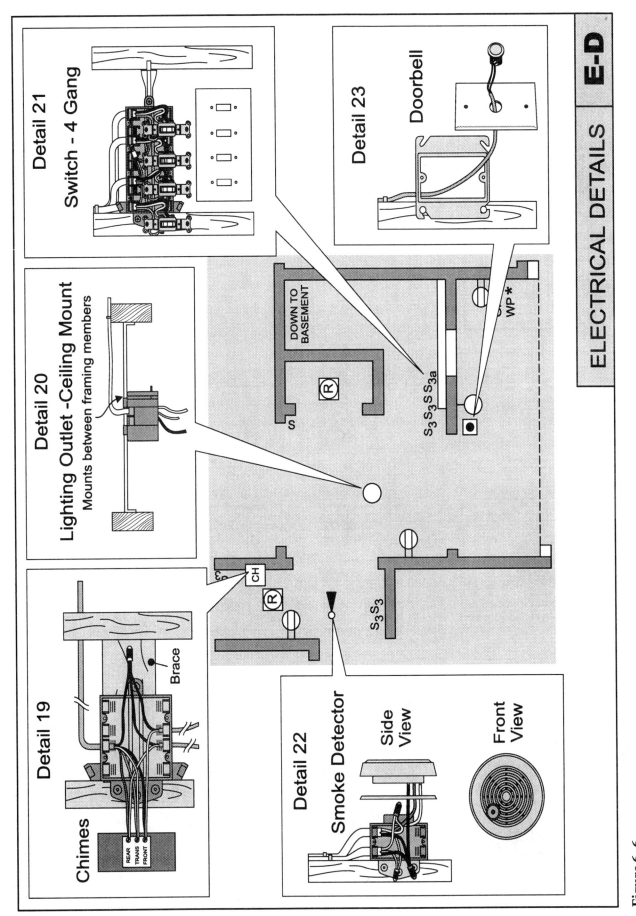

Figure 6–6
Blueprint Symbol Graphics

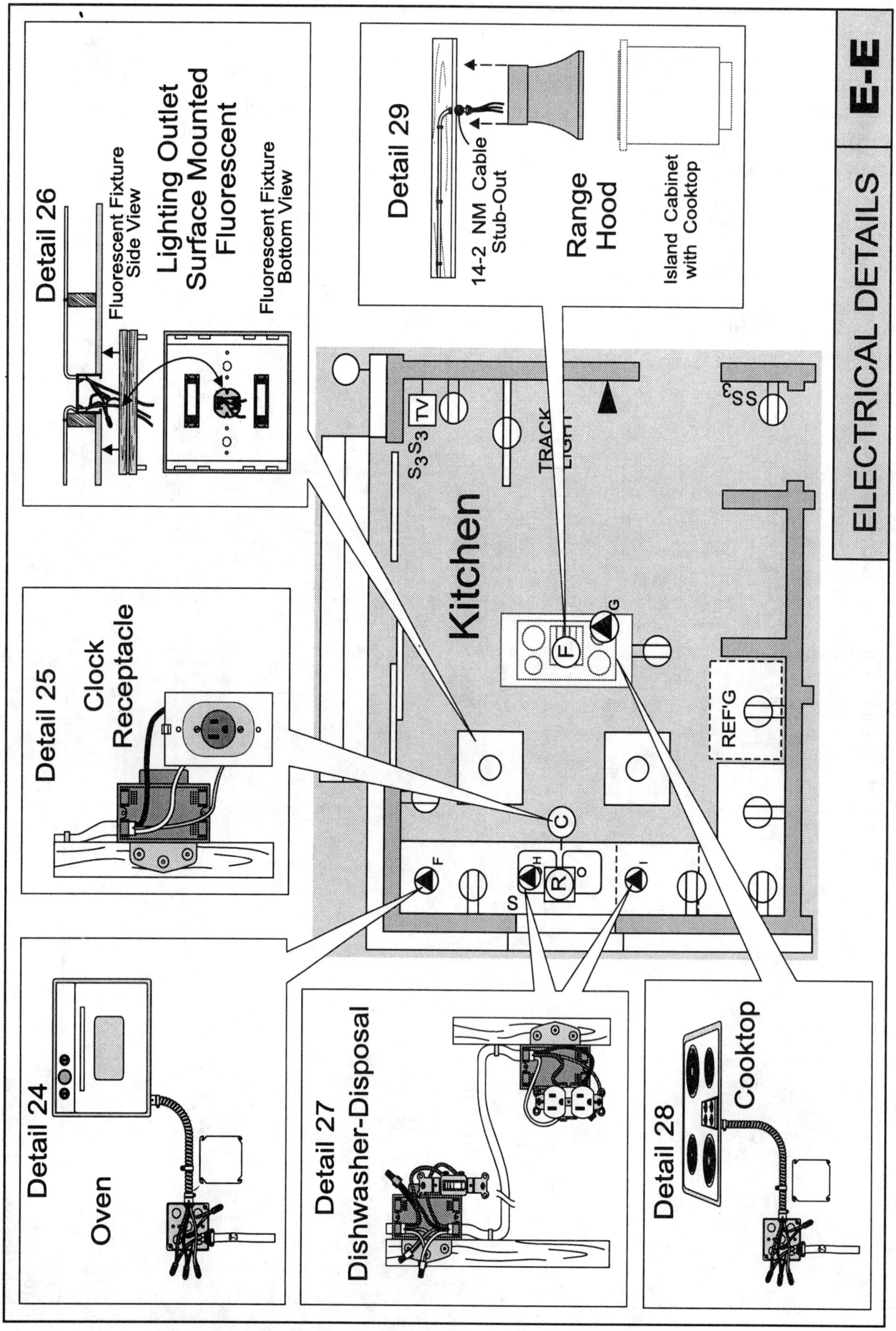

Figure 6-7
Blueprint Symbol Graphics

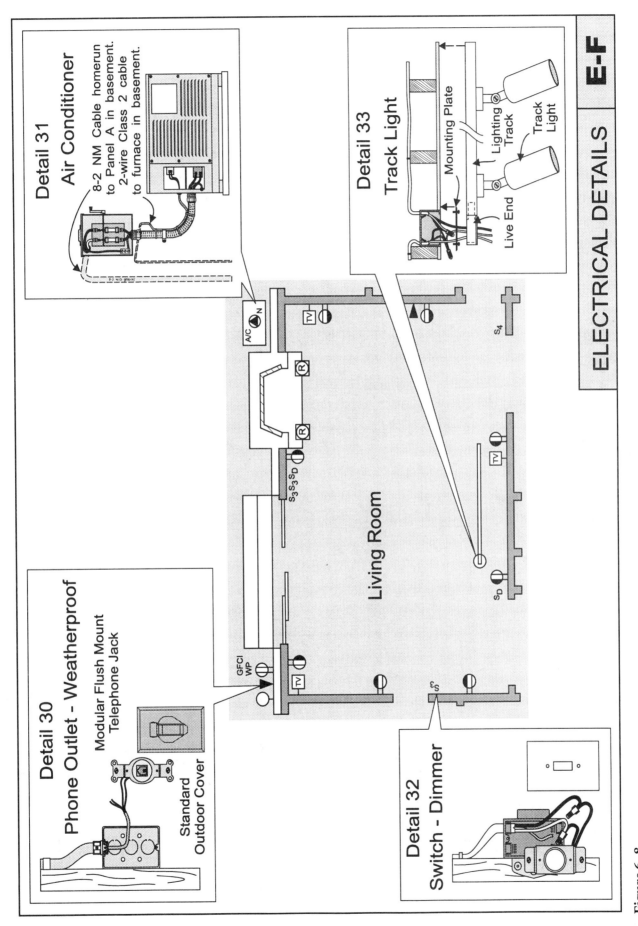

Figure 6-8
Blueprint Symbol Graphics

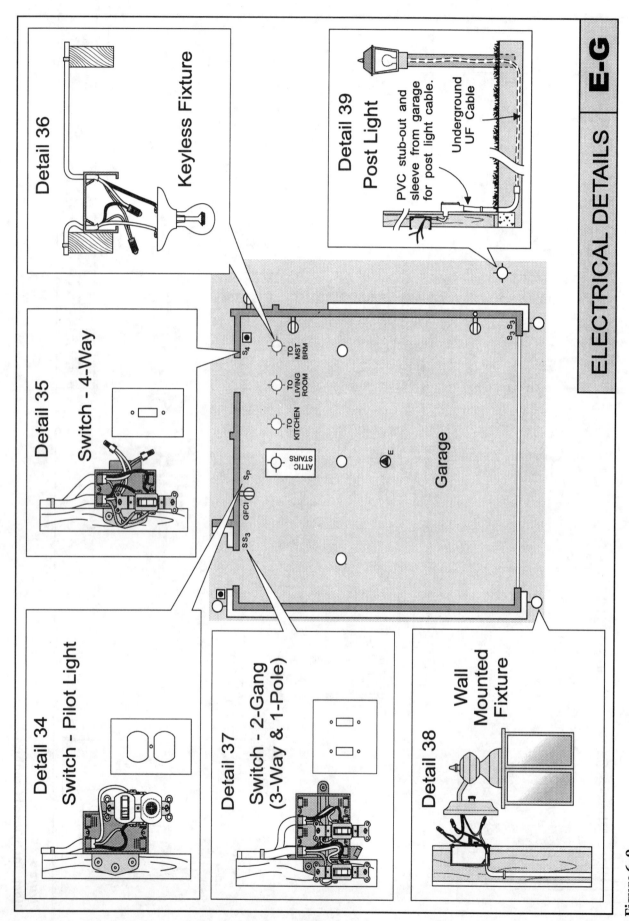

Figure 6–9
Blueprint Symbol Graphics

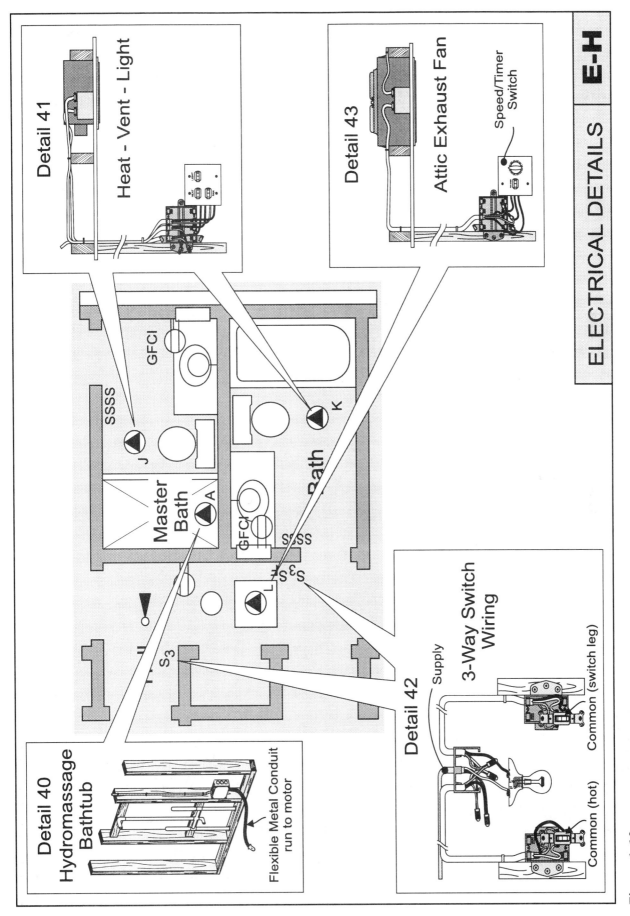

Figure 6-10
Blueprint Symbol Graphics

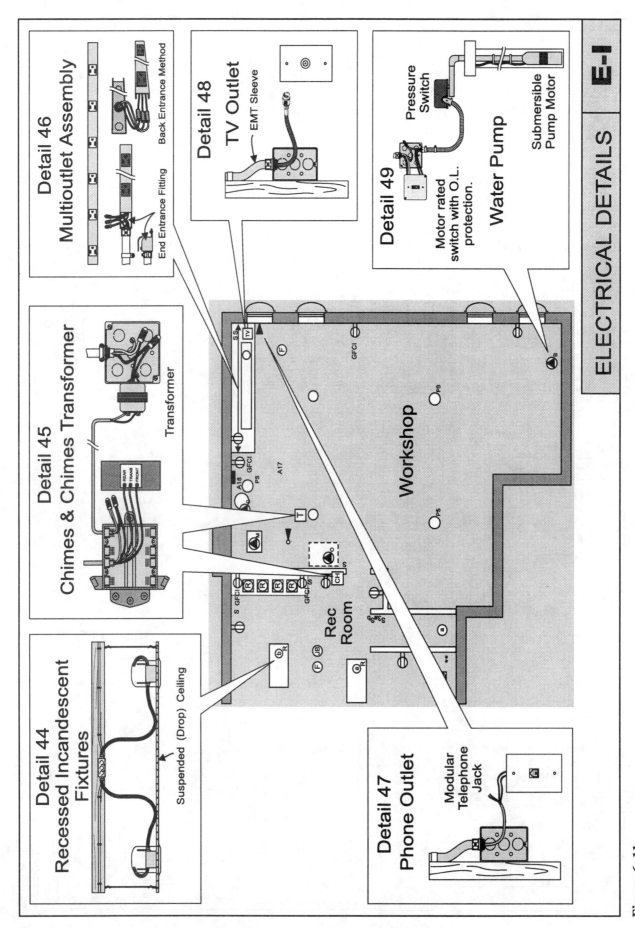

Figure 6-11
Blueprint Symbol Graphics

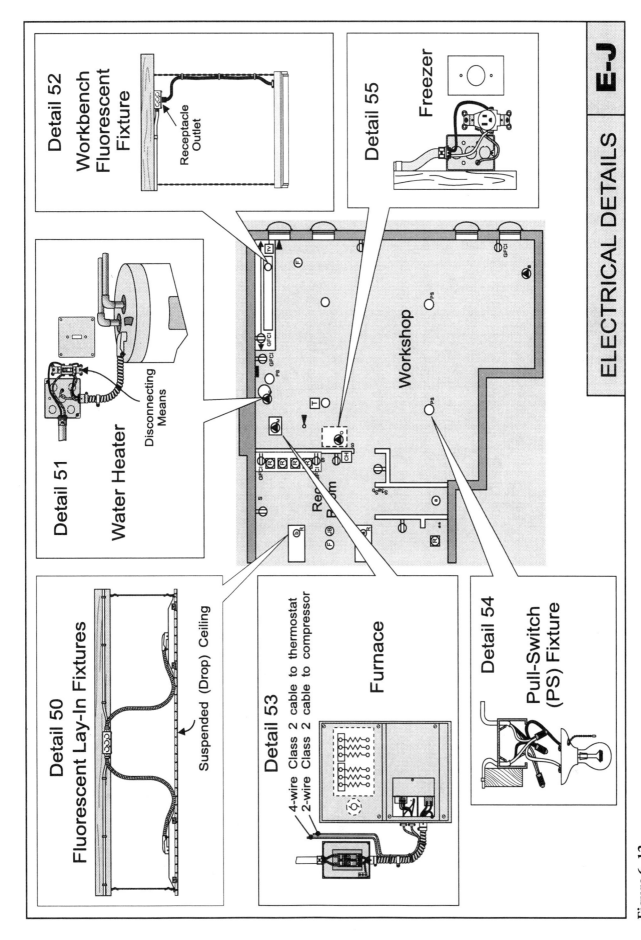

Figure 6–12
Blueprint Symbol Graphics

PART B – MANUAL ESTIMATE AND BID

Before you begin the actual take-off, you must be sure you have as much information about the job as possible. The first thing you should do when you receive the blueprints is to be sure they are readable and clear. You must have a complete and current set of blueprints and specifications to properly estimate the job.

Preparing The Estimate

You need to create a job folder to contain the estimate notes and worksheets. Then complete the Estimate Record Worksheet and hang it up on the wall over your take-off bench. See Figure 6–13 for a sample.

6.01 UNDERSTANDING SCOPE OF WORK

Before you get any deeper, you need to take some time in a quiet place to carefully read all notes on the blueprints, E–1 and E–2. Give the blueprints a thorough review to help you get a clear understanding of the scope of the project and the bid requirements before you begin the actual take-off. If applicable, be sure to visit the job site before you bid the job.

Take a close look at all blueprint legends, details, notations, and symbols. Watch for the electrical requirements for control wiring, underground wiring, area lighting fixtures, signs, and outdoor equipment. Become familiar with the entire installation and check for any special or unusual features such as unusual ceiling heights. Look to see if proper working space is provided for the electrical equipment and watch for the location of the utilities, etc. Once you have reviewed the plans, complete the Specification Check List. See Figure 6–14, page 110.

Estimate And Bid Notes

Keep a note pad handy so you can keep track of any questions you have about the estimate. If there is anything you don't understand, get the answer as soon as possible, don't wait until the last moment. Contact the architect or engineer, but be sure you get your questions answered. If not, don't bid the job. See the Estimate/Bid Notes form, Figure 6–15, page 111.

Note. A sample of the Specification Check List and Estimate/Bid Notes are contained in the Appendix.

ESTIMATE RECORD WORKSHEET

Job Information Detail   Estimate Job Number: 0601 - xxxxxA

Job Name: American Homes	Bid Due Date: June 5, xxxx
Job Location: 123 Main Street	City: Orlando, Florida
Job Owner: Mr. Ken Winston	
Fax: 1-444-432-2424	Phone: 1-444-432-4523

Plans And Specifications Detail

Date Of Plans: 1/1/xxxx   Blueprint Page Numbers: E-1 and E-2

Comments: _____

Contractor Information Detail

Contractor: US Builders, Inc.	Contact: Mike Seiple
Address: 1 Lake View	City: Clermont, Florida
	E-mail: usbuilt@gate.net
Phone: 1-800-444-1234	Mobil: 1-444-777-1234
Beeper: 1-444-777-1498	FAX: 1-444-777-1499

Important Contacts

Architect: Summit Architectural Group	Phone: 1-444-444-1234
Engineer: Allied Engineering, Inc.	Phone: 1-444-444-1235
Telephone Utility: John Strous	Phone: 1-444-444-1236
Electric Utility: Susan Billings	Phone: 1-444-444-1237
Electrical Inspector: Mr. Johnston	Phone: 1-444-444-1238

Notes: _____
_____
_____
_____
_____

Figure 6-13
Estimate Record Worksheet

Specification Check List

Labor-Unit Adjustment (See Section 3.09)
1. Building Conditions _____
2. Change Orders _____
3. Concealed And Exposed Wiring _____
4. Construction Schedule _____
5. Job Factors − 10%
6. Labor Skill (Efficiency) _____
7. Ladder And Scaffold _____
8. Management _____
9. Material (Miscellaneous) + 8%
10. Off Hours And Occupied _____
11. Overtime _____
12. Remodel (Old Work) _____
13. Repetitive Factor _____
14. Restrictive Working Conditions _____
15. Shift Work _____
16. Teamwork _____
17. Temperature _____
18. Weather And Humidity _____

Labor Adjustment

Additional Labor (See Section 5.01)
1. As-Built Plans _____
2. Demolition _____
3. Energized Parts _____
4. Environmental Hazards _____
5. Excavation, Trenching And Backfill _____
6. Fire Seals _____
7. Job Location _____
8. Match-Up Of Existing Equipment _____
9. Miscellaneous Material Items _____
10. Mobilization (Startup) _____
11. Nonproductive Labor _____
12. OSHA Compliance _____
13. Plans and Specifications _____
14. Public Safety _____
15. Security _____
16. Shop Time _____
17. Site Conditions _____
18. Sub-Contract Supervision _____
19. Temporary, Stand-By Power _____

Hour Adder

Direct Job Expenses (See Section 5.05)
1. As-Built Plans $_____
2. Bus. & Occupational Fees $_____
3. Engineering Drawings $_____
4. Equipment Rental $_____
5. Field Office Expenses $_____
6. Fire Seals $_____

7. Guarantee $_____
8. Insurance – Special $_____
9. Miscellaneous, Labels $_____
10. Mobilization $_____
11. OSHA Compliance $_____
12. Out of Town Expenses $_____
13. Parking Fees $_____
14. Permits/Inspection Fees $ 375
15. Public Safety $_____
16. Recycle Fees $_____
17. Storage & Handling $_____
18. Sub Contract:
    _____ $_____
19. Supervision Cost $_____
20. Temporary Wiring:
    Lighting $_____
    Power $ 250
    Maintenance $_____
21. Testing $_____
22. Trash $_____
23. Utility Cost $_____

Total Direct Cost $625

Other Final Cost
1. Allowances $_____
2. Back-Charges $_____
3. Bond $_____
4. Completion Penalty $_____
5. Finance Cost $_____
6. Gross Receipts Or Net Profit Tax $_____
7. Inspector Problems $_____
8. Retainage $_____

Total Other Cost

Other Considerations
1. Conductor Size - Minimum? 14
2. Raceway Size - Minimum? N/A
3. Control Wiring Responsibility? N/A
4. Cutting Responsibility? N/A
5. Demolition Responsibility? N/A
6. Excavation/Back Fill Responsibility? _____
7. Painting Responsibility? _____
8. Patching Responsibility? _____
9. Special Equipment? _____
10. Specification Grade Devices Or Fittings? _____

Figure 6–14
Specification Check List

Estimate And Bid Notes Page ____ of ____

Blueprint Questions or Comments:

Specification Questions or Comments:

Figure 6–15
Estimate And Bid Notes

6.02 THE TAKE-OFF

After you have analyzed the blueprints and you understand the electrical wiring requirements, you can begin the take-off. Use a medium-soft pencil and write as clearly as possible to reduce the risk of error. The take-off must be quick and detailed enough so it can be used to determine the project's bill-of-material. The following take-off sequence is probably the most efficient method for manually estimating residential wiring. This take-off sequence generally follows:

1. Count lighting fixtures and develop quantities for lighting fixture quotes.
2. Count switches.
3. Count receptacles.
4. Count and measure special systems such as phone and CATV.
5. Measure general branch circuit wiring.
6. Count and measure individual branch circuits and home runs. Develop quantities for quotes.
7. Count panels and measure feeder.
8. Service runs, service equipment.

Step 1 – Count Lighting Fixture Symbols

See the Estimating Workbook for Blueprints E–3 and E–4 required for this section.

Prepare a take-off worksheet for all lighting fixtures, listed on the lighting fixture legend as shown in Figure 6–16.

Now count the first lighting fixture symbol that catches your eye and continue counting the same symbol until you have counted all similar symbols on blueprint E–3. As you count each lighting fixture symbol color them yellow, and click the counter. Once you have counted all of the same lighting fixture symbols, write this number on the take-off worksheet. Now count the next lighting fixture symbol that catches your eye and continue this process for all lighting fixtures on Blueprint E–3. Once you have completed counting all lighting fixtures symbols on Blueprint E–3, continue this process for Blueprint E–4.

Mark	Description	Symbol
A	Post Light	○
B	Outside Wall Light	○-
C	Keyless	○
D	8' Fluorescent Strip With Two 95 Watt Lamps	▭
E	Fluorescent 2 x 4 Lay-In Four 34 Watt Lamps	▭R
F	8' Fluorescent Strip With One 95 Watt Lamp	▭
G1	Fluorescent Wrap - 2' x 2', 2- 34W U-Lamps	▭
G2	Fluorescent Wrap - With Four 34W Lamps	▭
H	Paddle Fan - With Lighting Fixture	✕F
I	Pull Chain Fixture	○PS
J	Recessed Light	Ⓡ
K1	Track Light - 4 Foot	⊙—
K2	Track Light - 8 Foot	⊙——
L	Wall Bath Strip	○-
M	Ceiling Light	○
Lighting Schedule	Revised 1-1-xxxx	💡

Figure 6–16
Lighting Fixture Schedule

See the lighting fixture count listed in Figure 6–17.

Note. Some items not listed on the lighting fixture schedule should be taken-off with the lighting fixtures. They include: smoke detectors, exhaust fans, and range hood fan.

Once you have counted all of the lighting fixture symbols for Blueprints E–3 and E–4, complete the Lighting Fixture Quote Worksheet. Fax the Lighting Fixture Quote Worksheet, Figure 6–18 (page 114), with a copy of the lighting fixture schedule, Figure 6–16, to at least three suppliers for pricing.

TAKE-OFF WORKSHEET Figure 6-17			
LIGHTING FIXTURE COUNT			
Fixtures	E-3	E-4	Total
Post Light – Type A	1	*	1
Outside Wall Light – Type B	6	*	6
Keyless (attic/workshop) – Type C	4	2	6
Fluorescent – Workbench – Type D	*	1	1
Fluorescent – Lay-In – Type E	*	6	6
Fluorescent – Strip Type F	1	*	1
Fluorescent Wrap (Kitchen) – Type G1	2	*	2
Fluorescent Wrap (Laundry) – Type G2	1	*	1
Paddle Fan – Type H	3	*	3
Pull Chain Fixture – Type I	*	3	3
Recessed Light – Type J	10	6	16
Track Light 4' – Type K1	1	*	1
Track Light 8' - Type K2	1	*	1
Wall Fixture (Bath) – Type L	3	*	3
Ceiling Fixture – Type M	6	*	6
Smoke Detector	2	2	4
Exhaust Fan	2	2	4
Range Hood Fan	1	*	1

Step 2 – Count Switch Symbols

See the *Estimating Workbook* for Blueprints E–5 and E–6 required for this section.

This process can be divided into three sections: count the individual switch symbols, trace the switch legs, and count the number of ganged boxes.

1. The first switch symbol to count is the single pole switch. As you count your single pole switch symbols, mark them with the color blue. Transfer the single pole switch count to the worksheet, then count the three-way switch symbols, the four-way switch symbols, dimmer symbols, fan control switch symbols, etc.

2. Once you have counted all of the switch symbols, lightly trace the switch legs from lighting fixture to lighting fixture with a blue pencil. Tracing the switch legs should help you find other switches or lighting fixtures you might have missed.

3. The final step is to count the number of ganged switch boxes. This step is not required, but it's recommended so that when you order the material, you have the correct number of ganged boxes.

See the switch count listed in Figure 6–19.

Step 3 – Count Convenience Receptacle Symbols

See the *Estimating Workbook* for Blueprints E–7 and E–8 required for this section.

The counting of convenience 15-ampere receptacle symbols is very quick and there's no set sequence to follow. Use a light green pencil to mark the convenience receptacle symbols that have been counted. It's okay to count the receptacles for the clock receptacle and the garage door at this time, since they are on a general purpose circuit.

Note. Do not count special purpose receptacles on a separate circuit, such as the washer or freezer outlet.

FIXTURE QUOTE WORKSHEET Figure 6-18

Type	Description	Qnty.	Supplier - 1 Unit	Extension	Supplier - 2 Unit	Extension	Supplier - 3 Unit	Extension
A	Post Light	1						
B	Outside Wall Light	6						
C	Keyless	6						
D	Fluorescent - Workbench	1						
E	Fluorescent - Lay-In	6						
F	Fluorescent - Strip 8'	1						
G1	Fluorescent Wrap	2						
G2	Fluorescent Wrap	1						
H	Paddle Fan	3						
I	Pull Chain Fixture	3						
J	Recessed Light	16						
K1	Track Light 4' - 2 heads	1						
K2	Track Light 8' - 5 heads	1						
L	Wall Fixture (Bath)	3						
M	Ceiling Fixture	6						
	Miscellaneous							
	Smoke Detector	4						
	Exhaust Fan - By Owner							
	Total							

1. Include all associated cost, such as delivery, freight, and any fixture accessories listed on the blueprints or specifications.
2. Comply with Lighting Fixture Schedule enclosed with this fax.

SWITCH COUNT WORKSHEET Figure 6-19

Switches	E-5	E-6	Total
Single Pole Switch	21	5	26
Three Way Switch	21	1	22
Four Way Switch	2	*	2
Dimmer	2	*	2
Fan Control Switch	3	*	3
Pilot Switch	1	*	1
1 Gang Switch Box	17	4	21
2 Gang Switch Box	8	1	9
3 Gang Switch Box	3	*	3
4 Gang Switch Box	2	*	2

Author's Comment. According to the National Electrical Code, Section 210-8, GFCI protection is required for receptacles located outdoors, supplying kitchen counter top surfaces, bathrooms, garages, within 6 feet of wet bar sinks, and unfinished basements. According to Section 210-52, 20-ampere circuits are required for receptacles located in the kitchen, laundry and bathroom areas. 15-ampere circuits are required in all other areas such as bedrooms, living rooms, basements, and garages.

Note. Blueprints pages E-7 and E-8 indicate which receptacle outlets are of the GFCI type and which are GFCI protected by GFCI receptacles. See the convenience receptacle symbol count listed in Figure 6-20.

RECEPTACLE COUNT WORKSHEET Figure 6-20			
Receptacles - 15 Ampere Circuit	E-7	E-8	Total
Weatherproof - GFCI Receptacle	5	*	5
Convenience Receptacle	6	9	15
Switched Receptacle	21	*	21
Garage/Basement Wet Bar - GFCI Receptacle	2	2	4
Clock Outlet	1	*	1
Garage Door Receptacle	1	*	1
Receptacles - 20 Ampere Circuit			
Kitchen/Bath - GFCI Receptacle	4	*	4
Kitchen/Bath/Laundry Area - Convenience Receptacle	10	*	10

Note. The feed-thru GFCI protected receptacles in the garage, bathroom, kitchen and basement are counted as convenience receptacles, and not as a GFCI receptacle.

Count EMT Wiring Device Symbols

See the *Estimating Workbook* for Blueprint E-8 required for this section.

Since the receptacles on the workshop walls are required to be in EMT, you need to count them separately and mark them with a color to indicate that they have been counted. See the miscellaneous device count in Figure 6-21.

Step 4 – Special Systems

See the *Estimating Workbook* for Blueprints E-9 and E-10 required for this section.

For this house, special systems include the chime, CATV system, and telephone. The first step is to list the components for each system on the Take-Off Worksheet. Then count the device symbols, mark them with a red pencil to indicate that they have been counted, and transfer the count to the take-off worksheet.

EMT DEVICE COUNT WORKSHEET Figure 6-21		
EMT Wired Devices	E-6/E-8	Total
2 Gang – Switch/Switch	1	1
Receptacle Convenience – GFCI	2	2
Receptacle Convenience	1	1
Receptacle – Multioutlet Assembly	1	1

Once you have counted each of the device symbols, measure the circuit run lengths of the cables and transfer this value to the take-off worksheet. See the special system take-off listed in Figure 6-22.

Step 5 – Branch Circuit Wiring

See the *Estimating Workbook* for Blueprints E-11 and the E-12 required for this section.

There are two generally accepted methods for determining the branch circuit wiring for the lighting fixtures, switches, and general use receptacles. You can either measure the circuit wiring, or use an average run length per outlet.

Measuring Circuit Wiring

Before you measure the circuits, test the scale and verify that you have the correct scale. Do not measure the home runs at this time.

Begin by measuring the two-wire circuits, then the three-wire circuits. Draw a scaled line on the blueprint to represent the distance of the drops for the switches and receptacles. When you come across a drop for a switch or a receptacle, simply roll the distance from the pre-scaled line.

As you measure the wiring, trace the line that represents the wiring with a color pen or pencil. It doesn't matter what color you use, just be consistent. See the circuit wiring measurement listed in Figure 6-23.

SPECIAL SYSTEMS TAKE-OFF Figure 6-22			
Special Systems (Low Voltage)	E-9	E-10	Total
Chime System With Two Push-Buttons	*	*	1
Chime Wire 18/2	50 feet	*	50 Feet
Chime Wire 18/3	*	8 feet	8 feet
Chime Wire 18/5	42 feet	*	42 feet
Television Outlet	8	4	12
Television Wire	308 feet	156 feet	464 feet
Phone Outlet	6	2	8
Phone Outlet Weatherproof	1	*	1
Phone Wire	227 feet	54 feet	281 feet

CIRCUIT WIRING MEASUREMENT – ACTUAL METHOD Figure 6-23			
Branch Circuits	E-11	E-12	Total
14/2 NM Cable	646 feet	262 feet	908 feet
14/3 NM Cable	891 feet	91 feet	982 feet
12/2 NM Cable	134 feet	*	134 feet
14/2 UF Cable	40 feet	*	40 feet

CIRCUIT WIRING MEASUREMENT – AVERAGE METHOD Figure 6-24			
No. 14 Wire Branch Circuits	E-11	E-12	Total
Outlets	98	36	134
Total 14/2 NM Cable @ 7 Feet	686 ft.	252 ft.	938 ft.
Total 14/3 NM Cable @ 7 Feet	735 ft.	270 ft.	1,005 ft.
No. 12 Wire Branch Circuits			
Outlets	18		18
12/2 NM Cable (outlets × 8 feet)	144 ft.		144 ft.

Average Run Length

Another method of determining the circuit wiring is to use the average run length method. This method is much simpler than measuring each circuit. You simply multiply the number of outlet boxes times the average run length. For this estimate you will assume that past experience indicated that the average run length per outlet is about seven feet for 14/2 NM cable, seven one-half feet for 14/3 NM cable, and eight feet for 12/2 NM cable. The reason the 14/2 NM cable is low and the 14/3 NM cable is high is because of the great number of switch receptacles.

Note. The average run length for 14/2 NM cable is actually a little more than seven feet, and a little less than 7 1/2 feet for 14/3 NM.

Author's Comment. You must have historical experience to use an average run length per outlet.

See the average circuit wiring measurements listed in Figure 6-24.

Step 6 – Separate Circuits And Home Runs

See the *Estimating Workbook* for Blueprints E-13 and E-14 required for this section.

Taking-off the separate circuits and home runs requires the following steps:

1. Locate the termination symbol for each circuit in the blueprints and highlight it. No particular color is required, and don't worry about which circuit termination you highlight first.
2. Draw a line to represent the circuit wiring from the circuit termination symbol to the panel symbol.

   Note. Blueprint pages E-13 and E-14 use dotted lines to represent this step.

3. Measure the distance of the line that represents the circuit wiring and be sure to include the rise and drops at the termination. Make a note of the circuit length on the blueprints next to the circuit termination.
4. Make a copy of the panel schedule and transfer the circuit lengths from the blueprints to the panel schedule. If you don't have a copy machine or if there is no panel schedule, than make a worksheet that lists the circuit number, ampacity and length. See Figures 6-25, 6-25a, 6-26, and 6-26a.

Quotes

As you transfer the circuit lengths to the panel schedule or worksheet, determine the items that require special pricing, such as breakers, panel or disconnects. Make a note of these items on to the Gear Quote Worksheet (Figure 6-28 on page 119).

Subcontracting

If any subcontracting is required or if the job requires special equipment, be sure to make a note of this on your notepad.

Step 7 – Feeders

See the *Estimating Workbook* for Blueprint E-14 required for this section.

1. Locate the symbols that indicate where each feeder originates and where it terminates. Highlight these symbols in a bright color and draw a line to represent the feeder run.

   Note. Blueprint E-14 uses dotted lines to represent this.

PANEL A HOME RUN TAKE-OFF (E-13) Figure 6-25			
Panel A - 200 Amperes	Circuit #	Ampere	Length
Electric Heat	1 + 3	70 Ampere	10 feet
Water Pump	5 + 7	20 Ampere	60 feet
Hydromassage Bathtub	9	15 Ampere	51 feet
Bath Heat/Vent/Light Equipment	11	20 Ampere	48 feet
Freezer Receptacle Outlet	13	15 Ampere	29 feet
Entry/Porch Light Circuit	15	15 Ampere	57 feet
Workshop Light Circuit	17	15 Ampere	10 feet
Master Bedroom Circuit	19	15 Ampere	15 feet
Front Study/Bedroom Circuit	21	15 Ampere	40 feet
Sub-Feed to Panel B	27 + 29	100 Ampere	50 feet
Air Conditioning Equipment	2 + 4	40 Ampere	10 feet
Water Heater	6 + 8	20 Ampere	5 feet
Attic Fan Equipment	10	15 Ampere	41 feet
Bath Heat/Vent/Light Equipment	12	20 Ampere	40 feet
Bath and Hall Light Circuit	14	15 Ampere	41 feet
Front Bedroom Circuit	16	15 Ampere	56 feet
Work Bench Receptacle Circuit	18	15 Ampere	5 feet
Workshop Receptacle Circuit	20	15 Ampere	40 feet
Bath Receptacle Circuit - Master Bedroom	22	20 Ampere	52 feet

Panel A - Directory

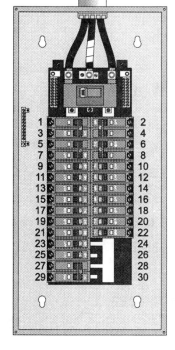

#		#	
1	Electric	2	A/C
3	Heat	4	Compressor
5	Water	6	Water
7	Pump	8	Heater
9	Hydromassage	10	Attic Fan
11	Bath / Heat / Vent	12	Heat Vent - M.B.
13	Freezer	14	Bath / Hall
15	Entry / Porch Lgt.	16	Front Bedroom
17	Workshop Lights	18	Work Bench Recp.
19	Master Bedroom	20	Workshop Recp.
21	Study / Bedroom	22	Mstr/Frt Bath Recp.
23	Spare	24	Space
25	Spare	26	Space
27	Sub-Feed	28	Space
29	Panel B	30	Space

Figure 6–25a
Panel A Directory

HOME RUN TAKE-OFF (E-13) Figure 6-26			
Panel B – 100 Amperes	Circuit #	Ampere	Length
Dryer	1 + 3	30 Ampere	35 feet
Dishwasher	5	20 Ampere	15 feet
Kitchen Light Circuit	7	15 Ampere	15 feet
Wet Bar Circuit	9	15 Ampere	5 feet
Recreation Room Receptacle Circuit	11	15 Ampere	5 feet
Kitchen Receptacle Circuit	13	20 Ampere	15 feet
Kitchen Receptacle Circuit	15	20 Ampere	15 feet
Living Room Circuit	17	15 Ampere	40 feet
Disposal	19	20 Ampere	20 feet
Powder Room Receptacle Circuit	21	20 Ampere	46 feet
Cook Top	2 + 4	40 Ampere	20 feet
Oven	6 + 8	30 Ampere	12 feet
Laundry/Powder Room Lighting Circuit	10	15 Ampere	35 feet
Recreation Room Lighting Circuit	12	15 Ampere	23 feet
Garage Circuit	14	15 Ampere	45 feet
Kitchen Receptacle Circuit	16	20 Ampere	35 feet
Washing Machine Receptacle Circuit	18	20 Ampere	40 feet
Laundry Receptacle Circuit	20	20 Ampere	45 feet

Panel B - Directory

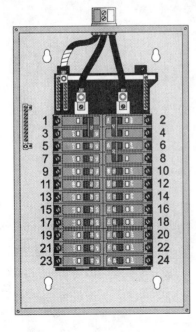

1 3	Dryer	2 4	Range
5	Dishwasher	6	Oven
7	Kitchen Lights	8	
9	Wet Bar	10	Laundry/Powd'r Rm
11	Rec Room Recp.	12	Rec Room Lights
13	Kitchen Recp.	14	Garage
15	Kitchen Recp.	16	Kitchen Recp./Ref
17	Living Room	18	Washer
19	Disposal	20	Laundry
21	Powder Rm Recp.	22	Spare
23	Spare	24	Spare

Figure 6–25a
Panel B Directory

| FEEDER TAKE-OFF Figure 6–27 |||||
From	To	Raceway	Wire	Length
Service	Panel A	2" PVC	Two No. 2/0 and One No. 1	10 feet
Panel A	Panel B	1¼" EMT	Two No. 3 and One No. 6	50 feet

2. Now measure the lines that represent each feeder and mark the length on the blueprints next to the feeder termination. Careful, be sure you verify the scale by using a ruler against a known distance.
3. Transfer the length for each of the feeders to the take-off worksheet. See Figure 6–27.

Quotes

As you transfer the length of the feeders to the panel schedule and take-off worksheet, determine if any items require special pricing: panels, circuit breakers, disconnects, fuses, etc. Transfer these items to the Gear Quote Worksheet. See Figure 6–28.

Subcontracting

If any subcontracting is required or if the job requires special equipment, be sure to make a note of this on your notepad.

Step 8 – Service Conductors And Equipment

Take-off as quickly as possible the main components that represent enough detail for you to determine the bill-of-material. For this estimate, you only need to note that the service is 200 amperes back-to-back.

Gear Quote

As you take-off the service equipment, determine which items require special pricing, such as meters and disconnects, fuses, etc. Transfer these items to the Gear Quote Worksheet and fax this to at least three suppliers for pricing. See Figure 6–28.

Subcontracting

If any subcontracting is required or if the job requires special equipment, be sure to make a note of this on your note pad.

| GEAR QUOTE WORKSHEET Figure 6–28 ||||||||
| Description | Qnty. | Supplier – 1 || Supplier – 2 || Supplier – 3 ||
		Unit	Extension	Unit	Extension	Unit	Extension
Breaker 1 Pole (15–20 Ampere) *	29						
Breaker 2 Pole (15–50 Ampere)	6						
Breaker 2 Pole 70 Ampere	1						
Breaker 2 Pole 100 Ampere	1						
Disconnect 60 Ampere	1						
Disconnect 70 Ampere	1						
Multioutlet Assembly 8'	1						
Meter - 200 Ampere	1						
Panel 100 Ampere	1						
Panel 200 Ampere	1						
Ground Rod Copper	2						
Total							

* **Note.** GFCI protection provided by GFCI feed-thru receptacles.

6.03 COMPLETED RESIDENTIAL TAKE-OFF

To help you see the big picture, review the take-off worksheets Figures 6–29, 6–30 and 6–31.

COMPLETED RESIDENTIAL TAKE-OFF (1 of 3) Figure 6-29			
Lighting Fixtures, Switches and Miscellaneous			1 of 3
	E-3	E-4	Total
Fixtures			
Post Light – Type A	1	*	1
Outside Wall Light – Type B	6	*	6
Keyless (Attic/Workshop) – Type C	4	2	6
Fluorescent – Workbench – Type D	*	1	1
Fluorescent – Lay-In – Type E	*	6	6
Fluorescent – Strip Type F	1	*	1
Fluorescent Wrap (Kitchen) – Type G1	2	*	2
Fluorescent Wrap (laundry) – Type G2	1	*	1
Paddle Fan – Type H	3	*	3
Pull Chain Fixture – Type I	*	3	3
Recessed Light – Type J	10	6	16
Track Light 4' – Type K1	1	*	1
Track Light 8' – Type K2	1	*	1
Wall Fixture (bath) – Type L	3	*	3
Ceiling Fixture – Type M	6	*	6
Miscellaneous Ceiling			
Smoke Detector	2	2	4
Exhaust Fan	2	2	4
Range Hood Fan	1	*	1
Switches	E-5	E-6	Total
Single Pole Switch	21	5	26
Three Way Switch	21	1	22
Four Way Switch	2	*	2
Dimmer	2	*	2
Fan Controlled Switch	3	*	3
Pilot Switch	1	*	1
Switch - 1 Gang Box	17	4	21
Switch - 2 Gang Box	9	*	9
Switch - 3 Gang Box	3	*	3
Switch - 4 Gang Box	2	*	2

COMPLETED RESIDENTIAL TAKE-OFF (2 of 3) Figure 6-30			
Receptacles, Low Voltage, And Branch Circuit Wire			2 of 3
Receptacles – 15 ampere Circuit	E-7	E-8	Total
Weatherproof	5	*	5
Convenience	6	9	15
Switched	21	*	21
Garage/Recreation Room – GFCI	2	2	4
Clock	1	*	1
Garage Door	1	*	1
Receptacles – 20 Ampere Circuit			
Kitchen/Bath – GFCI	4	*	4
Kitchen/Bath/Laundry – Convenience	10	*	10
EMT Wired Devices			
2 Gang - Switch/Switch	*	1	1
Receptacle Convenience – GFCI	*	2	2
Receptacle Convenience	*	1	1
Receptacle – Multioutlet	*	1	1
Low Voltage Systems	E-9	E-10	Total
Chime System With Two Push-buttons	*	*	1
Television Outlet	8	4	12
Phone Outlet	6	2	8
Phone Outlet Weatherproof	1	*	1
Phone Outlet (EMT)	*	1	1
Television Outlet (EMT)	*	1	1
Branch Circuits	E-11	E-12	Total
14/2 NM Cable	646 Feet	262 Feet	908 Feet
14/3 NM Cable	891 Feet	91 Feet	982 Feet
12/2 NM Cable	134 Feet	*	134 Feet
14/2 UF Cable	40 Feet	*	40 Feet

COMPLETED RESIDENTIAL TAKE-OFF (3 of 3) Figure 6-31			
Home Runs And Separate Circuits			3 of 3
Panel A Circuits	Circuit #	Ampere	Length
Electric Heat	1-3	70 Ampere	10 feet
Water Pump	5-7	20 Ampere	60 feet
Hydromassage Bathtub	9	15 Ampere	51 feet
Bath Heat/Vent/Light Equipment	11	20 Ampere	48 feet
Freezer Receptacle Outlet	13	15 Ampere	29 feet
Entry/Porch Light Circuit	15	15 Ampere	57 feet
Workshop Light Circuit	17	15 Ampere	10 feet
Master Bedroom Circuit	19	15 Ampere	15 feet
Front Study/Bedroom Circuit	21	15 Ampere	40 feet
Sub-Feed To Panel B	27-29	100 Ampere	50 feet
Air Conditioning Equipment	2-4	40 Ampere	10 feet
Water Heater	6-8	20 Ampere	5 feet
Attic Fan Equipment	10	15 Ampere	41 feet
Bath Heat/Vent/Light Equipment	12	20 Ampere	40 feet
Bath and Hall Light Circuit	14	15 Ampere	41 feet
Front Bedroom Circuit	16	15 Ampere	56 feet
Work Bench Receptacle Circuit	18	15 Ampere	5 feet
Workshop Receptacle Circuit	20	15 Ampere	40 feet
Bath Receptacle Circuit (Master Bedroom)	22	20 Ampere	52 feet
Panel B Circuits			
Dryer	1-3	30 Ampere	35 feet
Dishwasher	5	20 Ampere	15 feet
Kitchen Light Circuit	7	15 Ampere	15 feet
Wet Bar Circuit	9	15 Ampere	5 feet
Recreation Room Receptacle Circuit	11	15 Ampere	5 feet
Kitchen Receptacle Circuit	13	20 Ampere	15 feet
Kitchen Receptacle Circuit	15	20 Ampere	15 feet
Living Room Circuit	17	15 Ampere	40 feet
Disposal	19	20 Ampere	20 feet
Powder Room Receptacle Circuit	21	20 Ampere	46 feet
Cooktop	2-4	40 Ampere	20 feet
Oven	6-8	30 Ampere	12 feet
Laundry/Powder Room Lighting Circuit	10	15 Ampere	35 feet
Recreation Room Lighting Circuit	12	15 Ampere	23 feet
Garage Circuit	14	15 Ampere	45 feet
Kitchen Receptacle Circuit	16	20 Ampere	35 feet
Washing Machine Receptacle Circuit	18	20 Ampere	40 feet
Laundry Receptacle Circuit	20	20 Ampere	45 feet
Service Back to Back		200 Ampere	Utility

6.04 DETERMINING THE BILL-OF-MATERIAL

Since you are estimating this job manually, you need to manually determine the bill-of-material. Your take-off should provide enough information so you can determine the type, size, and quality of all material items required to complete the installation.

Some parts of the take-off lend themselves to the use of a spreadsheet to determine the quantity of the most common items. This would include, but is not limited to lighting fixtures, switches, convenience receptacles and their circuit wiring, see Figures 6–32 through 6–37.

The bill-of-material for individual circuits cannot be determined by the use of a spreadsheet. For these items, you must list the material requirement directly on to the Price/Labor Worksheet.

Note. Labor unit for lighting fixtures are listed in the labor unit manual in Chapter 1 of the Electrical Estimating Workbook.

From Spreadsheet To Price/Labor Sheet

Once you have determined the material items and their quantities by the use of a spreadsheet, transfer this information on to the Pricing/Labor Worksheet. See Figures 6–38 and 6–39.

LIGHTING FIXTURES WORKSHEET Figure 6-32					
Lighting Fixtures		Boxes			Installation
	Qnty.	Round	Fan	4 x 4	Hours
Post Light At 1 Hour	1				2.00
Outside Wall Light At .25 Hour	6	6			1.50
Keyless At .25 Hour	6	6			1.50
Fluorescent - Workbench	1	Receptacle For Fixture Connection			
Fluorescent - Lay-In At .75 Hour	6			3	4.50
Fluorescent - Strip At .75 Hour	1				0.75
Fluorescent Wrap - Kitchen At .75 Hour	2	2			1.50
Fluorescent Wrap - Laundry At .75 Hour	1	1			0.75
Paddle Fan/Light At .75 Hour	3		3		2.25
Pull Chain Fixture At .25 Hour	3	3			0.75
Recessed Light At 1 Hour	16				16.00
Track Light 4' At .75 Hour	1	1			0.75
Track Light 8' At 1 Hour	1	1			1.00
Wall Fixture (Bath) At .25 Hour	3	3			0.75
Ceiling Fixture At .25 Hour	6	6			1.50
Miscellaneous					
Smoke Detector At .25 Hour	4	4			1.00
Exhaust Fan At 1 Hour	4				4.00
Range Exhaust Fan At .5 Hour	1				0.50
Total	65	33	3	3	41.00

SWITCHES WORKSHEET Figure 6-33

Switch	Qnty.	Box				Switch						Plate			
		1G	2G	3G	4G	1P	3W	4W	Dim	Fan	Plt	1G	2G	3G	4G
1 Pole	26					26									
3 Way	22						22								
4 Way	2							2							
Dimmer	2								2						
Fan Switch	3									3					
Pilot Switch	1										1				
1 Gang Box	21	21										21			
2 Gang Box	9		9										9		
3 Gang Box	3			3										3	
4 Gang Box	2				2										2
Total	91	21	9	3	2	26	22	2	2	3	1	21	9	3	2

RECEPTACLES WORKSHEET Figure 6-34

Receptacle	Quantity	Box	Receptacle		Plate		
		1G	Standard	GFCI	Std	GFCI	WP
15 Ampere Circuit							
Weatherproof Receptacle	5	5	*	5	*	*	5
Convenience Receptacle	5	5	5	*	5	*	*
Garage Receptacle	9	9	9	*	9	*	*
Switched Receptacle	21	21	21	*	21	*	*
Garage/Recreation Room - GFCI	4	4	*	4	*	4	*
Clock Receptacle	1	1	1	*	1	*	*
Garage Door Receptacle	1	1	1	*	1	*	*
Workbench Light Outlet	1	1	1	*	1	*	*
20 Ampere Circuit							
Kitchen - GFCI Receptacle	2	2	*	2	*	2	*
Bath - GFCI Receptacle	2	2	*	2	*	2	*
Kitchen - Convenience Receptacle	7	7	7	*	7	*	*
Bath - Convenience Receptacle	1	1	1	*	1	*	*
Total	59	59	46	13	46	8	5

EMT WIRED DEVICES WORKSHEET Figure 6-35

EMT Devices	Qnty.	Box	Receptacle		Plate		Switch	Plate	EMT	Wire
		4" × 4"	Reg.	GFCI	Rec.	GFCI	1Pole	2Gang	1/2"	No. 12
2 Gang - Switch	1	1	*	*	*	*	2	1	15	38
Receptacle - GFCI	2	2	*	2	*	2	*	*	30	75
Receptacle	1	1	1	*	1	*	*	*	15	38
Multioutlet	1	*	*	*	*	*	*	*	15	38
Total	5	4	1	2	1	2	2	1	75	189

SPECIAL SYSTEMS (LOW VOLTAGE) WORKSHEET Figure 6-36

	Qnty.	Ring	Box	Plate			Cable			EMT
			4 × 4	TV	Phone	18/2	18/5	TV	Phone	1/2"
Low Voltage										
Chime With Push-buttons	2					50	52			
Television Outlet	11	11		11				494		
Phone Outlet	7	7			7				284	
Phone Outlet WP	1	1			1 (WP)					
Phone Outlet (EMT)	1		1		1					20
Television Outlet (EMT)	1		1	1						20
Total	22	19	2	12	8 + 1	50	52	494	284	40

GENERAL CIRCUIT AND HOME RUNS WORKSHEET Figure 6-37

Home Runs		Breaker	NM Cable	
	Amps	1 Pole	14/2	12/2
Panel A				
Entry/Porch Light Circuit	15	1	57	
Workshop Light Circuit	15	1	10	
Master Bedroom Circuit	15	1	15	
Front Study/Bedroom Circuit	15	1	40	
Bath and Hall Light Circuit	15	1	41	
Front Bedroom Circuit	15	1	56	
Work Bench Receptacle Circuit	15	1	5	
Workshop Receptacle Circuit	15	1	40	
Bath Receptacle Circuit (Master Br)	20	1		52
Panel B				
Kitchen Light Circuit	15	1	15	
Wet Bar Circuit	15	1	5	
Recreation Room Receptacle Circuit	15	1	5	
Kitchen Receptacle Circuit	20	1		15
Kitchen Receptacle Circuit	20	1		15
Living Room Circuit	15	1	40	
Powder Room Receptacle Circuit	20	1		46
Laundry/Powder Room Lighting Circuit	15	1	35	
Recreation Room Lighting Circuit	15	1	23	
Garage Circuit	15	1	45	
Kitchen Receptacle Circuit	20	1		35
Laundry Receptacle Circuit	20	1	35	
Total		21	467	163

PRICING/LABORING WORKSHEET (1 OF 7) Figure 6-38							
	Quantity	Cost	Unit	Extension	Labor	Unit	Extension
Fixtures							
Round Plastic Box	33						
Paddle Fan Box	3						
Box 4" x 4"	3						
Fixture Labor	1						
Switches							
1 Gang Box	21						
2 Gang Box	9						
3 Gang Box	3						
4 Gang Box	2						
1 Pole Switch	26						
3 Way Switch	22						
4 Way Switch	2						
Dimmer Switch	2						
Fan Switch	3						
Pilot Switch	1						
1 Gang Plate	21						
2 Gang Plate	9						
3 Gang Plate	3						
4 Gang Plate	2						
Receptacles							
Box 1 Gang	59						
Receptacle	46						
Receptacle GFCI	13						
Plate - Receptacle	46						
Plate - GFCI (Decorator)	8						
Plate - WP Receptacle	5						

PRICING/LABORING WORKSHEET (2 OF 7) Figure 6-39							
	Quantity	Cost	Unit	Extension	Labor	Unit	Extension
Miscellaneous							
Box - 4" × 4"	4						
Receptacle	1						
Receptacle GFCI	2						
Plate Receptacle - Raised	1						
Plate GFCI - Raised	2						
Switch 1 Pole	2						
Plate 2 G Switch - Raised	1						
1/2" EMT	75						
EMT - Connector	8						
EMT - Coupling	6						
EMT - Strap	6						
No. 12 Wire	189						
Special Systems							
Ring - 1 Gang	19						
Box - Metal 4" × 4"	2						
TV Outlet	12						
Phone Outlet	9						
18/2 Cable	50						
18/3 Cable	8						
18/5 Cable	42						
TV Cable	464						
Phone Cable	281						
1/2" EMT	40						
1/2" EMT Connector	2						
1/2" EMT Strap	2						
Chime	2						
Push Button	2						
Ring 1 Gang	2						
Branch Circuits And Home Runs							
14/2 NM Cable	908						
14/3 NM Cable	982						
12/2 NM Cable	134						
14/2 UF Cable	40						
1 Pole Breaker	21						
14/2 NM Cable	467						
14/3 NM Cable	5						
12/2 NM Cable	163						

Bill-Of-Material Individual Circuits, Feeders And Service

A spreadsheet cannot be used to determine the bill-of-material for the remaining items that have been taken-off. Therefore you must determine the individual items separately. This is accomplished by simply listing the material items and their quantities on to the Price/Labor Worksheet. See Figures 6–40 through 6–44.

PRICING/LABORING WORKSHEET (3 OF 7) Figure 6-40							
	Quantity	Cost	Unit	Extension	Labor	Unit	Extension
Electric Heat (A1-3)							
Breaker - 70 ampere	1						
Disconnect - 70 ampere	1						
1" EMT	10						
1" EMT Connector	2						
1" EMT Coupling	2						
1" EMT Strap	2						
1" Flex	6						
1" Flex Straight Connector	1						
1" Flex 90 Connector	1						
No. 4 Wire	40						
No. 8 Wire	8						
Water Pump (A5-7)							
Breaker - 20 Ampere (2-Pole)	1						
Metal Box 2" X 4"	1						
Switch - Single Pole 15 Ampere	1						
Switch Plate - Raised 2" X 4"	1						
1/2" EMT	60						
1/2" EMT Connector	2						
1/2" EMT Coupling	7						
1/2" Flex	3						
1/2" Flex Straight Connector	1						
1/2" Flex 90 Connector	1						
1/2" Strap	8						
No. 14 Wire	125						
Hydromassge Bathtub (A9)							
Breaker 15 Ampere (1 Pole)	1						
Box - Metal 4" X 4"	1						
Switch - 15 Ampere (1 Pole)	1						
Cover - Switch Raised	1						
1/2" Flex	6						
No. 14 Wire	20						
14/2 NM Cable	51						

PRICING/LABORING WORKSHEET (4 OF 7) Figure 6-41

	Quantity	Cost	Unit	Extension	Labor	Unit	Extension
Heat, Vent, Light, And Switch (A11)							
Breaker - 20 Ampere (1-Pole)	1						
Equipment Installation	1						
12/2 NM Cable	50						
NM Cable Connector	1						
Box - Plastic 1 Gang	1						
Switch - (Special)	1						
Freezer Receptacle (EMT Supply) (A13)							
Breaker - 15 Ampere (1 Pole)	1						
Box - Metal 2" × 4"	1						
Receptacle - 15 A (Single)	1						
Plate - Receptacle Raised	1						
1/2" EMT	30						
1/2" EMT Connector	2						
1/2" EMT Coupling	4						
1/2" EMT Strap	4						
No. 14 Wire	90						
Air Conditioning Equipment (A2-4)							
Breaker - 40 Ampere (2-Pole)	1						
Disconnect - 60 Ampere	1						
8/2 NM Cable	10						
NM Cable Connector	1						
1/2" Liquidtight	6						
1/2" Liquidtight - Straight Connector	1						
1/2" Liquidtight - 90 Connector	1						
No. 8 Wire	15						
No. 10 Wire	8						
Water Heater (A6-8)							
Breaker - 20 Ampere (2 Pole)	1						
Box - Metal 4" × 4"	1						
Switch - 20 Ampere (2 Pole)	1						
Cover - Switch Raised	1						
No. 12 Wire	20						
1/2" Flex	6						
1/2" Flex - Straight Connector	1						
1/2" Flex - 90 Connector	1						
1/2" EMT	5						
1/2" EMT Connector	2						
1/2" EMT Coupling	1						
1/2" EMT Strap	1						

PRICING/LABORING WORKSHEET (5 OF 7) Figure 6-42							
	Qnty.	Cost	Unit	Extension	Labor	Unit	Extension
Attic Fan (A10)							
Breaker - 15 Ampere (1 Pole)	1						
Box - Plastic 1 Gang	1						
14/2 NM Cable	45						
NM Cable Connector	1						
Switch - Attic Fan	1						
Heat, Vent, Light, And Switch (A11)							
Breaker - 20 Ampere (1 Pole)	1						
Equipment Installation	1						
12/2 NM Cable	40						
NM Cable - Connector	1						
Box - Plastic 1 Gang	1						
Switch - Fan/Vent/Light	1						
Panel B (A21-29)							
Panel - 100 Ampere	1						
Breaker - 100 Ampere (2 Pole)	1						
1¼" EMT	50						
1¼" EMT Elbow	3						
1¼" EMT Connector	2						
1¼" EMT Coupling	8						
1¼" EMT Strap	6						
1¼" Plastic Bushing	2						
No. 3 Wire	125						
No. 6 Wire	65						
Dryer (B2-4)							
Breaker - 30 Ampere (2 Pole)	1						
Box - Plastic 2 Gang	1						
Receptacle - 30 Ampere	1						
Plate (Plastic) - 30 Ampere	1						
10/3 NM Cable	35						
NM Cable Connector	1						
Dryer Cord 4 Wire	1						

PRICING/LABORING WORKSHEET (6 OF 7) Figure 6-43							
	Quantity	Cost	Unit	Extension	Labor	Unit	Extension
Dishwasher/Disposal (B5-19)							
Breaker - 20 Ampere (2 Pole)	1						
Box - Plastic 1 Gang	2						
Receptacle - 15 Ampere	1						
Plate (Plastic) - Receptacle	1						
Switch - 1 Pole	1						
Plate (Plastic) - Switch	1						
12/3 NM Cable	20						
NM Cable Connector	2						
Cord - 3'	1						
Cord - 6'	1						
Cooktop (B2-4)							
Breaker - 40 Ampere (2 Pole)	1						
Box - Metal 4" X 4"	1						
Cover - 4" X 4" Raised	1						
8/3 NM Cable	20						
NM Cable - Connector	1						
Oven (B6-8)							
Breaker - 30 Ampere (2 Pole)	1						
Box - Metal 4" X 4"	1						
Cover - 4" X 4" Metal	1						
1/2" Flex	6						
1/2" Flex - Straight Connector	1						
1/2" Flex - 90 Connector	1						
10/3 NM Cable	15						
NM Cable - Connector	1						
Washing Machine (B18)							
Breaker - 20 Ampere (1 Pole)	1						
Box - Plastic 1 Gang	1						
Receptacle Single 20 Ampere	1						
Plate (Plastic) - Receptacle	1						
12/2 NM Cable	40						
NM Cable Connector	1						

PRICING/LABORING WORKSHEET (7 OF 7) Figure 6-44							
	Quantity	Cost	Unit	Extension	Labor	Unit	Extension
Meter							
Meter/Main	1						
2" Rigid Conduit	10						
Rigid Conduit - Strap	2						
Weatherhead	1						
No. 2/0 Wire	30						
No. 1 Wire	15						
Grounding							
Ground Rod - ½" Copper	2						
Ground Clamp - Direct Burial	2						
No. 6 Copper Wire (Bare)	15						
½" PVC	10						
½" PVC Male Adapter	1						
½" Locknut	1						
Bonding							
No. 4 Wire	10						
Ground Clamp - Water Pipe	1						
Panel A							
Panel - 200 Ampere	1						
2" PVC	10						
2" PVC - LB	2						
2" PVC - Coupling	1						
2" PVC - Male Adapter	2						
2" Locknut	2						
2" Bushing	2						
No. 2/0 Wire	30						
No. 1 Wire	15						

6.05 PRICING, LABOR, EXTENDING, AND TOTALS

Now that you have determined the bill-of-material, you need to price and labor each item from the catalog in Chapter 1 of the Electrical Estimating Workbook. Then you can extend each item and determine the total material cost and total labor hours per worksheet. See Figures 6–45 through 6–50 for pricing, laboring, extending and totals for the residence.

Note. The previous seven Pricing/Laboring Worksheets have been consolidated into the following six pages.

PRICING/LABORING WORKSHEET (1 OF 6) Figure 6-45							
	Quantity	Cost	Unit	Extension	Labor	Unit	Extension
Fixtures							
Round Plastic Boxes	33	$89.00	C	$29.37	10.00	C	3.30
Paddle Fan Boxes	3	$399.00	C	$11.97	18.00	C	0.54
Box 4" X 4"	3	$59.00	C	$1.77	18.00	C	0.54
Fixture Labor	1	*	E	*	1.00	E	41.00
Switches							
Switch 1 Pole	26	$57.00	C	$14.82	20.00	C	5.20
3 Way Switch	22	$127.00	C	$27.94	25.00	C	5.50
4 Way Switch	2	$719.00	C	$14.38	28.00	C	0.56
Dimmer Switch	2	$610.00	C	$12.20	25.00	C	0.50
Fan Switch	3	$930.00	C	$27.90	25.00	C	0.75
Pilot Switch	1	$507.00	C	$5.07	25.00	C	0.25
1 Gang Switch Plate	21	$47.00	C	$9.87	2.50	C	0.53
2 Gang Switch Plate	9	$81.00	C	$7.29	4.00	C	0.36
3 Gang Switch Plate	3	$99.00	C	$2.97	6.00	C	0.18
4 Gang Switch Plate	2	$129.00	C	$2.58	7.00	C	0.14
1 Gang Switch Box	21	$97.00	C	$20.37	10.00	C	2.10
2 Gang Switch Box	9	$167.00	C	$15.03	12.00	C	1.08
3 Gang Switch Box	3	$174.00	C	$5.22	16.00	C	0.48
4 Gang Switch Box	2	$225.00	C	$4.50	20.00	C	0.40
Receptacles							
Box 1 Gang	59	$97.00	C	$57.23	10.00	C	5.90
Receptacle	46	$52.00	C	$23.92	18.00	C	8.28
Receptacle GFCI	13	$12.00	E	$156.00	0.30	E	3.90
Plate - Receptacle	46	$47.00	C	$21.62	2.50	C	1.15
Plate - GFCI (Decorator)	8	$59.00	C	$4.72	2.50	C	0.20
Plate - WP Receptacle	5	$389.00	C	$19.45	4.00	C	0.20
Miscellaneous							
Box - Metal 4" X 4"	4	$59.00	C	$2.36	18.00	C	0.72
Receptacle	1	$52.00	C	$0.52	18.00	C	0.18
Receptacle GFCI	2	$12.00	E	$24.00	0.30	E	0.60
Plate Receptacle - Raised	1	$99.00	C	$0.99	6.00	C	0.06
Plate GFCI - Raised	2	$99.00	C	$1.98	6.00	C	0.12
Switch 1 Pole	2	$57.00	C	$1.14	20.00	C	0.40
Plate 2 G Switch - Raised	1	$99.00	C	$0.99	6.00	C	0.06
½" EMT	75	$13.00	C	$9.75	2.25	C	1.69
½" EMT - Connector	8	$21.00	C	$1.68	2.00	C	0.16
½" EMT - Coupling	6	$24.00	C	$1.44	2.00	C	0.12
½" EMT - Strap	6	$6.00	C	$0.36	2.50	C	0.15
No. 12 Wire	189	$48.00	M	$9.07	4.25	M	0.80
Total				$550.42			88.10

PRICING/LABORING WORKSHEET (2 OF 6) Figure 6-46							
	Quantity	Cost	Unit	Extension	Labor	Unit	Extension
Special Systems							
Ring - 1 Gang	19	$39.00	C	$7.41	4.50	C	0.86
Box - Metal 4" X 4"	2	$59.00	C	$1.18	18.00	C	0.36
TV Outlet	12	$300.00	C	$36.00	25.00	C	3.00
Phone Outlet	9	$300.00	C	$27.00	25.00	C	2.25
18/2 Cable	50	$70.00	M	$3.50	13.00	M	0.65
18/3 Cable	8	$120.00	M	$0.96	13.00	M	0.10
18/5 Cable	42	$120.00	M	$5.04	13.00	M	0.55
TV Cable	464	$180.00	M	$83.52	12.00	M	5.57
Phone Cable	281	$90.00	M	$25.29	12.00	M	3.37
½" EMT	40	$13.00	C	$5.20	2.25	C	0.90
½" EMT Connector	2	$21.00	C	$0.42	2.00	C	0.04
½" EMT Strap	2	$6.00	C	$0.12	2.50	C	0.05
Chime	2	*	*	*	1.00	E	2.00
Push Buttons	2	$3.00	C	$0.06	2.50	C	0.05
Ring - 1 Gang	2	$39.00	C	$0.78	4.50	C	0.09
Branch Circuits And Home Runs							
14/2 MN Cable	908	$108.00	M	$98.06	10.00	M	9.08
14/3 NM Cable	982	$186.00	M	$182.65	12.00	M	11.78
12/2 NM Cable	134	$148.00	M	$19.83	12.00	M	1.61
14/2 UF Cable	40	$133.00	M	$5.32	5.00	M	0.20
Breaker - 1 Pole	21	*	*	*	0.10	E	2.10
14/2 NM Cable Homerun	467	$108.00	M	$50.44	10.00	M	4.67
14/3 NM Cable Homerun	5	$186.00	M	$0.93	12.00	M	0.06
12/2 NM Cable Homerun	163	$148.00	M	$24.12	12.00	M	1.96
Electric Heat							
Breaker - 70 Ampere (2 Pole)	1	*	*	*	0.20	E	0.20
Disconnect - 70 Ampere (2 Pole)	1	*	*	*	1.00	E	1.00
1" EMT	10	$31.00	C	$3.10	3.75	C	0.38
1" EMT Connector	2	$64.00	C	$1.28	5.00	C	0.10
1" EMT Coupling	2	$62.00	C	$1.24	5.00	C	0.10
1" EMT Strap	2	$17.00	C	$0.34	2.70	C	0.05
1" Flex	6	$82.00	C	$4.92	4.00	C	0.24
1" Flex Straight Connector	1	$95.00	C	$0.95	9.00	C	0.09
1" Flex 90 Connector	1	$495.00	C	$4.95	10.00	C	0.10
No. 4 Wire	40	$350.00	M	$14.00	7.25	M	0.29
No. 8 Wire	8	$146.00	M	$1.17	6.00	M	0.05
Total				$609.78			53.90

PRICING/LABORING WORKSHEET (3 OF 6) Figure 6-47							
	Quantity	Cost	Unit	Extension	Labor	Unit	Extension
Water Pump							
Breaker - 20 Ampere (1 Pole)	1	*	*	*	0.10	E	0.10
Box - Metal 2" × 4"	1	$59.00	C	$0.59	18.00	C	0.18
Switch - Single Pole 15 Ampere	1	$57.00	C	$0.57	20.00	C	0.20
Switch Plate - Raised 2" × 4"	1	$47.00	C	$0.47	4.00	C	0.04
1/2" EMT	60	$13.00	C	$7.80	2.25	C	1.35
1/2" EMT Connector	2	$21.00	C	$0.42	2.00	C	0.04
1/2" EMT Coupling	7	$24.00	C	$1.68	2.00	C	0.14
1/2" Flex	3	$31.00	C	$0.93	3.25	C	0.10
1/2" Flex Connector Straight	1	$55.00	C	$0.55	6.25	C	0.06
1/2" Flex Connector 90	1	$76.00	C	$0.76	7.50	C	0.08
1/2" Rigid Strap	8	$6.00	C	$0.48	2.50	C	0.20
No. 12 Wire	125	$46.00	M	$6.00	4.25	M	0.53
Hydromassge Bathtub							
Breaker 15 Ampere (1 Pole)	1	*	*	*	0.10	E	0.10
Box - Metal 4" × 4"	1	$59.00	C	$0.59	18.00	C	0.18
Switch - 15 Ampere (1 Pole)	1	$57.00	C	$0.57	20.00	C	0.20
Cover - Switch Raised	1	$99.00	C	$0.99	6.00	C	0.06
1/2" Flex	6	$31.00	C	$1.86	3.25	C	0.20
1/2" Flex Straight Connector	1	$55.00	C	$0.55	6.25	C	0.06
1/2" Flex 90 Connector	1	$76.00	C	$0.76	7.50	C	0.08
No. 14 Wire	20	$38.00	M	$0.76	3.60	M	0.07
14/2 NM Cable	51	$108.00	M	$5.51	10.00	M	0.51
Heat, Vent, Light And Switch							
Breaker - 20 Ampere (1 Pole)	1	*	*	*	0.10	E	0.10
Equipment Installation	1	*	*	*	1.00	E	1.00
12/2 NM Cable	50	$148.00	M	$7.40	12.00	M	0.60
NM Cable Connector	1	$16.00	C	$0.16	1.00	C	0.01
Box - Plastic 1 Gang	1	$97.00	C	$0.97	10.00	C	0.10
Switch - (Special)	1	$17.00	E	$17.00	0.50	E	0.50
Freezer Receptacle							
Breaker - 15 Ampere (1 Pole)	1	*	*	*	0.10	E	0.10
Box - Metal 2" × 4"	1	$59.00	C	$0.59	18.00	C	0.18
Receptacle - 15 A (Single)	1	$5.00	E	$5.00	0.20	E	0.20
Plate - Receptacle Raised	1	$99.00	C	$0.99	6.00	C	0.06
1/2" EMT	30	$13.00	C	$3.90	2.25	C	0.68
1/2" EMT Connector	2	$21.00	C	$0.42	2.00	C	0.04
1/2" EMT Coupling	4	$24.00	C	$0.96	2.00	C	0.08
1/2" EMT Strap	4	$6.00	C	$0.24	2.50	C	0.10
No. 14 Wire	90	$38.00	M	$3.42	3.60	M	0.32
Total				$72.89			8.55

PRICING/LABORING WORKSHEET (4 OF 6) Figure 6-48							
	Qnty.	Cost	Unit	Extension	Labor	Unit	Extension
Air Conditioning Equipment							
Breaker - 40 Ampere (2 Pole)	1	*	*	*	0.15	E	0.15
Disconnect - 60 Ampere	1	*	*	*	1.00	E	1.00
8/2 NM Cable	10	$583.00	M	$5.83	16.00	M	0.16
NM Cable Connector	1	$24.00	C	$0.24	2.00	C	0.02
1/2" Liquidtight	6	$47.00	C	$2.82	3.25	C	0.20
1/2" Liquidtight Straight Connector	1	$155.00	C	$1.55	6.00	C	0.06
1/2" Liquidtight 90 Connector	1	$205.00	C	$2.05	9.00	C	0.09
No. 8 Wire	15	$146.00	M	$2.19	6.00	M	0.09
No. 10 Wire	8	$78.00	M	$0.62	5.10	M	0.04
Water Heater							
Breaker - 20 Ampere (2 Pole)	1	*	*	*	0.15	E	0.15
Box - Metal 4" X 4"	1	$59.00	C	$0.59	18	C	0.18
Switch - 20 Ampere (2 Pole)	1	$1,200.00	C	$12.00	32	C	0.32
Cover - Switch Raised	1	$99.00	C	$0.99	6	C	0.06
No. 12 Wire	20	$48.00	M	$0.96	4.25	M	0.09
1/2" Flex	6	$31.00	C	$1.86	3.25	C	0.20
1/2" Flex Straight Connector	1	$55.00	C	$0.55	6.25	C	0.06
1/2" Flex 90 Connector	1	$76.00	C	$0.76	7.5	C	0.08
1/2" EMT	5	$13.00	C	$0.65	2.25	C	0.11
1/2" EMT Connector	2	$21.00	C	$0.42	2	C	0.04
1/2" EMT Coupling	1	$24.00	C	$0.24	2	C	0.02
1/2" EMT Strap	1	$6.00	C	$0.06	2.5	C	0.03
Attic Fan							
Breaker - 15 Ampere (1 Pole)	1	*	*	*	0.1	E	0.10
Box - Plastic 1 Gang	1	$97.00	C	$0.97	10	C	0.10
14/2 NM Cable	45	$108.00	M	$4.86	10	M	0.45
NM Cable Connector	1	$16.00	C	$0.16	1	C	0.01
Attic Fan Switch (Special)	1	$15.00	E	$15.00	0.5	E	0.50
Heat, Vent, Light And Switch							
Breaker - 20 Ampere (1 Pole)	1	*	*	*	0.1	E	0.1
Equipment Installation	1	*	*	*	1	E	1.00
12/2 NM Cable	40	$148.00	M	$5.92	12	M	0.48
NM Cable Connector	1	$16.00	C	$0.16	1	C	0.01
Box - Plastic 1 Gang	1	$97.00	C	$0.97	10	C	0.10
Switch - Fan/Vent/Light	1	$17.00	E	$17.00	0.5	E	0.50
Total				$79.42			6.50

PRICING/LABORING WORKSHEET (5 OF 6) Figure 6-49							
	Qnty.	Cost	Unit	Extension	Labor	Unit	Extension
Panel B							
Panel - 100 Ampere	1	*	*	*	1.00	E	1.00
Breaker - 100 Ampere (2 Pole)	1	*	*	*	0.20	E	0.20
1¼" EMT	50	$50.00	C	$25.00	4.00	C	2.00
1¼" EMT Elbows	3	$311.00	C	$9.33	15.00	C	0.45
1¼" EMT Connectors	2	$108.00	C	$2.16	7.00	C	0.14
1¼" EMT Coupling	8	$99.00	C	$7.92	7.00	C	0.56
1¼" EMT Straps	6	$19.00	C	$1.14	2.90	C	0.17
1¼" Plastic Bushing	2	$89.00	C	$1.78	9.00	C	0.18
No. 3 Wire	125	$375.00	M	$46.88	7.50	M	0.94
No. 6 Wire	65	$250.00	M	$16.25	7.00	M	0.46
Dryer							
Breaker - 30 Ampere (2 Pole)	1	*	*	*	0.15	E	0.15
Box - Plastic 2 Gang	1	$167.00	C	$1.67	12.00	C	0.12
Receptacle - 30 Ampere	1	$7.00	E	$7.00	0.25	E	0.25
Plate (Plastic) - 30 Ampere	1	$99.00	C	$0.99	5.00	C	0.05
10/3 NM Cable	35	$397.00	M	$13.90	16.00	M	0.56
NM Cable Connector	1	$16.00	C	$0.16	1.00	C	0.01
Dryer Cord 4 Wire	1	$12.00	E	$12.00	0.30	E	0.30
Dishwasher/Disposal							
Breaker - 20 Ampere (2 Pole)	1	*	*	*	0.15	E	0.15
Box - Plastic 1 Gang	2	$97.00	C	$1.94	10.00	C	0.20
Receptacle - 15 Ampere	1	$52.00	C	$0.52	18.00	C	0.18
Plate (Plastic) - Receptacle	1	$47.00	C	$0.47	2.50	C	0.03
Switch - Single Pole	1	$57.00	C	$0.57	20.00	C	0.20
Plate (Plastic) - Switch	1	$47.00	C	$0.47	2.50	C	0.03
12/3 NM Cable	20	$267.00	M	$5.34	15.00	M	0.30
NM Cable Connector	2	$16.00	C	$0.32	1.00	C	0.02
Cord - 3'	1	$400.00	C	$4.00	25.00	C	0.25
Cord - 6'	1	$400.00	C	$4.00	25.00	C	0.25
Cooktop							
Breaker - 40 Ampere (2 Pole)	1	*	*	*	0.15	E	0.15
Box - Metal 4" X 4"	1	$59.00	C	$0.59	18.00	C	0.18
Cover - 4" X 4" Raised	1	$99.00	C	$0.99	6.00	C	0.06
8/3 NM Cable	20	$820.00	M	$16.40	20.00	M	0.40
NM Cable - Connector	1	$23.62	C	$0.24	2.00	C	0.02
Total				$182.03			9.96

PRICING/LABORING WORKSHEET (6 OF 6) Figure 6-50							
	Quantity	Cost	Unit	Extension	Labor	Unit	Extension
Oven							
Breaker - 30 Ampere (2 Pole)	1	*	*	*	0.15	E	0.15
Box - Metal 4" × 4"	1	$59.00	C	$0.59	18.00	C	0.18
Cover - Metal 4" × 4"	1	$99.00	C	$0.99	6.00	C	0.06
1/2" Flex	6	$31.00	C	$1.86	3.25	C	0.20
1/2" Flex - Straight Connector	1	$55.00	C	$0.55	6.25	C	0.06
1/2" Flex - 90 Connector	1	$76.00	C	$0.76	7.50	C	0.08
10/3 NM Cable	15	$397.00	M	$5.96	16.00	M	0.24
NM Cable - Connector	1	$24.00	C	$0.24	2.00	C	0.02
Washing Machine							
Breaker - 20 Ampere (1 Pole)	1	*	*	*	0.10	E	0.10
Box - Plastic 1 Gang	1	$97.00	C	$0.97	10.00	C	0.10
Receptacle Single 20 Ampere	1	$500.00	C	$5.00	25.00	C	0.25
Plate (Plastic) - Receptacle	1	$47.00	C	$0.47	2.50	C	0.03
12/2 NM Cable	40	$148.00	M	$5.92	12.00	M	0.48
NM Cable Connector	1	$16.00	C	$0.16	1.00	C	0.01
Meter							
Meter/Main	1	*	*	*	2.00	E	2.00
2" Rigid Conduit	10	$228.00	C	$22.80	7.00	C	0.70
Rigid Conduit Strap	2	$28.00	C	$0.56	3.25	C	0.07
Weatherhead	1	$7.00	E	$7.00	0.40	E	0.40
No. 2/0 Wire	30	$816.00	M	$24.48	15.25	M	0.46
No. 1 Wire	15	$591.00	M	$8.87	8.00	M	0.12
Grounding							
Ground Rod - 1/2" Copper	2	$1,000.00	C	$20.00	50.00	C	1.00
Ground Clamp - Direct Burial	2	$174.00	C	$3.48	13.50	C	0.27
No. 6 Copper Wire (Bare)	15	$203.00	M	$3.05	7.25	M	0.11
1/2" PVC	10	$13.00	C	$1.30	1.90	C	0.19
1/2" PVC Male Adapter	1	$22.00	C	$0.22	4.50	C	0.05
1/2" Locknut	1	$15.00	C	$0.15	4.00	C	0.04
Bonding							
No. 4 Wire	10	$350.00	M	$3.50	7.25	M	0.07
Ground Clamp - Water Pipe	1	$164.00	C	$1.64	22.50	C	0.23
Panel A							
Panel - 200 Ampere	1	*	*	*	1.00	E	1.00
2" PVC	10	$49.00	C	$4.90	4.50	C	0.45
2" PVC - LB	2	$466.00	C	$9.32	27.00	C	0.54
2" PVC - Coupling	1	$68.00	C	$0.68	6.00	C	0.06
2" PVC - Male Adapter	2	$72.00	C	$1.44	12.50	C	0.25
2" Locknut	2	$69.00	C	$1.38	6.00	C	0.12
2" Bushing	2	$139.00	C	$2.78	13.00	C	0.26
No. 2/0 Wire	30	$816.00	M	$24.48	15.25	M	0.46
No. 1 Wire	15	$591.00	M	$8.87	8.00	M	0.12
Total				$174.37			10.93

6.06 ESTIMATE AND BID SUMMARY

One simple mistake at this point can be costly. This phase of the estimating process requires you to make judgments on intangibles such as job conditions, labor productivity, miscellaneous material requirements, waste, theft, small tools, direct job expenses, and overhead.

The Summary Worksheet is used to help organize the different costs so you don't make a mistake. The summary worksheet contains eight major sections:

Step A – Total Labor-Hours
Step B – Labor Cost
Step C – Adjusted Material Cost
Steps D, E, and F – Direct Job Cost
Step G – Estimated Prime Cost
Step H – Overhead
Steps I, J, K, L – Estimated Cost and Bid Price

Use the information contained on the Specification Check List, Figure 6–14, and the Estimate and Bid Notes, Figure 6–15 to complete the Estimate/Bid Summary worksheet, Figure 6–51.

Step A – Total Labor-Hours, 173 Hours

Once you have determined the estimated man-hours (178 hours), using a labor-unit manual, you must adjust those hours to account for specific job conditions. There are eighteen job site factors. See Section 3.09 that might impact the labor-unit total.

For this job, apply a labor-unit adjustment of −10%, and additional labor of +8%. See Section 5.01.

Author's Comment. For this job, we did not include the labor-hours for the following material items:

LABOR HOUR CALCULATION	
Labor-Unit Estimated Hours	177.94
Labor-Unit Adjustment, − 10%	− 17.79
Total Estimated Hours	160.14
Additional Labor, + 8%	12.81
Total Hours	172.95

MISCELLANEOUS ITEMS	
Item	Hours
Covers	0.25
NM Cable Nail Plates	1.60
Push buttons	0.50
PVC and LB	0.20
Multioutlet Assembly	1.00
NM Cable Staples	3.50
Wire Nuts	5.00
Other Items	0.76
Total	12.81 Hours

Step B – Labor Cost, $1,842

The labor cost for the job is determined by multiplying the labor man-hours by the labor rate per man-hour. See Section 5.02.

For this bid assume the following:
1 Foreman $12.45
1 Journeyman $11.00
1 Helper   + $8.50
Labor Rate $31.95
Labor Cost = $31.95/3 = $10.65

Note. Don't apply a labor burden factor at this time, because labor burden costs will be recovered when overhead is applied. See Section 5.03.

Labor Cost Calculation
Labor Cost = Labor-Hours × Labor Rate
Labor Cost = 179.95 hours × $10.65
Labor Cost = $1,841.96

Step C – Adjusted Material Cost, $1,969

Once you have determined the estimated material cost ($1,669), you must adjust this cost for the following factors (see Section 5.04):

1. Miscellaneous items not accounted for, +10%
2. Waste and theft, +5%
3. Small tools, +3%

ADJUSTED MATERIAL COST CALCULATION	
Price Sheet Total Cost	$1,668.97
Miscellaneous Items, + 10%	$166.90
Waste and Theft, + 5%	$83.45
Small Tool Allowance, + 3%	$50.07
Total Adjusted Material Cost	$1,969.39

Step D – Quotes, $425

Written quotes are included in the estimate at cost so be sure your supplier includes the cost for all accessories, freight, and delivery. Careful, don't make a mistake when transfering the supplier prices to the estimate summary worksheet.

Gear Cost — $425
Fixtures supplied by owner

Step E – Sales Tax, $168

For this bid, figure sales tax at 7% of total taxable material cost.

SALES TAX CALCULATION	
Pricing Worksheet Material Cost	$1,668.97
Miscellaneous Material Items, + 10%	$166.90
Waste And Theft, + 5%	$83.45
Small Tool Allowance, + 3%	$50.07
Total Adjusted Material Cost	$1,969.39
Quote - Gear	$425.00
Total Taxable Material Cost	$2,394.39
Sales Tax, + 7%	$167.61
Total Cost Of Material	$2,561.99

Step F – Direct Cost, $705

Direct job expenses are often not shown on the blueprints, but are probably indicated in the specifications. See Section 5.05 for details.

Permit cost — $375
Temporary service — $330

Step G – Estimated Prime Cost, $5,109

Estimated prime cost is the sum of labor, material, and direct job costs. This value is used in determining overhead. See Section 5.06.

ESTIMATED PRIME COST CALCULATION	
Labor Cost at $10.65 per hour	$1,841.96
Total Material Cost, including Tax	$2,561.99
Direct Job Cost	$705.00
Total Prime Cost	$5,108.95

Step H – Overhead Expenses, $2,044

Overhead cost should be the lesser of 40% of prime cost or $13 per man-hour. See Section 5.09.

Overhead Calculation
Percentage Method = $5,108.95 × 40% = $2,043.58
Hour Method = 172.95 hours × $13 = $2,249

Step I – Break Even Cost, $7,153

The estimated break even cost is the sum of prime cost plus overhead. See Section 5.10.

BREAK EVEN COST CALCULATION	
Total Prime Cost	$5,108.95
Overhead $5,108.95 at 40%	$2,043.58
Estimated Break Even Cost	$7,152.53

Step J – Profit, $1,073

To properly apply profit, the estimate must cover all contingencies fairly and correctly. For this job, figure profit to be 15% of break even cost. See Section 5.11.

PROFIT CALCULATION	
Total Prime Cost	$5,108.95
Overhead $5,075 at 40%	$2,043.58
Estimated Break Even Cost	$7,152.53
Profit at 15% of $7,152.53	$1,072.88

Step K – Other Final Costs (None)

Just before you submit your bid, you must consider if there are any other final costs such as bid bonds, finance cost, inspector problems, or other expenses that must be taken into consideration. See Section 5.12.

Step L – Bid Price, $8,225

The bid price is the sum of the estimated break even cost, plus profit, plus other final costs.

BID PRICE CALCULATION	
Estimated Break Even Cost	$7,152.53
Profit, + 15%	$1,072.88
Other Cost: None	$0
Bid Price	$8,225.41

Figure 6–51 is the summary worksheet for the residential bid.

	SUMMARY WORKSHEET Figure 6-51		
	Description		
	Labor Calculations	Hours	
	Labor Worksheet Page 1 of 6	88.10	
	Labor Worksheet Page 2 of 6	53.90	
	Labor Worksheet Page 3 of 6	8.55	
	Labor Worksheet Page 4 of 6	6.50	
	Labor Worksheet Page 5 of 6	9.96	
	Labor Worksheet Page 6 of 6	10.93	
	Pricing/Laboring Worksheet Labor-Hours	177.94	
	Labor-Unit Adjustment, − 10%	− 17.79	
	Total Adjusted Hours	160.14	
	Additional Labor Adjustment, + 8%	12.81	
Step A	Total Final Adjusted Hours	172.95	
Step B	Labor Cost - $10.65 Per Hour × 172.72 Hours		$1,841.96
	Material Cost	Dollars	
	Price Worksheet Page 1 of 6	$550.47	
	Price Worksheet Page 2 of 6	$609.79	
	Price Worksheet Page 3 of 6	$72.89	
	Price Worksheet Page 4 of 6	$79.42	
	Price Worksheet Page 5 of 6	$182.03	
	Price Worksheet Page 6 of 6	$174.37	
	Pricing Worksheet Material Cost	$1,668.97	
	Miscellaneous Material Items, + 10%	$166.90	
	Waste And Theft, + 5%	$83.45	
	Small Tool Allowance, + 3%	$50.07	
Step C	Total Adjusted Material Cost	$1,969.39	
Step D	Quote - Gear	$425.00	
	Total Taxable Material Cost	$2,394.39	
Step E	Sales Tax, + 7%	$167.61	
	Total Material Cost		$2,561.99
Step F	Direct Cost: Permit And Temporary		$705.00
Step G	Total Prime Cost ($1,839.36 + $2,558.96 + $705.00)		$5,108.95
Step H	Overhead $5,103 At 40%		$2,043.58
Step I	Break Even Cost		$7,152.53
Step J	Profit, + 15%		$1,072.88
Step K	Other Final Cost: None		$0.00
Step L	Bid Selling Price		$8,225.41

6.07 BID ACCURACY AND ANALYSIS

Before submitting the bid, review the Estimate/Bid Notes (Figure 6–14), and be sure you have included all costs. Now you must verify the bid's accuracy and perform a bid analysis.

Bid Accuracy

Make sure you have not made any of the following mistakes:

- Errors in multiplication or addition.
- Failing to include outside or underground work.
- Failure to visit site to determine job conditions.
- Forgetting a major item, such as gear quotes.
- Forgetting to include subcontract cost or equipment rental requirements.
- Forgetting to include the changes to the original specifications or blueprints.
- Hurrying and rushing the bid.
- Improper estimating forms.
- Not double checking all figures.
- Not transferring totals to the summary worksheet properly, transposing numbers.
- Omitting a section of the bid.
- Using supplier take-offs for quotes or depending on verbal quotes.
- Wrong extensions.
- Wrong scale on reduced blueprints.
- Wrong unit for labor-unit.
- Wrong unit for material cost.

Bid Analysis

It's impossible to perform extensive bid analysis if you estimate manually. However, you can determine the following bench-marks and compare them against other similar bids you have completed.

Cost Distribution

Labor	$1,842/$8,225 =	22.39%
Material	$2,562/$8,225 =	31.15%
Other	$705/$8,225 =	8.57%
Overhead	$2,044/$8,225 =	24.84%
Profit	$1,073/$8,225 =	13.04%

Cost Per Square Foot = $2.19

Cost Per Sq. Ft. = $8,216/3,744 square foot
Cost Per Sq. Ft. = $2.19 per square foot

Labor Analysis

Total Hours = . . .   172.72 hours
Total 8 Hour. Days (2 men)
(172.72 hours/16 hours) = 10.79 days
Total 5 Day Weeks
(10.79 days/5 days) =   2.16 wk

6.08 THE PROPOSAL

Now that you have completed the estimate and have established a bid price, you must submit a written proposal. The proposal must clarify what the bid includes and what is not included. Check the Estimate/Bid Notes Worksheet 6–15 for details. Since this bid is based on a manual estimate, use a preprinted wiring contract and apply the necessary corrections.

PART C – COMPUTER ASSISTED ESTIMATE AND BID

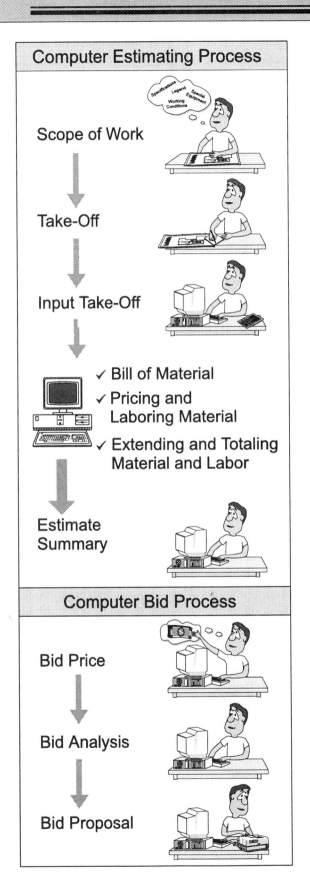

This part explains the computer assisted estimating process for residential wiring and gives you the approximate time to complete each phase of the bid. The total amount of time required from start to finish is about 1 hour.

6.09 BID PREPARATION AND TAKE-OFF

The bid preparation and the take-off using a computer is about the same as with the manual estimate method. With experience and proper take-off forms you should be able to review the blueprints and complete the take-off in about 30 minutes. Figures 6–52 and 6–53 contain the take-off details for the residence.

6.10 INPUT TAKE-OFF INTO COMPUTER

The next step is to input the take-off quantities in the computer. Once you are up to speed, this should take less than 15 minutes.

6.11 BILL-OF-MATERIAL, PRICING, LABORING, EXTENDING AND TOTALING

The computer automatically determines the bill-of-material, price and labor for each material item, determines the extended cost and labor-hour for each material item, as well as totaling in about 10 seconds.

6.12 ESTIMATE AND BID SUMMARY

It only takes about 3 minutes to input labor-hour adjustments, labor-rate, material adjustments, overhead and profit in the computer to produce the bid price.

6.13 BID ACCURACY AND ANALYSIS

Before you submit your bid you must verify its accuracy and perform a bid analysis.

TAKE-OFF WORKSHEET 1 OF 2 Figure 6-52

Fixtures	Total	Receptacles - 15 ampere Circuit	Total
Post Light - Type A	1	Weatherproof	5
Outside Wall Light - Type B	6	Convenience	14
Keyless (Attic/Workshop) - Type C	6	Switched	21
Fluorescent - Workbench - Type D	1	Garage/Recreation Room - GFCI	4
Fluorescent - Lay-In - Type E	6	Clock	1
Fluorescent - Strip Type F	1	Garage Door	1
Fluorescent Wrap (Kitchen)-Type G1	2		
Fluorescent Wrap (Laundry) - Type G2	1		
Paddle Fan - Type H	3	**Receptacles - 20 Ampere Circuit**	
Pull Chain Fixture - Type I	3	Kitchen/Bath - GFCI	4
Recessed Light - Type J	16	Kitchen/Bath/Laundry - Convenience	10
Track Light 4' - Type K1	1		
Track Light 8' - Type K2	1		
Wall Fixture (Bath) - Type I	3	**EMT Wired Devices**	
Ceiling Fixture - Type M	6	2 Gang - Switch/Switch	1
Miscellaneous Ceiling		Receptacle Convenience - GFCI	2
Smoke Detector	4	Receptacle Convenience	1
Exhaust Fan	4	Receptacle - Multioutlet	1
Range Hood Fan	1		
Switches			
Single Pole Switch	26	**Special Systems (Low Voltage)**	
Three Way Switch	22	Chime With Push-buttons	1
Four Way Switch	2	Television Outlet	12
Dimmer	2	Phone Outlet	8
Fan Controlled Switch	3	Phone Outlet Weatherproof	1
Pilot Switch	1	Phone Outlet (EMT Protected)	1
Switch - 1 Gang Box	21	Television Outlet (EMT Protected)	1
Switch - 2 Gang Box	9		
Switch - 3 Gang Box	3		
Switch - 4 Gang Box	2		

TAKE-OFF WORKSHEET 2 OF 2 Figure 6-53			
Panel A Circuits	Circuit #	Ampere	Length
Electric Heat	1-3	70 Amp	10 feet
Water Pump	5-7	15 Amp	60 feet
Hydromassage Bathtub	9	15 Amp	51 feet
Bath Heat/Vent/Light Equipment	11	20 Amp	48 feet
Freezer Receptacle Outlet	13	15 Amp	29 feet
Entry/Porch Light Circuit	15	15 Amp	57 feet
Workshop Light Circuit	17	15 Amp	10 feet
Master Bedroom Circuit	19	15 Amp	15 feet
Front Study/Bedroom Circuit	21	15 Amp	40 feet
Sub-Feed to Panel B	27-29	100 Amp	50 feet
Air Conditioning Equipment	2-4	40 Amp	10 feet
Water Heater	6-8	30 Amp	5 feet
Attic Fan Equipment	10	15 Amp	41 feet
Bath Heat/Vent/Light Equipment	12	20 Amp	40 feet
Bath and Hall Light Circuit	14	15 Amp	41 feet
Front Bedroom Circuit	16	15 Amp	56 feet
Work Bench Receptacle Circuit	18	15 Amp	5 feet
Workshop Receptacle Circuit	20	15 Amp	40 feet
Bath Receptacle Circuit (Master Bedroom)	22	20 Amp	52 feet
Panel B Circuits			
Dryer	1-3	30 Amp	35 feet
Dishwasher	5	20 Amp	15 feet
Kitchen Light Circuit	7	15 Amp	15 feet
Wet Bar Circuit	9	15 Amp	5 feet
Recreation Room Receptacle Circuit	11	15 Amp	5 feet
Kitchen Receptacle Circuit	13	20 Amp	15 feet
Kitchen Receptacle Circuit	15	20 Amp	15 feet
Living Room Circuit	17	15 Amp	40 feet
Disposal	19	20 Amp	20 feet
Bathroom Receptacle Circuit	21	20 Amp	46 feet
Cook Top	2-4	40 Amp	20 feet
Oven	6-8	30 Amp	12 feet
Laundry Lighting Circuit	10	15 Amp	35 feet
Recreation Room Lighting Circuit	12	15 Amp	23 feet
Garage Circuit	14	15 Amp	45 feet
Kitchen Receptacle Circuit	16	20 Amp	35 feet
Washing Machine Receptacle Circuit	18	20 Amp	40 feet
Laundry Receptacle Circuit	20	15 Amp	45 feet
Service Requirements Back to Back		200 Amp	Utility
Other Cost			
Quotes: Gear	$425		
Direct Costs: Permit	$375		
Temporary Wiring Allowance	$330		

Bid Accuracy

Be sure you have not made any of the following common errors:

- Assuming standard grade devices, where specification grades are required.
- Failing to include outside or underground work.
- Failure to visit site to determine job conditions.
- Forgetting a major item such as quotes.
- Forgetting to include subcontract cost or equipment rental requirements.
- Forgetting to include the changes to the original specifications or blueprints.
- Hurrying the bid.
- Improper estimating forms.
- Omitting a section of the bid.
- Using supplier take-offs for quotes or depending on verbal quotes.
- Wrong scale on reduced blueprints.

Bid Analysis

The following is a description of typical computer analysis reports that can help you insure that your bid is correct.

Bid Summary Analysis – Report No. 1

This report permits you the opportunity to analyze the following:

1. Estimated hours, man-days and man-weeks.
2. Ratio of labor cost to selling price.
3. Material ratio to selling price.
4. Ratio of overhead to selling price.

Detailed Summary Analysis – Report No. 2

This report permits a detailed analysis of each take-off item, such as labor-hours, cost, sales tax, overhead and profit. If you made a mistake in your estimate, you'll see the error very quickly.

Labor Analysis – Report No. 3

This report analyzes which material items are most labor intensive and which are not. If you made a mistake on the quantity in your take-off, you will catch it here, or if you did not include sufficient labor or forgot to include the labor completely, you'll know about it.

Material Quantity Analysis – Report No. 4

You'll be able to analyze the quantity of material items estimated, their individual and extended cost as well as their unit and extended labor-hours.

Material Cost Analysis – Report No. 5

With this report you'll know which material items cost you the most. If you made a mistake on the quantity in your take-off, you will catch it here, or if you did not include sufficient cost or forgot to include the cost at all, you'll know about it.

Work Phase Analysis – Report No. 6

This report analyzes the distribution of the material for the slab, the rough, the trim, etc. With this information you will be able to detect any material items that you might have missed.

Job Proposal – Report No. 7

You will be able to analyze the unit price for each take-off item, as well as the total cost for selected groups of the take-off, such as lighting fixtures, switches, receptacles, special systems, equipment and service, etc.

Draw Schedule – Report No. 8

Before you submit your bid, you want to compare the bid's cost per square foot against similar bids. In addition, you want to compare the bid cost distribution percentage for the slab, rough, trim, etc. This report gives you the information to help you manage your cash flow.

Report Contents

Report 1	Bid Summary Analysis	149
Report 2	Detailed Summary Analysis	151
Report 3	Labor Analysis	155
Report 4	Material Quantity Analysis	159
Report 5	Material Cost Analysis	163
Report 6	Work Phase Analysis	167
Report 7	Job Proposal	171
Report 8	Draw Schedule	175

Report #1

```
                    [5.1.1] Brief Summary
                    MIKE HOLT ENTERPRISES                        Page:    1
                  Bid Summary Analysis - Report No. 1
                      Job Name: Residential Bid
        Job #: 000000002   Type:    Square footage:     3,744
```

```
=========================================================================
                         Labor Hour Adjustments
=========================================================================

             Job Duration........ :    0.00 (months)
             Labor Hours (Fixed). :                       0.00
             ***********........ :    0.00%              0.00
             Labor (Adjustable).. :                    191.99
                Experience....... :  - 10.00%    -      19.20
             Labor Hours Subtotal :                    172.79
                ADJUSTMENT....... :    0.00%              0.00
             -------------------------------------------------
             Total Labor Hours... :                    172.79
             Total Labor Days.... : 8.00 Hours/Day     10.79    2.00 Men/Day
             Total Labor Weeks... : 5.00 Days/Week      2.16    2.00 Men/Day

=========================================================================
                         Labor Cost Adjustments
=========================================================================

 Rate   Count
$14.00   1.00     Average Labor Rate   :    $10.65
$12.25   1.00     Labor Cost (Base)... :               1,840.21    22.39%
$ 9.00   3.00     ***********........ :    0.00%           0.00     0.00%
$ 0.00   0.00     Inflation Adj...... :    0.00%           0.00     0.00%
$ 0.00   0.00     -----------------------------------------------------
$ 0.00   0.00     Labor Cost Subtotal. :               1,840.21    22.39%
$ 0.00   0.00
$ 0.00   0.00     ***********........ :    0.00%           0.00     0.00%
$ 0.00   0.00     ***********........ :    0.00%           0.00     0.00%
 -------------    -----------------------------------------------------
$10.65 100.00%    Total Labor Cost.... :               1,840.21    22.39%

=========================================================================
                          Material Adjustments
=========================================================================

             Material Cost (Fixed):                       0.00     0.00%
                ***********....... :    0.00%             0.00     0.00%
             Total Fixed Material :                       0.00     0.00%

             Material (Adjustable):                   1,800.38    21.91%
                ADJUSTMENT........ :    0.00%             0.00     0.00%
                Inflation Adj..... :    0.00%             0.00     0.00%
                Misc.............. :    1.00%            18.00     0.22%
             Total Adj. Material. :                   1,818.38    22.13%
             -------------------------------------------------
             Material Subtotal... :                   1,818.38    22.13%

                Theft/Waste....... :    5.00%            90.92     1.11%
                SMALL TOOLS....... :    3.00%            57.28     0.70%
             -------------------------------------------------
             Material Cost Total. :                   1,966.58    23.93%


                 MIKE HOLT ENTERPRISESVersion [091889]
                                                              [8008812580]
```

Report #1

```
                    [5.1.1] Brief Summary
                     MIKE HOLT ENTERPRISES                      Page:   2
                  Bid Summary Analysis - Report No. 1
                      Job Name: Residential Bid
       Job #: 000000002  Type:    Square footage:      3,744


:===============================================================================
                         Prime Cost Calculations
:===============================================================================

        Total Material Cost. :                 1,966.58        23.93%
        Quotes.............. :                   425.00         5.17%
          HANDLING.......... :       0.00%         0.00         0.00%
        Sales tax........... :       7.000%      167.41         2.04%
        Total Labor Cost.... :                 1,840.21        22.39%
                              ------------------------------------------
        Subtotal of above... :                 4,399.20        53.53%
                              ------------------------------------------

        Sub-Contracts....... :                     0.00         0.00%
          ************...... :       0.00%         0.00         0.00%
        Direct Costs........ :                   705.00         8.58%
          HANDLING.......... :       0.00%         0.00         0.00%
                              ------------------------------------------
        Total Prime Cost.... :                 5,104.20        62.11%
                              ------------------------------------------


:===============================================================================
                              Final Summary
:===============================================================================

        Total Prime Cost.... :                 5,104.20        62.11%
          OVERHEAD.......... :      40.00%    2,041.68        24.84%
        Total Net Cost...... :                 7,145.88        86.96%
          PROFIT............ :      15.00%    1,071.88        13.04%
                              ------------------------------------------
        Subtotal............ :                 8,217.76       100.00%
                              ------------------------------------------

          BONDING........... :       0.00%         0.00         0.00%
          INSURANCE......... :       0.00%         0.00         0.00%
          CONTINGENT........ :       0.00%         0.00         0.00%
          PERMIT............ :       0.00%         0.00         0.00%
                              ==========================================
        Total............... :                 8,217.76       100.00%
                              ==========================================
```

MIKE HOLT ENTERPRISESVersion [091889]

[8008812580]

Report #2

```
                    Detailed Summary Analysis - Report No. 2
                              MIKE HOLT ENTERPRISES                               Page: 1
                  Based on ELECTRICAL WIRING - Residential, Ray Mullin
                  Job #: 000000002  Type:    Square footage:    3,744
                              Hours    Total    Labor       Cost     Cost      Sales                            Total
Tran I.D. Description        Quantity (each)   Hours  Extension  (Each) Extension  Tax OVERHEAD PROFIT          Cost
```

Tran	I.D.	Description	Quantity	Hours (each)	Total Hours	Labor Extension	Cost (Each)	Cost Extension	Sales Tax	OVERHEAD	PROFIT	Total Cost
0	B 0	==> START OF BID <==	1.00	*****.**	*****.**	*****.**	*****.**	*****.**	*****.**	*****.**	*****.**	*****.**
1	B 1	Computer Estimate	1.00	*****.**	*****.**	*****.**	*****.**	*****.**	*****.**	*****.**	*****.**	*****.**
2	C 1	Fixtures	1.00	*****.**	*****.**	*****.**	*****.**	*****.**	*****.**	*****.**	*****.**	*****.**
3	A 600	Post Light	1.00	2.04	2.04	21.76	2.72	2.72	0.19	9.87	5.18	39.72
4	A 602	Outside Wall Light	6.00	0.37	2.22	23.68	1.43	8.55	0.60	13.13	6.89	52.85
5	A 610	Attic Light	6.00	0.37	2.22	23.68	1.43	8.55	0.60	13.13	6.89	52.85
6	A 614	Flourscent Workbench	1.00	0.33	0.33	3.55	2.60	2.60	0.18	2.53	1.33	10.19
7	A 615	Flourscent Lay-In	6.00	0.81	4.87	51.85	1.09	6.54	0.46	23.54	12.36	94.75
8	A 616	Flourscent Strip	1.00	0.74	0.74	7.86	0.63	0.63	0.04	3.42	1.79	13.74
9	A 617	Flourscent Wrap Fix.	2.00	0.83	1.66	17.64	1.60	3.20	0.22	8.42	4.42	33.90
10	A 617	Flourscent Wrap Fix.	1.00	0.83	0.83	8.82	1.61	1.61	0.11	4.21	2.21	16.96
11	A 618	Paddle Fan	3.00	0.89	2.68	28.56	4.81	14.44	1.01	17.61	9.24	70.86
12	A 619	Pull Chain Fixture	3.00	0.28	0.85	9.01	0.46	1.37	0.10	4.19	2.20	16.87
13	A 621	Recessed Light	16.00	1.04	16.70	177.90	1.94	31.11	2.18	84.47	44.35	340.01
14	A 624	Track Light 4'	1.00	0.91	0.91	9.68	2.41	2.41	0.17	4.91	2.58	19.75
15	A 625	Track Light 8'	1.00	1.13	1.13	12.08	2.41	2.41	0.17	5.86	3.08	23.60
16	A 626	Wall Fixture	3.00	0.37	1.12	11.89	1.43	4.28	0.30	6.59	3.46	26.52
17	A 627	Ceiling Fixture	6.00	0.37	2.22	23.68	1.43	8.55	0.60	13.13	6.89	52.85
18	A 629	Smoke Detector	4.00	0.37	1.49	15.82	1.51	6.05	0.42	8.92	4.68	35.89
19	A 630	Exhaust Fan	4.00	0.97	3.86	41.12	0.63	2.51	0.18	17.52	9.20	70.53
20	A 728	Range Hood Fan	1.00	0.46	0.46	4.89	0.17	0.17	0.01	2.03	1.07	8.17
	C 1	Fixtures Unit 2.89 days	1.00	*****.**	46.33	493.47	*****.**	107.70	7.54	243.48	127.82	980.01
	C 1	Fixtures Total 2.89 days	1.00	*****.**	46.33	493.47	*****.**	107.70	7.54	243.48	127.82	980.01
21	C 2	Switches	1.00	*****.**	*****.**	*****.**	*****.**	*****.**	*****.**	*****.**	*****.**	*****.**
22	2091	SW 15,125V 1POLE	26.00	0.18	4.68	49.84	0.62	16.19	1.13	26.87	14.10	108.13
23	2092	SW 15,125V 3WAY	22.00	0.23	4.95	52.72	1.39	30.52	2.14	34.15	17.93	137.46
24	2093	SW 15,125V 4WAY	2.00	0.25	0.50	5.37	7.86	15.71	1.10	8.87	4.66	35.71
25	235	DIM 600W 120V 1POLE	2.00	0.23	0.45	4.79	6.67	13.33	0.93	7.62	4.00	30.67
26	2110	SW 15,125V 1POLE FAN	3.00	0.23	0.68	7.19	10.16	30.48	2.13	15.92	8.36	64.08
27	2104	SW 15 AMP PILOT 125V	1.00	0.23	0.23	2.40	5.54	5.54	0.39	3.33	1.75	13.41
28	A 637	1G Switch Assembly	21.00	0.17	3.54	37.67	2.03	42.56	2.98	33.28	17.47	133.96
29	A 638	2G Switch Assembly	8.00	0.20	1.60	17.06	3.16	25.30	1.77	17.65	9.27	71.05
30	A 639	3G Switch Assembly	3.00	0.26	0.77	8.15	3.44	10.31	0.72	7.67	4.03	30.88
31	A 640	4G Switch Assemblly	2.00	0.30	0.59	6.33	4.32	8.64	0.60	6.23	3.27	25.07
	C 2	Switches Unit 1.12 days	1.00	*****.**	17.99	191.52	*****.**	198.58	13.90	161.59	84.84	650.43
	C 2	Switches Total 1.12 days	1.00	*****.**	17.99	191.52	*****.**	198.58	13.90	161.59	84.84	650.43
32	C 3	Receptacles	1.00	*****.**	*****.**	*****.**	*****.**	*****.**	*****.**	*****.**	*****.**	*****.**
33	A 645	GFCI Weather Proof	5.00	0.45	2.26	24.06	18.87	94.35	6.60	50.01	26.25	201.27
34	A 646	Duplex Receptacle	5.00	0.33	1.66	17.64	2.60	12.98	0.91	12.61	6.62	50.76
35	A 646	Duplex Receptacle	9.00	0.33	2.97	31.63	2.59	23.35	1.63	22.65	11.89	91.15
36	A 647	Switched Receptacle	21.00	0.33	6.94	73.90	2.59	54.48	3.81	52.88	27.76	212.83
37	A 648	GFCI Recept 15A Ckt	4.00	0.44	1.76	18.69	15.27	61.06	4.27	33.61	17.65	135.28
38	A 649	Clock Receptacle	1.00	0.55	0.55	5.85	6.19	6.19	0.43	4.99	2.62	20.08

MIKE HOLT ENTERPRISES Version [091889]

[8008812580]

Report #2

```
                           Detailed Summary Analysis - Report No. 2
                                     MIKE HOLT ENTERPRISES                                    Page:  2
                        Based on ELECTRICAL WIRING - Residential, Ray Mullin
                        Job #: 000000002  Type:   Square footage:   3,744
                            Hours    Total    Labor     Cost     Cost    Sales                         Total
Tran I.D. Description    Quantity  (each)   Hours  Extension  (Each) Extension   Tax OVERHEAD PROFIT    Cost
 39 A 650 Garage Door Recept.   1.00   0.59    0.59    6.33      6.30    6.30    0.44    5.23    2.74   21.04
 40 A 651 GFCI Kitchen Recept.  2.00   0.49    0.98   10.45     16.04   32.07    2.24   17.91    9.40   72.07
 41 A 652 GFCI Bath Receptacle  2.00   0.49    0.98   10.45     16.04   32.07    2.24   17.91    9.40   72.07
 42 A 653 Kitchen Receptacle    7.00   0.38    2.69   28.66      3.37   23.56    1.65   21.55   11.31   86.73
 43 A 654 Bath Receptacle       1.00   0.39    0.39    4.12      3.36    3.36    0.24    3.09    1.62   12.43
                              ----------------------------------------------------------------------------
     C    3 Receptacles         1.00 *****.**  21.77  231.78 *****.**  349.77   24.49  242.44  127.26  975.74
          Unit      1.36 days
   ========================================================================================================
     C    3 Receptacles         1.00 *****.**  21.77  231.78 *****.**  349.77   24.49  242.44  127.26  975.74
          Total     1.36 days
   ========================================================================================================
 44 C     4 EMT Devices         1.00 *****.** *****.** *****.** *****.** *****.** *****.** *****.** *****.** *****.**
 45 A 657 2G Switch (EMT)       1.00   0.93    0.93    9.87      6.18    6.18    0.43    6.60    3.46   26.54
 46 A 658 GFCI Receptacle EMT   2.00   0.55    1.10   11.69     15.54   31.07    2.17   17.97    9.44   72.34
 47 A 659 Duplex Recept. EMT    1.00   0.44    0.44    4.70      2.99    2.99    0.21    3.16    1.66   12.72
 48 A 660 Multioutlet Assembly  1.00   1.13    1.13   12.08      2.01    2.01    0.14    5.69    2.99   22.91
                              ----------------------------------------------------------------------------
     C    4 EMT Devices         1.00 *****.**   3.60   38.34 *****.**   42.25    2.96   33.42   17.55  134.52
          Unit      0.22 days
   ========================================================================================================
     C    4 EMT Devices         1.00 *****.**   3.60   38.34 *****.**   42.25    2.96   33.42   17.55  134.52
          Total     0.22 days
   ========================================================================================================
 49 C     5 Branch Circuit      1.00 *****.** *****.** *****.** *****.** *****.** *****.** *****.** *****.** *****.**
 50  1740 ROMEX 14/2 COPPER   908.00   0.01    8.17   87.03      0.12  107.11    7.50   80.66   42.34  324.64
 51  1741 ROMEX 14/3 COPPER   982.00   0.01   10.60  112.91      0.20  199.51   13.97  130.55   68.54  525.48
 52  1742 ROMEX 12/2 COPPER   134.00   0.01    1.45   15.43      0.16   21.66    1.52   15.44    8.11   62.16
 53  2290 UF CABLE 14/2 COPPER 40.00   0.00    0.18    1.92      0.15    5.81    0.41    3.25    1.71   13.10
                              ----------------------------------------------------------------------------
     C    5 Branch Circuit      1.00 *****.**  20.40  217.29 *****.**  334.09   23.39  229.90  120.70  925.37
          Unit      1.27 days
   ========================================================================================================
     C    5 Branch Circuit      1.00 *****.**  20.40  217.29 *****.**  334.09   23.39  229.90  120.70  925.37
          Total     1.27 days
   ========================================================================================================
 54 C     6 Special Systems     1.00 *****.** *****.** *****.** *****.** *****.** *****.** *****.** *****.** *****.**
 55 A 664 Chime W/Push Buttons  1.00   2.60    2.60   27.70     17.78   17.78    1.24   18.69    9.81   75.22
 56 A 666 TV Outlet            11.00   0.72    7.91   84.25     11.96  131.57    9.21   90.01   47.26  362.30
 57 A 667 Phone Outlet          7.00   0.61    4.28   45.53      6.85   47.94    3.36   38.73   20.33  155.89
 58 A 668 Phone Weather Proof   1.00   0.61    0.61    6.52     12.31   12.31    0.86    7.88    4.13   31.70
 59 A 669 Phone EMT             1.00   0.99    0.99   10.54      9.14    9.14    0.64    8.13    4.27   32.72
 60 A 670 TV Outlet EMT         1.00   1.10    1.10   11.69     14.25   14.25    1.00   10.78    5.66   43.38
 61 A 672 Thermostat            1.00   1.50    1.50   16.01     12.86   12.86    0.90   11.91    6.25   47.93
 62 A 675 Home Run 15 Amp Ckt. 12.00   0.38    4.54   48.31      3.83   45.99    3.22   39.01   20.48  157.01
 63 A 676 Home Run 20 Amp Ckt.  5.00   0.47    2.33   24.83      5.67   28.36    1.98   22.07   11.58   88.82
 64 A 678 Home Run EMT 20 Amp   2.00   0.73    1.46   15.53      5.53   11.05    0.77   10.94    5.74   44.03
                              ----------------------------------------------------------------------------
     C    6 Special Systems     1.00 *****.**  27.32  290.91 *****.**  331.25   23.19  258.15  135.51 1039.01
          Unit      1.70 days
   ========================================================================================================

                              MIKE HOLT ENTERPRISESVersion [091889]

                                                                                              [8008812580]
```

Report #2

```
                              Detailed Summary Analysis - Report No. 2
                                       MIKE HOLT ENTERPRISES                              Page:   3
                              Based on ELECTRICAL WIRING - Residential, Ray Mullin
                              Job #: 000000002  Type:   Square footage:    3,744
                                        Hours     Total    Labor    Cost     Cost    Sales                         Total
Tran I.D. Description         Quantity  (each)    Hours  Extension (Each)  Extension  Tax  OVERHEAD PROFIT          Cost
         C   6 Special Systems     1.00 *****.**   27.32   290.91 ******.**  331.25 23.19   258.15  135.51        1039.01
             Total     1.70 days        ================================================================================

 65 C   7 Separate Circuits       1.00 *****.** ******.** ******.** ******.** ******.** ******.** ******.** ******.** *******.**
 66 A 722 Heat 70 Ampere EMT      1.00   1.80     1.80    19.17    21.86     21.86     1.53   17.02    8.94      68.52
 67 A 251 FEEDER #4 CU 1PH EMT   15.00   0.05     0.68     7.19     0.79     11.83     0.83    7.94    4.17      31.96
 68 A 735 Water Pump              1.00   0.68     0.68     7.19     4.82      4.82     0.34    4.94    2.59      19.88
 69 A   1 1/2 EMT 2#14 THHN      60.00   0.03     1.73    18.40     0.26     15.60     1.09   14.04    7.37      56.50
 70 A 724 Hydromassage Bathtub    1.00   0.79     0.79     8.43     5.71      5.71     0.40    5.82    3.06      23.42
 71   1740 ROMEX 14/2 COPPER     51.00   0.01     0.46     4.89     0.12      6.02     0.42    4.53    2.38      18.24
 72 A 713 Bath Heat/Vent/Light    1.00   1.59     1.59    16.97    20.26     20.26     1.42   15.46    8.12      62.23
 73   1742 ROMEX 12/2 COPPER     50.00   0.01     0.54     5.75     0.16      8.08     0.57    5.76    3.02      23.18
 74 A 717 Freezer EMT             1.00   0.55     0.55     5.85     7.21      7.21     0.50    5.42    2.85      21.83
 75 A   1 1/2 EMT 2#14 THHN      30.00   0.03     0.86     9.20     0.26      7.80     0.55    7.02    3.68      28.25
 76 A 711 Air Condition 40 Amp    1.00   1.94     1.94    20.61    10.72     10.72     0.75   12.83    6.74      51.65
 77   1746 ROMEX 8/2 COPPER      10.00   0.01     0.14     1.53     0.66      6.56     0.46    3.42    1.80      13.77
 78 A 734 Water Heater 20 Amp     1.00   1.08     1.08    11.50    19.37     19.37     1.36   12.89    6.77      51.89
 79 A  11 1/2 EMT 2#10 THHN       5.00   0.03     0.16     1.73     0.36      1.78     0.12    1.45    0.76       5.84
 80 A 712 Attic Fan & Switch      1.00   0.69     0.69     7.38    18.08     18.08     1.27   10.69    5.61      43.03
 81   1740 ROMEX 14/2 COPPER     45.00   0.01     0.41     4.31     0.12      5.31     0.37    4.00    2.10      16.09
 82 A 713 Bath Heat/Vent/Light    1.00   1.59     1.59    16.97    20.26     20.26     1.42   15.46    8.12      62.23
 83   1742 ROMEX 12/2 COPPER     40.00   0.01     0.43     4.60     0.16      6.47     0.45    4.61    2.42      18.55
 84 A 754 Panel "B" 100 Ampere EMT 1.00  2.25     2.25    23.96    33.86     33.86     2.37   24.08   12.64      96.91
 85    109 BREAKER 70AMP 2 POLE   1.00   0.18     0.18     1.92   ***.**      0.00     0.00    0.77    0.40       3.09
 86 A 755 Feeder Panel "B" 100 Amp 50.00 0.07     3.38    35.94     2.03    101.69     7.12   57.90   30.40     233.05
 87 A 716 Dryer                   1.00   0.80     0.80     8.53    23.92     23.92     1.67   13.65    7.17      54.94
 88   1745 ROMEX 10/3 COPPER     35.00   0.01     0.50     5.37     0.43     15.18     1.06    8.65    4.54      34.80
 89 A 715 Dishwasher/Disposal     1.00   1.26     1.26    13.42    14.30     14.30     1.00   11.49    6.03      46.24
 90   1741 ROMEX 14/3 COPPER     20.00   0.01     0.22     2.30     0.20      4.06     0.28    2.66    1.40      10.70
 91 A 714 Cook Top 40 Ampere      1.00   0.46     0.46     4.89     3.98      3.98     0.28    3.66    1.92      14.73
 92   1747 ROMEX 8/3 COPPER      20.00   0.02     0.36     3.83     0.90     17.91     1.25    9.20    4.83      37.02
 93 A 725 Oven 30 Ampere          1.00   0.62     0.62     6.61     4.86      4.86     0.34    4.73    2.48      19.02
 94   1745 ROMEX 10/3 COPPER     15.00   0.01     0.22     2.30     0.43      6.51     0.46    3.71    1.95      14.93
 95 A 733 Washing Machine         1.00   0.54     0.54     5.75     8.26      8.26     0.58    5.83    3.06      23.48
 96   1742 ROMEX 12/2 COPPER     40.00   0.01     0.43     4.60     0.16      6.47     0.45    4.61    2.42      18.55
 97 A 750 Meter/Main 200 Amp      1.00   5.13     5.13    54.63   105.35    105.35     7.37   66.95   35.15     269.45
 98 A 751 Panel Back/Back 200A    1.00   2.93     2.93    31.25    58.81     58.81     4.12   37.67   19.78     151.63
 99 Q   1 Switchgear              1.00   0.00     0.00     0.00   425.00    425.00    29.75  181.90   95.50     732.15
100 D   1 Permit                  1.00   0.00     0.00     0.00   375.00    375.00 *****.**  150.00   78.75     603.75
101 D   2 Temporary               1.00   0.00     0.00     0.00   330.00    330.00 *****.**  132.00   69.30     531.30
                                       ----------------------------------------------------------------------------------
         C   7 Separate Circuits  1.00 *****.**   35.40   376.97 ******.** 1732.90    71.95  872.76  458.22    3512.80
             Unit      2.21 days


         C   7 Separate Circuits  1.00 *****.**   35.40   376.97 ******.** 1732.90    71.95  872.76  458.22    3512.80
             Total     2.21 days        ================================================================================


         B   1 Computer Estimate  1.00 *****.**  172.81  1840.28 ******.** 3096.54   167.41 2041.74 1071.90    8217.87
             Total    10.80 days         ==============================================================================
```

MIKE HOLT ENTERPRISESVersion [091889]

[8008812580]

Report #2

```
                     Detailed Summary Analysis - Report No. 2
                                MIKE HOLT ENTERPRISES                                    Page:   4
                      Based on ELECTRICAL WIRING - Residential, Ray Mullin
                      Job #: 000000002  Type:   Square footage:    3,744
                               Hours    Total    Labor     Cost     Cost    Sales                      Total
Tran I.D. Description  Quantity (each)  Hours  Extension  (Each)  Extension  Tax OVERHEAD PROFIT        Cost

****************************************====================================================================
***      Sub-Total         ***   1.00 *****.**  172.81  1840.28 ******.**  3096.54  167.41  2041.74  1071.90   8217.87
****************************************====================================================================


****************************************                                                          **********
***      BONDING           ***                                                                          0.00
****************************************                                                          **********


****************************************                                                          **********
***      INSURANCE         ***                                                                          0.00
****************************************                                                          **********


****************************************                                                          **********
***      CONTINGENT        ***                                                                          0.00
****************************************                                                          **********


****************************************                                                          **********
***      PERMIT            ***                                                                          0.00
****************************************                                                          **********


****************************************                                                          **********
***      Bid Total         ***                                                                       8217.87
****************************************                                                          **********
```

MIKE HOLT ENTERPRISESVersion [091889]

[8008812580]

Report #3

```
                              Labor Analysis - Report No. 3
                                    MIKE HOLT ENTERPRISES                                          Page: 1
                          Based on ELECTRICAL WIRING - Residential, Ray Mullin
Residential Bid           000000002                        3,744   sqft
                                                                         Labor      Labor
Id Code Description           Quantity      Cost/Unit    Cost Ext   Labor/Unit   Hour Ext   Cost Ext   Last Updt
I- 1740 ROMEX 14/2 COPPER     1,616.00   117.970 M/Feet    190.64    9.004 M/Feet   14.55    154.96    9/ 3/96
I-  478 RECESS FIXTURE           16.00           E/Each              0.900 E/Each   14.40    153.36   11/ 7/96
I- 1741 ROMEX 14/3 COPPER     1,002.00   203.164 M/Feet    203.57   10.798 M/Feet   10.82    115.23    9/ 3/96
I-   50 BOX PLASTIC 1 GANG       90.00   105.978 C/Pcs.     95.38    9.000 C/Pcs.    8.10     86.27    9/ 2/96
I- 1490 REC DUPLEX 15A 125V      47.00    56.830 C/Pcs.     26.71   16.170 C/Pcs.    7.60     80.94    9/ 3/96
I- 2091 SW 15,125V 1POLE         31.00    62.258 C/Pcs.     19.30   18.000 C/Pcs.    5.58     59.43    9/ 3/96
I- 2223 WIRE COAX TV CABLE      504.00   196.627 M/FEET     99.10   10.794 M/FEET    5.44     57.94    9/ 3/96
I- 2092 SW 15,125V 3WAY          22.00   138.727 C/Pcs.     30.52   22.500 C/Pcs.    4.95     52.72    9/ 3/96
I- 1742 ROMEX 12/2 COPPER       434.00   161.659 M/Feet     70.16   10.806 M/Feet    4.69     49.95    9/ 3/96
I-  583 FLUO LAYIN 2X4' 120V      6.00           E/Each              0.675 E/Each    4.05     43.13   11/ 7/96
I- 1492 REC GFCI 15A 125V        15.00    13.109 E/Each    196.63    0.270 E/Each    4.05     43.13    9/ 3/96
I-  474 FAN, EXHAUST              4.00           E/Each              0.900 E/Each    3.60     38.34    1/ 3/91
I-  305 EMT 1/2"                170.00    14.200 C/Feet     24.14    2.041 C/Feet    3.47     36.96    9/ 2/96
I- 2228 TELEPHONE WIRE TWIST    288.00    98.333 M/Feet     28.32   10.764 M/Feet    3.10     33.02    9/ 3/96
I- 1760 ROMEX STAPLE SMALL      312.00     1.093 C/Pcs.      3.41    0.904 C/Pcs.    2.82     30.03    9/ 6/96
I- 2222 TV RECEPTACLE            12.00     3.278 E/Each     39.33    0.226 E/Each    2.71     28.86    9/ 3/96
I- 2442 WIRE NUTS RED           285.00     8.709 C/Pcs.     24.82    0.916 C/Pcs.    2.61     27.80    9/ 6/96
I-   55 BOX PLASTIC OCTAGON      24.00    97.167 C/Pcs.     23.32    9.000 C/Pcs.    2.16     23.00    9/ 2/96
I-   31 BOX MET 4*4 REGULAR      13.00    64.154 C/Pcs.      8.34   16.077 C/Pcs.    2.09     22.26    9/ 2/96
I-  473 FAN AND LIGHT             3.00           E/Each              0.677 E/Each    2.03     21.62    9/ 3/96
I-  492 INSIDE FIXTURE            9.00           E                   0.226 E         2.03     21.62    9/ 3/96
I-  579 FLUO WRAP 2-2' 120V       3.00           E/Each              0.677 E/Each    2.03     21.62    1/ 3/91
I- 2229 PHONE JACK INDOOR         8.00     3.278 E/Each     26.22    0.226 E/Each    1.81     19.28    9/ 3/96
I-  308 EMT 1 1/4"               50.00    54.620 C/Feet     27.31    3.600 C/Feet    1.80     19.17    9/ 2/96
I-  501 BATH HEAT/FAN/LIGHT       2.00           E/                  0.900 E/        1.80     19.17    2/19/96
I-  505 POST LIGHT                1.00           E/                  1.800 E/        1.80     19.17    9/ 3/96
I- 1005 METER/MAIN 200 AMP        1.00           E/Each              1.800 E/Each    1.80     19.17   10/ 7/96
I-  100 BREAKER 15AMP 1 POLE     15.00           E/Each              0.090 E/Each    1.35     14.38    1/ 3/91
I-  491 OUTSIDE FIXTURE           6.00           E/                  0.225 E/        1.35     14.38    2/19/96
I-  502 KEYLESS FIXTURE           6.00           E/                  0.225 E/        1.35     14.38    2/22/96
I- 2441 WIRE NUTS YELLOW        139.00     6.597 C/Pcs.      9.17    0.935 C/Pcs.    1.30     13.85    9/ 3/96
I- 2426 THERMO WIRE 18/5        100.00   131.000 M/Feet     13.10   11.800 M/Feet    1.18     12.57    9/ 3/96
I- 1263 PLATE PLAS 1G DPX RE     46.00    51.304 C/Pcs.     23.60    2.370 C/Pcs.    1.09     11.61    9/ 3/96
I- 1711 RING 1 GANG              24.00    42.667 C          10.24    4.250 C         1.02     10.86    9/ 3/96
I- 2396 WIRE #3 CU 600V         150.00   409.733 M/Feet     61.46    6.800 M/Feet    1.02     10.86    9/ 3/96
I-   51 BOX PLASTIC 2 GANG        9.00   182.333 C/Pcs.     16.41   10.778 C/Pcs.    0.97     10.33    9/ 2/96
I-  101 BREAKER 20AMP 1 POLE     10.00           E                   0.090 E         0.90      9.59    2/21/96
I-  251 DISC. 60 AMP 1PHASE       1.00           E                   0.900 E         0.90      9.59    2/21/96
I-  252 DISC. 70 AMP 1PHASE       1.00           E                   0.900 E         0.90      9.59    2/21/96
I-  500 SMOKE DETECTOR            4.00           E/                  0.225 E/        0.90      9.59    2/19/96
I-  621 FIX TRACK 8'              1.00           E/Each              0.900 E/Each    0.90      9.59    1/ 3/91
I-  762 GROUND ROD 1/2X8' CU      2.00  1,092.500 C/Pcs.    21.85   45.000 C/Pcs.    0.90      9.59    9/ 3/96
I-  941 CHIME                     1.00           E/Each              0.900 E/Each    0.90      9.59    9/ 3/96
I- 1190 PANEL 100 AMP 1PHASE      1.00           E/Each              0.900 E/Each    0.90      9.59    1/ 3/91
I- 1191 PANEL 200 AMP 1PHASE      1.00           E/Each              0.900 E/Each    0.90      9.59    2/22/96
I- 1521 MULTIOUTLET ASSEMBLY      1.00           E                   0.900 E         0.90      9.59    9/ 3/96
I- 2108 SW BATH/FAN/LIGHT         2.00    18.570 E          37.14    0.450 E         0.90      9.59    9/ 3/96
```

MIKE HOLT ENTERPRISESVersion [091889]

[8008812580]

Report #3

Labor Analysis - Report No. 3
MIKE HOLT ENTERPRISES
Based on ELECTRICAL WIRING - Residential, Ray Mullin

Residential Bid 000000002 3,744 sqft

Page: 2

Id Code Description	Quantity	Cost/Unit	Cost Ext	Labor/Unit	Labor Hour Ext	Labor Cost Ext	Last Updt
I- 1762 KICK PLATES	78.00	5.538 C/Pcs.	4.32	1.090 C/Pcs.	0.85	9.05	9/ 3/96
I- 2405 WIRE 2/0 CU 600V	60.00	891.333 M/Feet	53.48	13.667 M/Feet	0.82	8.73	9/ 3/96
I- 2425 THERMO WIRE 18/2	70.00	76.429 M/Feet	5.35	11.714 M/Feet	0.82	8.73	9/ 3/96
I- 1745 ROMEX 10/3 COPPER	50.00	433.800 M/Feet	21.69	14.400 M/Feet	0.72	7.67	9/ 3/96
I- 2445 WIRE NUTS GREEN	64.00	10.844 C/Pcs.	6.94	1.125 C/Pcs.	0.72	7.67	9/ 6/96
I- 2390 WIRE #14 CU 600V	218.00	41.514 M/Feet	9.05	3.257 M/Feet	0.71	7.56	9/ 3/96
I- 480 PULL CHAIN	3.00	E/Each		0.227 E/Each	0.68	7.24	1/ 3/91
I- 576 FLUO STR 1-8' 120V	1.00	E/Each		0.680 E/Each	0.68	7.24	1/ 3/91
I- 620 FIX TRACK 4'	1.00	E		0.680 E	0.68	7.24	2/26/96
I- 948 PUSH BUTTON	3.00	3.277 E/Each	9.83	0.227 E/Each	0.68	7.24	9/ 3/96
I- 2110 SW 15,125V 1POLE FAN	3.00	1,016.000 C	30.48	22.667 C	0.68	7.24	9/ 3/96
I- 640 FLEX STEEL 1/2"	21.00	33.857 C/Feet	7.11	3.000 C/Feet	0.63	6.71	9/ 3/96
I- 1595 RIGID 2"	10.00	249.000 C/Feet	24.90	6.300 C/Feet	0.63	6.71	9/ 3/96
I- 1743 ROMEX 12/3 COPPER	42.00	291.429 M/Feet	12.24	14.286 M/Feet	0.60	6.39	9/ 3/96
I- 358 EMT COUP SS 1 1/4"	9.00	108.222 C/Pcs.	9.74	6.333 C/Pcs.	0.57	6.07	9/ 2/96
I- 1261 PLATE PLAS 1G SWITCH	22.00	51.318 C/Pcs.	11.29	2.318 C/Pcs.	0.51	5.43	9/ 3/96
I- 307 EMT 1"	15.00	33.867 C/Feet	5.08	3.333 C/Feet	0.50	5.33	9/ 3/96
I- 2093 SW 15,125V 4WAY	2.00	785.500 C/Pcs.	15.71	25.000 C/Pcs.	0.50	5.33	9/ 3/96
I- 59 BOX PLASTIC FAN	3.00	436.000 C	13.08	16.333 C	0.49	5.22	9/ 3/96
I- 1415 PVC LB 2"	2.00	509.000 C/Pcs.	10.18	24.500 C/Pcs.	0.49	5.22	9/ 3/96
I- 2394 WIRE #6 CU 600V	78.00	273.077 M/Feet	21.30	6.282 M/Feet	0.49	5.22	9/ 3/96
I- 235 DIM 600W 120V 1POLE	2.00	6.665 E/Each	13.33	0.225 E/Each	0.45	4.79	9/ 2/96
I- 476 FAN, HOOD	1.00	E/Each		0.450 E/Each	0.45	4.79	1/ 3/91
I- 946 THERMOSTAT	1.00	E		0.450 E	0.45	4.79	9/ 3/96
I- 1751 ROMEX CONN 1/2"	48.00	17.417 C/Pcs.	8.36	0.938 C/Pcs.	0.45	4.79	9/ 6/96
I- 2109 ATTIC FAN SWITCH	1.00	16.380 E	16.38	0.450 E	0.45	4.79	4/22/96
I- 2391 WIRE #12 CU 600V	111.00	52.432 M/Feet	5.82	3.964 M/Feet	0.44	4.69	9/ 3/96
I- 52 BOX PLASTIC 3 GANG	3.00	190.000 C/Pcs.	5.70	14.333 C/Pcs.	0.43	4.58	9/ 2/96
I- 315 EMT CONN SS 1/2"	22.00	23.000 C/Pcs.	5.06	1.909 C/Pcs.	0.42	4.47	9/ 2/96
I- 398 EMT ELLS 90'S 1 1/4"	3.00	339.667 C/Pcs.	10.19	13.667 C/Pcs.	0.41	4.37	9/ 2/96
I- 1355 PVC 2" SCHEDULE 40	10.00	53.500 C/Feet	5.35	4.100 C/Feet	0.41	4.37	9/ 3/96
I- 1504 CLOCK OUTLET & PLATE	1.00	468.000 C/Pcs.	4.68	41.000 C/Pcs.	0.41	4.37	9/ 3/96
I- 2450 WIRE TERM # 12-6	3.00	E/Each		0.137 E/Each	0.41	4.37	1/ 3/91
I- 53 BOX PLASTIC 4 GANG	2.00	246.000 C/Pcs.	4.92	18.000 C/Pcs.	0.36	3.83	9/ 2/96
I- 56 BOX PLASTIC OCT/BAR	2.00	196.000 C/Pcs.	3.92	18.000 C/Pcs.	0.36	3.83	9/ 2/96
I- 109 BREAKER 70AMP 2 POLE	2.00	E		0.180 E	0.36	3.83	2/21/96
I- 1747 ROMEX 8/3 COPPER	20.00	895.500 M/Feet	17.91	18.000 M/Feet	0.36	3.83	9/ 4/96
I- 1890 SEALTIGHT 1/2"	12.00	51.333 C/Feet	6.16	3.000 C/Feet	0.36	3.83	9/ 3/96
I- 2360 WEATHERHEADS 2"	1.00	7.650 E/Each	7.65	0.360 E/Each	0.36	3.83	9/ 3/96
I- 26 BOX MET 2*4 REGULAR	2.00	64.000 C/Pcs.	1.28	16.000 C/Pcs.	0.32	3.41	9/ 2/96
I- 415 EMT STRAPS 1H 1/2"	12.00	6.500 C/Pcs.	0.78	2.417 C/Pcs.	0.29	3.09	9/ 3/96
I- 1265 PLATE PLAS 2G SWITCH	8.00	88.500 C/Pcs.	7.08	3.625 C/Pcs.	0.29	3.09	9/ 3/96
I- 2097 SW 20,125V 2POLE	1.00	1,311.000 C/Pcs.	13.11	29.000 C/Pcs.	0.29	3.09	9/ 3/96
I- 105 BREAKER 30AMP 2 POLE	2.00	E		0.140 E	0.28	2.98	2/21/96
I- 106 BREAKER 40AMP 2 POLE	2.00	E		0.140 E	0.28	2.98	2/21/96
I- 228 CORD DRYER 30A 4WIRE	1.00	13.110 E/Each	13.11	0.270 E/Each	0.27	2.88	9/ 2/96
I- 651 FLEX ST CONN 1/2"	5.00	60.000 C/Pcs.	3.00	5.400 C/Pcs.	0.27	2.88	9/ 4/96

MIKE HOLT ENTERPRISES Version [091889]

[8008812580]

Report #3

Labor Analysis - Report No. 3
MIKE HOLT ENTERPRISES   Page: 3
Based on ELECTRICAL WIRING - Residential, Ray Mullin

Residential Bid   000000002   3,744   sqft

Id Code Description	Quantity	Cost/Unit	Cost Ext	Labor/Unit	Labor Hour Ext	Labor Cost Ext	Last Updt
I- 1041 GREEN GROMMENTS 1"	284.00	54.754 M/Pcs.	15.55	0.915 M/Pcs.	0.26	2.77	9/12/96
I- 1285 PLATE IND 4X4 BLANK	5.00	108.000 C	5.40	5.200 C	0.26	2.77	9/ 3/96
I- 1350 PVC 1/2" SCHEDULE 40	15.00	14.200 C/Feet	2.13	1.733 C/Feet	0.26	2.77	9/ 3/96
I- 755 GROUND CLAMP 1/2" DB	2.00	190.000 C/Pcs.	3.80	12.000 C/Pcs.	0.24	2.56	9/ 3/96
I- 195 BUSHING PLASTIC 2"	2.00	152.000 C/Pcs.	3.04	11.500 C/Pcs.	0.23	2.45	9/ 2/96
I- 222 CORD 15.250V 3'	1.00	4.370 E/Each	4.37	0.230 E/Each	0.23	2.45	9/ 2/96
I- 223 CORD 15,250V 6'	1.00	4.370 E/Each	4.37	0.230 E/Each	0.23	2.45	9/ 2/96
I- 1425 PVC MALE ADAPT 2"	2.00	78.500 E/Each	1.57	11.500 C/Pcs.	0.23	2.45	7/14/96
I- 1496 REC SINGLE 20, 125V	1.00	5.460 E/Each	5.46	0.230 E/Each	0.23	2.45	9/ 3/96
I- 1498 REC DRYER 30A 4 WIRE	1.00	7.650 E/Each	7.65	0.230 E/Each	0.23	2.45	9/ 3/96
I- 2104 SW 15 AMP PILOT 125V	1.00	554.000 C/Pcs.	5.54	23.000 C/Pcs.	0.23	2.45	9/ 3/96
I- 2230 PHONE JACK WP.	1.00	8.740 E/Each	8.74	0.230 E/Each	0.23	2.45	4/22/96
I- 642 FLEX STEEL 1"	6.00	89.500 C/Feet	5.37	3.667 C/Feet	0.22	2.34	9/ 3/96
I- 2398 WIRE #1 CU 600V	30.00	646.000 M/FEET	19.38	7.333 M/FEET	0.22	2.34	9/ 3/96
I- 757 GROUND CLAMP 1/2-1"	1.00	179.000 C/Pcs.	1.79	21.000 C/Pcs.	0.21	2.24	9/ 3/96
I- 2395 WIRE #4 CU 600V	31.00	382.258 M/Feet	11.85	6.774 M/Feet	0.21	2.24	9/ 3/96
I- 355 EMT COUP SS 1/2"	11.00	26.182 C/Pcs.	2.88	1.818 C/Pcs.	0.20	2.13	9/ 2/96
I- 1215 PLATE DECORA 1 GANG	8.00	64.500 C/Pcs.	5.16	2.375 C/Pcs.	0.19	2.02	9/ 3/96
I- 1303 PLATE WP 1G DPX REC	5.00	425.000 C/Pcs.	21.25	3.600 C/Pcs.	0.18	1.92	9/ 3/96
I- 1495 REC SINGLE 15A 125V	1.00	5.460 E/Each	5.46	0.180 E/Each	0.18	1.92	9/ 3/96
I- 2290 UF CABLE 14/2 COPPER	40.00	145.250 M/Feet	5.81	4.500 M/Feet	0.18	1.92	9/ 3/96
I- 193 BUSHING PLAST 1 1/4"	2.00	97.000 C/Pcs.	1.94	8.000 C/Pcs.	0.16	1.70	9/ 2/96
I- 1270 PLATE PLAS 3G SWITCH	3.00	108.000 C/Pcs.	3.24	5.333 C/Pcs.	0.16	1.70	9/ 3/96
I- 418 EMT STRAPS 1H 1 1/4"	6.00	20.833 C/Pcs.	1.25	2.500 C/Pcs.	0.15	1.60	9/ 3/96
I- 103 BREAKER 20AMP 2 POLE	1.00	E/Each		0.140 E/Each	0.14	1.49	2/21/96
I- 104 BREAKER 20AMP 2 POLE	1.00	E		0.140 E	0.14	1.49	2/21/96
I- 670 FLEX 90 CONN 1/2"	2.00	83.000 C/Pcs.	1.66	7.000 C/Pcs.	0.14	1.49	9/ 3/96
I- 1746 ROMEX 8/2 COPPER	10.00	656.000 M/Feet	6.56	14.000 M/Feet	0.14	1.49	9/ 3/96
I- 318 EMT CONN SS 1 1/4"	2.00	118.000 C/Pcs.	2.36	6.500 C/Pcs.	0.13	1.38	9/ 2/96
I- 1271 PLATE PLAS 4G SWITCH	2.00	141.000 C	2.82	6.500 C	0.13	1.38	9/ 3/96
I- 1761 ROMEX STAPLE LARGE	14.00	1.071 C/Pcs.	0.15	0.929 C/Pcs.	0.13	1.38	9/ 6/96
I- 2393 WIRE #8 CU 600V	22.00	159.545 M/Feet	3.51	5.455 M/Feet	0.12	1.28	9/ 3/96
I- 915 LOCKNUTS STEEL 2"	2.00	75.500 C/Pcs.	1.51	5.500 C/Pcs.	0.11	1.17	9/ 3/96
I- 1286 PLATE IND GFCI	2.00	108.000 C	2.16	5.500 C	0.11	1.17	9/ 3/96
I- 1900 SEAL CONN STR 1/2"	2.00	169.000 C/Pcs.	3.38	5.000 C/Pcs.	0.10	1.07	9/ 3/96
I- 2421 WIRE BARE #6 COPPER	15.00	222.000 M/Feet	3.33	6.667 M/Feet	0.10	1.07	9/ 3/96
I- 317 EMT CONN SS 1"	2.00	70.000 C/Pcs.	1.40	4.500 C/Pcs.	0.09	0.96	9/ 2/96
I- 672 FLEX 90 CONN 1"	1.00	541.000 C/Pcs.	5.41	9.000 C/Pcs.	0.09	0.96	9/ 4/96
I- 1410 PVC LB 1/2"	1.00	151.000 C/Pcs.	1.51	9.000 C/Pcs.	0.09	0.96	9/ 3/96
I- 1680 RIGID STR 1H 1/2"	4.00	17.250 C/Pcs.	0.69	2.250 C/Pcs.	0.09	0.96	9/ 3/96
I- 2392 WIRE #10 CU 600V	18.00	85.556 M/Feet	1.54	5.000 M/Feet	0.09	0.96	9/ 3/96
I- 653 FLEX ST CONN 1"	1.00	104.000 C/Pcs.	1.04	8.000 C/Pcs.	0.08	0.85	9/ 3/96
I- 1276 PLATE IND 2X4 SWITCH	2.00	51.000 C/Pcs.	1.02	4.000 C/Pcs.	0.08	0.85	9/ 3/96
I- 1920 SEAL CONN 90 1/2"	1.00	224.000 C/Pcs.	2.24	8.000 C/Pcs.	0.08	0.85	9/ 3/96
I- 2422 WIRE BARE #4 COPPER	10.00	331.000 M/Feet	3.31	8.000 M/Feet	0.08	0.85	9/ 3/96
I- 357 EMT COUP SS 1"	1.50	68.000 C/Pcs.	1.02	4.667 C/Pcs.	0.07	0.75	9/ 2/96
I- 671 FLEX 90 CONN 3/4"	1.00	168.000 C/Pcs.	1.68	7.000 C/Pcs.	0.07	0.75	9/ 3/96

MIKE HOLT ENTERPRISES Version [091889]

[8008812580]

Report #3

```
                            Labor Analysis - Report No. 3
                                 MIKE HOLT ENTERPRISES                              Page:  4
                    Based on ELECTRICAL WIRING - Residential, Ray Mullin
Residential Bid        000000002                  3,744   sqft
                                                              Labor      Labor
Id Code Description        Quantity    Cost/Unit   Cost Ext   Labor/Unit   Hour Ext   Cost Ext   Last Updt
I- 1752 ROMEX CONN 3/4"      3.00     26.000 C/Pcs.   0.78    2.000 C/Pcs.   0.06       0.64     9/ 6/96
I-  417 EMT STRAPS 1H 1"     2.00     18.500 C/Pcs.   0.37    2.500 C/Pcs.   0.05       0.53     9/ 3/96
I- 1266 PLATE PLAS 30/50 AMP 1.00    108.000 C/Pcs.   1.08    5.000 C/Pcs.   0.05       0.53     9/ 3/96
I- 1279 PLATE IND 4X4 SWITCH 1.00    108.000 C/Pcs.   1.08    5.000 C/Pcs.   0.05       0.53     9/ 4/96
I- 1280 PLATE IND 4X4 DB SW  1.00    108.000 C/Pcs.   1.08    5.000 C/Pcs.   0.05       0.53     9/ 3/96
I- 1282 PLATE IND 4X4 RECEPT 1.00    108.000 C/Pcs.   1.08    5.000 C/Pcs.   0.05       0.53     9/ 3/96
I- 1375 PVC COUPLINGS 2"     1.00     74.000 C/Pcs.   0.74    5.000 C/Pcs.   0.05       0.53     9/ 3/96
I- 1420 PVC MALE ADAPT 1/2"  1.00     24.000 C/Pcs.   0.24    5.000 C/Pcs.   0.05       0.53     9/ 3/96
I- 1682 RIGID STR 1H 1"      2.00     43.500 C/Pcs.   0.87    2.500 C/Pcs.   0.05       0.53     9/ 3/96
I- 2443 WIRE NUTS GREY       5.00     11.000 C/Pcs.   0.55    1.000 C/Pcs.   0.05       0.53     9/ 6/96
I-  910 LOCKNUTS STEEL 1/2"  1.00     16.000 C/Pcs.   0.16    4.000 C/Pcs.   0.04       0.43     9/ 3/96
I- 1277 PLATE IND 2X4 SNG RE 1.00     51.000 C/Pcs.   0.51    4.000 C/Pcs.   0.04       0.43     9/ 3/96
I- 1262 PLATE PLAS 1G SGL RE 1.00     51.000 C/Pcs.   0.51    3.000 C/Pcs.   0.03       0.32     9/ 3/96
I- 1685 RIGID STR 2H 2"      1.00     31.000 C/Pcs.   0.31    3.000 C/Pcs.   0.03       0.32     9/ 3/96
Q-    1 Switchgear           1.00    425.000 E      425.00           E                  0.00    11/ 6/96
D-    1 Permit               1.00    375.000 E      375.00           E                  0.00    11/ 6/96
D-    2 Temporary            1.00    330.000 E      330.00           E                  0.00    11/ 6/96
                                                $  3,096.41                   173.67  1,849.66
     NOTE: These Figures do NOT Include Tax.       ==========                 ====== ========
```

MIKE HOLT ENTERPRISES Version [091889]

Report #4

Material Quantity Analysis - Report No. 4
MIKE HOLT ENTERPRISES
Based on ELECTRICAL WIRING - Residential, Ray Mullin
Page: 1

Residential Bid   000000002   3,744 sqft

Id Code Description	Quantity	Cost/Unit	Cost Ext	Labor/Unit	Labor Hour Ext	Labor Cost Ext	Last Updt
I- 26 BOX MET 2*4 REGULAR	2.00	64.000 C/Pcs.	1.28	16.000 C/Pcs.	0.32	3.41	9/ 2/96
I- 31 BOX MET 4*4 REGULAR	13.00	64.154 C/Pcs.	8.34	16.077 C/Pcs.	2.09	22.26	9/ 2/96
I- 50 BOX PLASTIC 1 GANG	90.00	105.978 C/Pcs.	95.38	9.000 C/Pcs.	8.10	86.27	9/ 2/96
I- 51 BOX PLASTIC 2 GANG	9.00	182.333 C/Pcs.	16.41	10.778 C/Pcs.	0.97	10.33	9/ 2/96
I- 52 BOX PLASTIC 3 GANG	3.00	190.000 C/Pcs.	5.70	14.333 C/Pcs.	0.43	4.58	9/ 2/96
I- 53 BOX PLASTIC 4 GANG	2.00	246.000 C/Pcs.	4.92	18.000 C/Pcs.	0.36	3.83	9/ 2/96
I- 55 BOX PLASTIC OCTAGON	24.00	97.167 C/Pcs.	23.32	9.000 C/Pcs.	2.16	23.00	9/ 2/96
I- 56 BOX PLASTIC OCT/BAR	2.00	196.000 C/Pcs.	3.92	18.000 C/Pcs.	0.36	3.83	9/ 2/96
I- 59 BOX PLASTIC FAN	3.00	436.000 C	13.08	16.333 C	0.49	5.22	9/ 3/96
I- 100 BREAKER 15AMP 1 POLE	15.00	E/Each		0.090 E/Each	1.35	14.38	1/ 3/91
I- 101 BREAKER 20AMP 1 POLE	10.00	E		0.090 E	0.90	9.59	2/21/96
I- 103 BREAKER 20AMP 2 POLE	1.00	E/Each		0.140 E/Each	0.14	1.49	2/21/96
I- 104 BREAKER 20AMP 2 POLE	1.00	E		0.140 E	0.14	1.49	2/21/96
I- 105 BREAKER 30AMP 2 POLE	2.00	E		0.140 E	0.28	2.98	2/21/96
I- 106 BREAKER 40AMP 2 POLE	2.00	E		0.140 E	0.28	2.98	2/21/96
I- 109 BREAKER 70AMP 2 POLE	2.00	E		0.180 E	0.36	3.83	2/21/96
I- 193 BUSHING PLAST 1 1/4"	2.00	97.000 C/Pcs.	1.94	8.000 C/Pcs.	0.16	1.70	9/ 2/96
I- 195 BUSHING PLASTIC 2"	2.00	152.000 C/Pcs.	3.04	11.500 C/Pcs.	0.23	2.45	9/ 2/96
I- 222 CORD 15.250V 3'	1.00	4.370 E/Each	4.37	0.230 E/Each	0.23	2.45	9/ 2/96
I- 223 CORD 15.250V 6'	1.00	4.370 E/Each	4.37	0.230 E/Each	0.23	2.45	9/ 2/96
I- 228 CORD DRYER 30A 4WIRE	1.00	13.110 E/Each	13.11	0.270 E/Each	0.27	2.88	9/ 2/96
I- 235 DIM 600W 120V 1POLE	2.00	6.665 E/Each	13.33	0.225 E/Each	0.45	4.79	9/ 2/96
I- 251 DISC. 60 AMP 1PHASE	1.00	E		0.900 E	0.90	9.59	2/21/96
I- 252 DISC. 70 AMP 1PHASE	1.00	E		0.900 E	0.90	9.59	2/21/96
I- 305 EMT 1/2"	170.00	14.200 C/Feet	24.14	2.041 C/Feet	3.47	36.96	9/ 2/96
I- 307 EMT 1"	15.00	33.867 C/Feet	5.08	3.333 C/Feet	0.50	5.33	9/ 3/96
I- 308 EMT 1 1/4"	50.00	54.620 C/Feet	27.31	3.600 C/Feet	1.80	19.17	9/ 2/96
I- 315 EMT CONN SS 1/2"	22.00	23.000 C/Pcs.	5.06	1.909 C/Pcs.	0.42	4.47	9/ 2/96
I- 317 EMT CONN SS 1"	2.00	70.000 C/Pcs.	1.40	4.500 C/Pcs.	0.09	0.96	9/ 2/96
I- 318 EMT CONN SS 1 1/4"	2.00	118.000 C/Pcs.	2.36	6.500 C/Pcs.	0.13	1.38	9/ 2/96
I- 355 EMT COUP SS 1/2"	11.00	26.182 C/Pcs.	2.88	1.818 C/Pcs.	0.20	2.13	9/ 2/96
I- 357 EMT COUP SS 1"	1.50	68.000 C/Pcs.	1.02	4.667 C/Pcs.	0.07	0.75	9/ 2/96
I- 358 EMT COUP SS 1 1/4"	9.00	108.222 C/Pcs.	9.74	6.333 C/Pcs.	0.57	6.07	9/ 2/96
I- 398 EMT ELLS 90'S 1 1/4"	3.00	339.667 C/Pcs.	10.19	13.667 C/Pcs.	0.41	4.37	9/ 2/96
I- 415 EMT STRAPS 1H 1/2"	12.00	6.500 C/Pcs.	0.78	2.417 C/Pcs.	0.29	3.09	9/ 3/96
I- 417 EMT STRAPS 1H 1"	2.00	18.500 C/Pcs.	0.37	2.500 C/Pcs.	0.05	0.53	9/ 3/96
I- 418 EMT STRAPS 1H 1 1/4"	6.00	20.833 C/Pcs.	1.25	2.500 C/Pcs.	0.15	1.60	9/ 3/96
I- 473 FAN AND LIGHT	3.00	E/Each		0.677 E/Each	2.03	21.62	9/ 3/96
I- 474 FAN, EXHAUST	4.00	E/Each		0.900 E/Each	3.60	38.34	1/ 3/91
I- 476 FAN, HOOD	1.00	E/Each		0.450 E/Each	0.45	4.79	1/ 3/91
I- 478 RECESS FIXTURE	16.00	E/Each		0.900 E/Each	14.40	153.36	11/ 7/96
I- 480 PULL CHAIN	3.00	E/Each		0.227 E/Each	0.68	7.24	1/ 3/91
I- 491 OUTSIDE FIXTURE	6.00	E/		0.225 E/	1.35	14.38	2/19/96
I- 492 INSIDE FIXTURE	9.00	E		0.226 E	2.03	21.62	9/ 3/96
I- 500 SMOKE DETECTOR	4.00	E/		0.225 E/	0.90	9.59	2/19/96
I- 501 BATH HEAT/FAN/LIGHT	2.00	E/		0.900 E/	1.80	19.17	2/19/96
I- 502 KEYLESS FIXTURE	6.00	E/		0.225 E/	1.35	14.38	2/22/96

MIKE HOLT ENTERPRISES Version [091889]   [8008812580]

Report #4

```
                        Material Quantity Analysis - Report No. 4
                                MIKE HOLT ENTERPRISES                              Page:  2
                        Based on ELECTRICAL WIRING - Residential, Ray Mullin
Residential Bid         000000002                    3,744    sqft
                                                                    Labor     Labor
Id Code Description     Quantity      Cost/Unit     Cost Ext   Labor/Unit   Hour Ext   Cost Ext   Last Updt
```

Id Code	Description	Quantity	Cost/Unit		Cost Ext	Labor/Unit		Labor Hour Ext	Labor Cost Ext	Last Updt
I- 505	POST LIGHT	1.00		E/		1.800	E/	1.80	19.17	9/ 3/96
I- 576	FLUO STR 1-8' 120V	1.00		E/Each		0.680	E/Each	0.68	7.24	1/ 3/91
I- 579	FLUO WRAP 2-2' 120V	3.00		E/Each		0.677	E/Each	2.03	21.62	1/ 3/91
I- 583	FLUO LAYIN 2X4' 120V	6.00		E/Each		0.675	E/Each	4.05	43.13	11/ 7/96
I- 620	FIX TRACK 4'	1.00		E		0.680	E	0.68	7.24	2/26/96
I- 621	FIX TRACK 8'	1.00		E/Each		0.900	E/Each	0.90	9.59	1/ 3/91
I- 640	FLEX STEEL 1/2"	21.00	33.857	C/Feet	7.11	3.000	C/Feet	0.63	6.71	9/ 3/96
I- 642	FLEX STEEL 1"	6.00	89.500	C/Feet	5.37	3.667	C/Feet	0.22	2.34	9/ 3/96
I- 651	FLEX ST CONN 1/2"	5.00	60.000	C/Pcs.	3.00	5.400	C/Pcs.	0.27	2.88	9/ 4/96
I- 653	FLEX ST CONN 1"	1.00	104.000	C/Pcs.	1.04	8.000	C/Pcs.	0.08	0.85	9/ 3/96
I- 670	FLEX 90 CONN 1/2"	2.00	83.000	C/Pcs.	1.66	7.000	C/Pcs.	0.14	1.49	9/ 3/96
I- 671	FLEX 90 CONN 3/4"	1.00	168.000	C/Pcs.	1.68	7.000	C/Pcs.	0.07	0.75	9/ 3/96
I- 672	FLEX 90 CONN 1"	1.00	541.000	C/Pcs.	5.41	9.000	C/Pcs.	0.09	0.96	9/ 4/96
I- 755	GROUND CLAMP 1/2" DB	2.00	190.000	C/Pcs.	3.80	12.000	C/Pcs.	0.24	2.56	9/ 3/96
I- 757	GROUND CLAMP 1/2-1"	1.00	179.000	C/Pcs.	1.79	21.000	C/Pcs.	0.21	2.24	9/ 3/96
I- 762	GROUND ROD 1/2X8' CU	2.00	1,092.500	C/Pcs.	21.85	45.000	C/Pcs.	0.90	9.59	9/ 3/96
I- 910	LOCKNUTS STEEL 1/2"	1.00	16.000	C/Pcs.	0.16	4.000	C/Pcs.	0.04	0.43	9/ 3/96
I- 915	LOCKNUTS STEEL 2"	2.00	75.500	C/Pcs.	1.51	5.500	C/Pcs.	0.11	1.17	9/ 3/96
I- 941	CHIME	1.00		E/Each		0.900	E/Each	0.90	9.59	9/ 3/96
I- 946	THERMOSTAT	1.00		E		0.450	E	0.45	4.79	9/ 3/96
I- 948	PUSH BUTTON	3.00	3.277	E/Each	9.83	0.227	E/Each	0.68	7.24	9/ 3/96
I- 1005	METER/MAIN 200 AMP	1.00		E/Each		1.800	E/Each	1.80	19.17	10/ 7/96
I- 1041	GREEN GROMMENTS 1"	284.00	54.754	M/Pcs.	15.55	0.915	M/Pcs.	0.26	2.77	9/12/96
I- 1190	PANEL 100 AMP 1PHASE	1.00		E/Each		0.900	E/Each	0.90	9.59	1/ 3/91
I- 1191	PANEL 200 AMP 1PHASE	1.00		E/Each		0.900	E/Each	0.90	9.59	2/22/96
I- 1215	PLATE DECORA 1 GANG	8.00	64.500	C/Pcs.	5.16	2.375	C/Pcs.	0.19	2.02	9/ 3/96
I- 1261	PLATE PLAS 1G SWITCH	22.00	51.318	C/Pcs.	11.29	2.318	C/Pcs.	0.51	5.43	9/ 3/96
I- 1262	PLATE PLAS 1G SGL RE	1.00	51.000	C/Pcs.	0.51	3.000	C/Pcs.	0.03	0.32	9/ 3/96
I- 1263	PLATE PLAS 1G DPX RE	46.00	51.304	C/Pcs.	23.60	2.370	C/Pcs.	1.09	11.61	9/ 3/96
I- 1265	PLATE PLAS 2G SWITCH	8.00	88.500	C/Pcs.	7.08	3.625	C/Pcs.	0.29	3.09	9/ 3/96
I- 1266	PLATE PLAS 30/50 AMP	1.00	108.000	C/Pcs.	1.08	5.000	C/Pcs.	0.05	0.53	9/ 3/96
I- 1270	PLATE PLAS 3G SWITCH	3.00	108.000	C/Pcs.	3.24	5.333	C/Pcs.	0.16	1.70	9/ 3/96
I- 1271	PLATE PLAS 4G SWITCH	2.00	141.000	C	2.82	6.500	C	0.13	1.38	9/ 3/96
I- 1276	PLATE IND 2X4 SWITCH	2.00	51.000	C/Pcs.	1.02	4.000	C/Pcs.	0.08	0.85	9/ 3/96
I- 1277	PLATE IND 2X4 SNG RE	1.00	51.000	C/Pcs.	0.51	4.000	C/Pcs.	0.04	0.43	9/ 3/96
I- 1279	PLATE IND 4X4 SWITCH	1.00	108.000	C/Pcs.	1.08	5.000	C/Pcs.	0.05	0.53	9/ 4/96
I- 1280	PLATE IND 4X4 DB SW	1.00	108.000	C/Pcs.	1.08	5.000	C/Pcs.	0.05	0.53	9/ 3/96
I- 1282	PLATE IND 4X4 RECEPT	1.00	108.000	C/Pcs.	1.08	5.000	C/Pcs.	0.05	0.53	9/ 3/96
I- 1285	PLATE IND 4X4 BLANK	5.00	108.000	C	5.40	5.200	C	0.26	2.77	9/ 3/96
I- 1286	PLATE IND GFCI	2.00	108.000	C	2.16	5.500	C	0.11	1.17	9/ 3/96
I- 1303	PLATE WP 1G DPX REC	5.00	425.000	C/Pcs.	21.25	3.600	C/Pcs.	0.18	1.92	9/ 3/96
I- 1350	PVC 1/2" SCHEDULE 40	15.00	14.200	C/Feet	2.13	1.733	C/Feet	0.26	2.77	9/ 3/96
I- 1355	PVC 2" SCHEDULE 40	10.00	53.500	C/Feet	5.35	4.100	C/Feet	0.41	4.37	9/ 3/96
I- 1375	PVC COUPLINGS 2"	1.00	74.000	C/Pcs.	0.74	5.000	C/Pcs.	0.05	0.53	9/ 3/96
I- 1410	PVC LB 1/2"	1.00	151.000	C/Pcs.	1.51	9.000	C/Pcs.	0.09	0.96	9/ 3/96
I- 1415	PVC LB 2"	2.00	509.000	C/Pcs.	10.18	24.500	C/Pcs.	0.49	5.22	9/ 3/96
I- 1420	PVC MALE ADAPT 1/2"	1.00	24.000	C/Pcs.	0.24	5.000	C/Pcs.	0.05	0.53	9/ 3/96

MIKE HOLT ENTERPRISES Version [091889]

[8008812580]

Report #4

```
                        Material Quantity Analysis - Report No. 4
                                  MIKE HOLT ENTERPRISES                              Page:  3
                        Based on ELECTRICAL WIRING - Residential, Ray Mullin
Residential Bid         000000002                         3,744   sqft
                                                                      Labor      Labor
Id Code Description          Quantity       Cost/Unit     Cost Ext   Labor/Unit   Hour Ext   Cost Ext   Last Updt
I- 1425 PVC MALE ADAPT 2"         2.00     78.500 C/Pcs.      1.57   11.500 C/Pcs.    0.23      2.45    7/14/96
I- 1490 REC DUPLEX 15A 125V      47.00     56.830 C/Pcs.     26.71   16.170 C/Pcs.    7.60     80.94    9/ 3/96
I- 1492 REC GFCI 15A 125V        15.00     13.109 E/Each    196.63    0.270 E/Each    4.05     43.13    9/ 3/96
I- 1495 REC SINGLE 15A 125V       1.00      5.460 E/Each      5.46    0.180 E/Each    0.18      1.92    9/ 3/96
I- 1496 REC SINGLE 20, 125V       1.00      5.460 E/Each      5.46    0.230 E/Each    0.23      2.45    9/ 3/96
I- 1498 REC DRYER 30A 4 WIRE      1.00      7.650 E/Each      7.65    0.230 E/Each    0.23      2.45    9/ 3/96
I- 1504 CLOCK OUTLET & PLATE      1.00    468.000 C/Pcs.      4.68   41.000 C/Pcs.    0.41      4.37    9/ 3/96
I- 1521 MULTIOUTLET ASSEMBLY      1.00              E                 0.900 E         0.90      9.59    9/ 3/96
I- 1595 RIGID 2"                 10.00    249.000 C/Feet     24.90    6.300 C/Feet    0.63      6.71    9/ 3/96
I- 1680 RIGID STR 1H 1/2"         4.00     17.250 C/Pcs.      0.69    2.250 C/Pcs.    0.09      0.96    9/ 3/96
I- 1682 RIGID STR 1H 1"           2.00     43.500 C/Pcs.      0.87    2.500 C/Pcs.    0.05      0.53    9/ 3/96
I- 1685 RIGID STR 2H 2"           1.00     31.000 C/Pcs.      0.31    3.000 C/Pcs.    0.03      0.32    9/ 3/96
I- 1711 RING 1 GANG              24.00     42.667 C          10.24    4.250 C         1.02     10.86    9/ 3/96
I- 1740 ROMEX 14/2 COPPER     1,616.00    117.970 M/Feet    190.64    9.004 M/Feet   14.55    154.96    9/ 3/96
I- 1741 ROMEX 14/3 COPPER     1,002.00    203.164 M/Feet    203.57   10.798 M/Feet   10.82    115.23    9/ 3/96
I- 1742 ROMEX 12/2 COPPER       434.00    161.659 M/Feet     70.16   10.806 M/Feet    4.69     49.95    9/ 3/96
I- 1743 ROMEX 12/3 COPPER        42.00    291.429 M/Feet     12.24   14.286 M/Feet    0.60      6.39    9/ 3/96
I- 1745 ROMEX 10/3 COPPER        50.00    433.800 M/Feet     21.69   14.400 M/Feet    0.72      7.67    9/ 3/96
I- 1746 ROMEX 8/2 COPPER         10.00    656.000 M/Feet      6.56   14.000 M/Feet    0.14      1.49    9/ 3/96
I- 1747 ROMEX 8/3 COPPER         20.00    895.500 M/Feet     17.91   18.000 M/Feet    0.36      3.83    9/ 4/96
I- 1751 ROMEX CONN 1/2"          48.00     17.417 C/Pcs.      8.36    0.938 C/Pcs.    0.45      4.79    9/ 6/96
I- 1752 ROMEX CONN 3/4"           3.00     26.000 C/Pcs.      0.78    2.000 C/Pcs.    0.06      0.64    9/ 6/96
I- 1760 ROMEX STAPLE SMALL      312.00      1.093 C/Pcs.      3.41    0.904 C/Pcs.    2.82     30.03    9/ 6/96
I- 1761 ROMEX STAPLE LARGE       14.00      1.071 C/Pcs.      0.15    0.929 C/Pcs.    0.13      1.38    9/ 6/96
I- 1762 KICK PLATES              78.00      5.538 C/Pcs.      4.32    1.090 C/Pcs.    0.85      9.05    9/ 3/96
I- 1890 SEALTIGHT 1/2"           12.00     51.333 C/Feet      6.16    3.000 C/Feet    0.36      3.83    9/ 3/96
I- 1900 SEAL CONN STR 1/2"        2.00    169.000 C/Pcs.      3.38    5.000 C/Pcs.    0.10      1.07    9/ 3/96
I- 1920 SEAL CONN 90 1/2"         1.00    224.000 C/Pcs.      2.24    8.000 C/Pcs.    0.08      0.85    9/ 3/96
I- 2091 SW 15,125V 1POLE         31.00     62.258 C/Pcs.     19.30   18.000 C/Pcs.    5.58     59.43    9/ 3/96
I- 2092 SW 15,125V 3WAY          22.00    138.727 C/Pcs.     30.52   22.500 C/Pcs.    4.95     52.72    9/ 3/96
I- 2093 SW 15,125V 4WAY           2.00    785.500 C/Pcs.     15.71   25.000 C/Pcs.    0.50      5.33    9/ 3/96
I- 2097 SW 20,125V 2POLE          1.00  1,311.000 C/Pcs.     13.11   29.000 C/Pcs.    0.29      3.09    9/ 3/96
I- 2104 SW 15 AMP PILOT 125V      1.00    554.000 C/Pcs.      5.54   23.000 C/Pcs.    0.23      2.45    9/ 3/96
I- 2108 SW BATH/FAN/LIGHT         2.00     18.570 E          37.14    0.450 E         0.90      9.59    9/ 3/96
I- 2109 ATTIC FAN SWITCH          1.00     16.380 E          16.38    0.450 E         0.45      4.79    4/22/96
I- 2110 SW 15,125V 1POLE FAN      3.00  1,016.000 C          30.48   22.667 C         0.68      7.24    9/ 3/96
I- 2222 TV RECEPTACLE            12.00      3.278 E/Each     39.33    0.226 E/Each    2.71     28.86    9/ 3/96
I- 2223 WIRE COAX TV CABLE      504.00    196.627 M/FEET     99.10   10.794 M/FEET    5.44     57.94    9/ 3/96
I- 2228 TELEPHONE WIRE TWIST    288.00     98.333 M/Feet     28.32   10.764 M/Feet    3.10     33.02    9/ 3/96
I- 2229 PHONE JACK INDOOR         8.00      3.278 E/Each     26.22    0.226 E/Each    1.81     19.28    9/ 3/96
I- 2230 PHONE JACK WP.            1.00      8.740 E/Each      8.74    0.230 E/Each    0.23      2.45    4/22/96
I- 2290 UF CABLE 14/2 COPPER     40.00    145.250 M/Feet      5.81    4.500 M/Feet    0.18      1.92    9/ 3/96
I- 2360 WEATHERHEADS 2"           1.00      7.650 E/Each      7.65    0.360 E/Each    0.36      3.83    9/ 3/96
I- 2390 WIRE #14 CU 600V        218.00     41.514 M/Feet      9.05    3.257 M/Feet    0.71      7.56    9/ 3/96
I- 2391 WIRE #12 CU 600V        111.00     52.432 M/Feet      5.82    3.964 M/Feet    0.44      4.69    9/ 3/96
I- 2392 WIRE #10 CU 600V         18.00     85.556 M/Feet      1.54    5.000 M/Feet    0.09      0.96    9/ 3/96
I- 2393 WIRE #8 CU 600V          22.00    159.545 M/Feet      3.51    5.455 M/Feet    0.12      1.28    9/ 3/96


                              MIKE HOLT ENTERPRISES Version [091889]
                                                                                      [8008812580]
```

Report #4

```
                    Material Quantity Analysis - Report No. 4
                              MIKE HOLT ENTERPRISES                              Page:  4
                    Based on ELECTRICAL WIRING - Residential, Ray Mullin
Residential Bid     000000002                            3,744    sqft
                                                                  Labor      Labor
Id Code Description      Quantity     Cost/Unit   Cost Ext    Labor/Unit   Hour Ext   Cost Ext   Last Updt
I- 2394 WIRE #6 CU 600V     78.00   273.077 M/Feet   21.30   6.282 M/Feet     0.49      5.22      9/ 3/96
I- 2395 WIRE #4 CU 600V     31.00   382.258 M/Feet   11.85   6.774 M/Feet     0.21      2.24      9/ 3/96
I- 2396 WIRE #3 CU 600V    150.00   409.733 M/Feet   61.46   6.800 M/Feet     1.02     10.86      9/ 3/96
I- 2398 WIRE #1 CU 600V     30.00   646.000 M/FEET   19.38   7.333 M/FEET     0.22      2.34      9/ 3/96
I- 2405 WIRE 2/0 CU 600V    60.00   891.333 M/Feet   53.48  13.667 M/Feet     0.82      8.73      9/ 3/96
I- 2421 WIRE BARE #6 COPPER 15.00   222.000 M/Feet    3.33   6.667 M/Feet     0.10      1.07      9/ 3/96
I- 2422 WIRE BARE #4 COPPER 10.00   331.000 M/Feet    3.31   8.000 M/Feet     0.08      0.85      9/ 3/96
I- 2425 THERMO WIRE 18/2    70.00    76.429 M/Feet    5.35  11.714 M/Feet     0.82      8.73      9/ 3/96
I- 2426 THERMO WIRE 18/5   100.00   131.000 M/Feet   13.10  11.800 M/Feet     1.18     12.57      9/ 3/96
I- 2441 WIRE NUTS YELLOW   139.00     6.597 C/Pcs.    9.17   0.935 C/Pcs.     1.30     13.85      9/ 3/96
I- 2442 WIRE NUTS RED      285.00     8.709 C/Pcs.   24.82   0.916 C/Pcs.     2.61     27.80      9/ 6/96
I- 2443 WIRE NUTS GREY       5.00    11.000 C/Pcs.    0.55   1.000 C/Pcs.     0.05      0.53      9/ 6/96
I- 2445 WIRE NUTS GREEN     64.00    10.844 C/Pcs.    6.94   1.125 C/Pcs.     0.72      7.67      9/ 6/96
I- 2450 WIRE TERM # 12-6     3.00           E/Each           0.137 E/Each     0.41      4.37      9/ 6/96
Q-    1 Switchgear           1.00   425.000 E       425.00              E                0.00     1/ 3/91
D-    1 Permit               1.00   375.000 E       375.00              E                0.00    11/ 6/96
D-    2 Temporary            1.00   330.000 E       330.00              E                0.00    11/ 6/96

      NOTE: These Figures do NOT Include Tax.     $  3,096.41              173.67    1,849.66
```

MIKE HOLT ENTERPRISES Version [091889]

[8008812580]

Report #5

```
                        Material Cost Analysis - Report No. 5
                                 MIKE HOLT ENTERPRISES                              Page: 1
                        Based on ELECTRICAL WIRING - Residential, Ray Mullin
Residential Bid         000000002                       3,744   sqft
                                                                  Labor        Labor
                                                                  Hour Ext     Cost Ext   Last Updt
Id Code Description        Quantity     Cost/Unit    Cost Ext   Labor/Unit
Q-    1 Switchgear             1.00     425.000 E     425.00               E                          11/ 6/96
D-    1 Permit                 1.00     375.000 E     375.00               E                          11/ 6/96
D-    2 Temporary              1.00     330.000 E     330.00               E                          11/ 6/96
I- 1741 ROMEX 14/3 COPPER  1,002.00     203.164 M/Feet 203.57   10.798 M/Feet   10.82      115.23      9/ 3/96
I- 1492 REC GFCI 15A 125V     15.00      13.109 E/Each 196.63    0.270 E/Each    4.05       43.13      9/ 3/96
I- 1740 ROMEX 14/2 COPPER  1,616.00     117.970 M/Feet 190.64    9.004 M/Feet   14.55      154.96      9/ 3/96
I- 2223 WIRE COAX TV CABLE   504.00     196.627 M/FEET  99.10   10.794 M/FEET    5.44       57.94      9/ 3/96
I-   50 BOX PLASTIC 1 GANG    90.00     105.978 C/Pcs.  95.38    9.000 C/Pcs.    8.10       86.27      9/ 2/96
I- 1742 ROMEX 12/2 COPPER    434.00     161.659 M/Feet  70.16   10.806 M/Feet    4.69       49.95      9/ 3/96
I- 2396 WIRE #3 CU 600V      150.00     409.733 M/Feet  61.46    6.800 M/Feet    1.02       10.86      9/ 3/96
I- 2405 WIRE 2/0 CU 600V      60.00     891.333 M/Feet  53.48   13.667 M/Feet    0.82        8.73      9/ 3/96
I- 2222 TV RECEPTACLE         12.00       3.278 E/Each  39.33    0.226 E/Each    2.71       28.86      9/ 3/96
I- 2108 SW BATH/FAN/LIGHT      2.00      18.570 E       37.14    0.450 E         0.90        9.59      9/ 3/96
I- 2092 SW 15,125V 3WAY       22.00     138.727 C/Pcs.  30.52   22.500 C/Pcs.    4.95       52.72      9/ 3/96
I- 2110 SW 15,125V 1POLE FAN   3.00   1,016.000 C       30.48   22.667 C         0.68        7.24      9/ 3/96
I- 2228 TELEPHONE WIRE TWIST 288.00      98.333 M/Feet  28.32   10.764 M/Feet    3.10       33.02      9/ 3/96
I-  308 EMT 1 1/4"            50.00      54.620 C/Feet  27.31    3.600 C/Feet    1.80       19.17      9/ 2/96
I- 1490 REC DUPLEX 15A 125V   47.00      56.830 C/Pcs.  26.71   16.170 C/Pcs.    7.60       80.94      9/ 3/96
I- 2229 PHONE JACK INDOOR      8.00       3.278 E/Each  26.22    0.226 E/Each    1.81       19.28      9/ 3/96
I- 1595 RIGID 2"              10.00     249.000 C/Feet  24.90    6.300 C/Feet    0.63        6.71      9/ 3/96
I- 2442 WIRE NUTS RED        285.00       8.709 C/Pcs.  24.82    0.916 C/Pcs.    2.61       27.80      9/ 6/96
I-  305 EMT 1/2"             170.00      14.200 C/Feet  24.14    2.041 C/Feet    3.47       36.96      9/ 2/96
I- 1263 PLATE PLAS 1G DPX RE  46.00      51.304 C/Pcs.  23.60    2.370 C/Pcs.    1.09       11.61      9/ 3/96
I-   55 BOX PLASTIC OCTAGON   24.00      97.167 C/Pcs.  23.32    9.000 C/Pcs.    2.16       23.00      9/ 2/96
I-  762 GROUND ROD 1/2X8' CU   2.00   1,092.500 C/Pcs.  21.85   45.000 C/Pcs.    0.90        9.59      9/ 3/96
I- 1745 ROMEX 10/3 COPPER     50.00     433.800 M/Feet  21.69   14.400 M/Feet    0.72        7.67      9/ 3/96
I- 2394 WIRE #6 CU 600V       78.00     273.077 M/Feet  21.30    6.282 M/Feet    0.49        5.22      9/ 3/96
I- 1303 PLATE WP 1G DPX REC    5.00     425.000 C/Pcs.  21.25    3.600 C/Pcs.    0.18        1.92      9/ 3/96
I- 2398 WIRE #1 CU 600V       30.00     646.000 M/FEET  19.38    7.333 M/FEET    0.22        2.34      9/ 3/96
I- 2091 SW 15,125V 1POLE      31.00      62.258 C/Pcs.  19.30   18.000 C/Pcs.    5.58       59.43      9/ 3/96
I- 1747 ROMEX 8/3 COPPER      20.00     895.500 M/Feet  17.91   18.000 M/Feet    0.36        3.83      9/ 4/96
I-   51 BOX PLASTIC 2 GANG     9.00     182.333 C/Pcs.  16.41   10.778 C/Pcs.    0.97       10.33      9/ 2/96
I- 2109 ATTIC FAN SWITCH       1.00      16.380 E       16.38    0.450 E         0.45        4.79      4/22/96
I- 2093 SW 15,125V 4WAY        2.00     785.500 C/Pcs.  15.71   25.000 C/Pcs.    0.50        5.33      9/ 3/96
I- 1041 GREEN GROMMENTS 1"   284.00      54.754 M/Pcs.  15.55    0.915 M/Pcs.    0.26        2.77      9/12/96
I-  235 DIM 600W 120V 1POLE    2.00       6.665 E/Each  13.33    0.225 E/Each    0.45        4.79      9/ 2/96
I-  228 CORD DRYER 30A 4WIRE   1.00      13.110 E/Each  13.11    0.270 E/Each    0.27        2.88      9/ 2/96
I- 2097 SW 20,125V 2POLE       1.00   1,311.000 C/Pcs.  13.11   29.000 C/Pcs.    0.29        3.09      9/ 3/96
I- 2426 THERMO WIRE 18/5     100.00     131.000 M/Feet  13.10   11.800 M/Feet    1.18       12.57      9/ 3/96
I-   59 BOX PLASTIC FAN        3.00     436.000 C       13.08   16.333 C         0.49        5.22      9/ 3/96
I- 1743 ROMEX 12/3 COPPER     42.00     291.429 M/Feet  12.24   14.286 M/Feet    0.60        6.39      9/ 3/96
I- 2395 WIRE #4 CU 600V       31.00     382.258 M/Feet  11.85    6.774 M/Feet    0.21        2.24      9/ 3/96
I- 1261 PLATE PLAS 1G SWITCH  22.00      51.318 C/Pcs.  11.29    2.338 C/Pcs.    0.51        5.43      9/ 3/96
I- 1711 RING 1 GANG           24.00      42.667 C       10.24    4.250 C         1.02       10.86      9/ 3/96
I-  398 EMT ELLS 90'S 1 1/4"   3.00     339.667 C/Pcs.  10.19   13.667 C/Pcs.    0.41        4.37      9/ 2/96
I- 1415 PVC LB 2"              2.00     509.000 C/Pcs.  10.18   24.500 C/Pcs.    0.49        5.22      9/ 3/96
I-  948 PUSH BUTTON            3.00       3.277 E/Each   9.83    0.227 E/Each    0.68        7.24      9/ 3/96


                                     MIKE HOLT ENTERPRISESVersion [091889]
                                                                                              [8008812580]
```

Report #5

Material Cost Analysis - Report No. 5
MIKE HOLT ENTERPRISES
Based on ELECTRICAL WIRING - Residential, Ray Mullin

Residential Bid   000000002   3,744 sqft

Id Code Description	Quantity	Cost/Unit	Cost Ext	Labor/Unit	Labor Hour Ext	Labor Cost Ext	Last Updt
I- 358 EMT COUP SS 1 1/4"	9.00	108.222 C/Pcs.	9.74	6.333 C/Pcs.	0.57	6.07	9/ 2/96
I- 2441 WIRE NUTS YELLOW	139.00	6.597 C/Pcs.	9.17	0.935 C/Pcs.	1.30	13.85	9/ 3/96
I- 2390 WIRE #14 CU 600V	218.00	41.514 M/Feet	9.05	3.257 M/Feet	0.71	7.56	9/ 3/96
I- 2230 PHONE JACK WP.	1.00	8.740 E/Each	8.74	0.230 E/Each	0.23	2.45	4/22/96
I- 1751 ROMEX CONN 1/2"	48.00	17.417 C/Pcs.	8.36	0.938 C/Pcs.	0.45	4.79	9/ 6/96
I- 31 BOX MET 4*4 REGULAR	13.00	64.154 C/Pcs.	8.34	16.077 C/Pcs.	2.09	22.26	9/ 2/96
I- 1498 REC DRYER 30A 4 WIRE	1.00	7.650 E/Each	7.65	0.230 E/Each	0.23	2.45	9/ 3/96
I- 2360 WEATHERHEADS 2"	1.00	7.650 E/Each	7.65	0.360 E/Each	0.36	3.83	9/ 3/96
I- 640 FLEX STEEL 1/2"	21.00	33.857 C/Feet	7.11	3.000 C/Feet	0.63	6.71	9/ 3/96
I- 1265 PLATE PLAS 2G SWITCH	8.00	88.500 C/Pcs.	7.08	3.625 C/Pcs.	0.29	3.09	9/ 3/96
I- 2445 WIRE NUTS GREEN	64.00	10.844 C/Pcs.	6.94	1.125 C/Pcs.	0.72	7.67	9/ 6/96
I- 1746 ROMEX 8/2 COPPER	10.00	656.000 M/Feet	6.56	14.000 M/Feet	0.14	1.49	9/ 3/96
I- 1890 SEALTIGHT 1/2"	12.00	51.333 C/Feet	6.16	3.000 C/Feet	0.36	3.83	9/ 3/96
I- 2391 WIRE #12 CU 600V	111.00	52.432 M/Feet	5.82	3.964 M/Feet	0.44	4.69	9/ 3/96
I- 2290 UF CABLE 14/2 COPPER	40.00	145.250 M/Feet	5.81	4.500 M/Feet	0.18	1.92	9/ 3/96
I- 52 BOX PLASTIC 3 GANG	3.00	190.000 C/Pcs.	5.70	14.333 C/Pcs.	0.43	4.58	9/ 2/96
I- 2104 SW 15 AMP PILOT 125V	1.00	554.000 C/Pcs.	5.54	23.000 C/Pcs.	0.23	2.45	9/ 3/96
I- 1495 REC SINGLE 15A 125V	1.00	5.460 E/Each	5.46	0.180 E/Each	0.18	1.92	9/ 3/96
I- 1496 REC SINGLE 20, 125V	1.00	5.460 E/Each	5.46	0.230 E/Each	0.23	2.45	9/ 3/96
I- 672 FLEX 90 CONN 1"	1.00	541.000 C/Pcs.	5.41	9.000 C/Pcs.	0.09	0.96	9/ 4/96
I- 1285 PLATE IND 4X4 BLANK	5.00	108.000 C	5.40	5.200 C	0.26	2.77	9/ 3/96
I- 642 FLEX STEEL 1"	6.00	89.500 C/Feet	5.37	3.667 C/Feet	0.22	2.34	9/ 3/96
I- 1355 PVC 2" SCHEDULE 40	10.00	53.500 C/Feet	5.35	4.100 C/Feet	0.41	4.37	9/ 3/96
I- 2425 THERMO WIRE 18/2	70.00	76.429 M/Feet	5.35	11.714 M/Feet	0.82	8.73	9/ 3/96
I- 1215 PLATE DECORA 1 GANG	8.00	64.500 C/Pcs.	5.16	2.375 C/Pcs.	0.19	2.02	9/ 3/96
I- 307 EMT 1"	15.00	33.867 C/Feet	5.08	3.333 C/Feet	0.50	5.33	9/ 3/96
I- 315 EMT CONN SS 1/2"	22.00	23.000 C/Pcs.	5.06	1.909 C/Pcs.	0.42	4.47	9/ 2/96
I- 53 BOX PLASTIC 4 GANG	2.00	246.000 C/Pcs.	4.92	18.000 C/Pcs.	0.36	3.83	9/ 3/96
I- 1504 CLOCK OUTLET & PLATE	1.00	468.000 C/Pcs.	4.68	41.000 C/Pcs.	0.41	4.37	9/ 3/96
I- 222 CORD 15.250V 3'	1.00	4.370 E/Each	4.37	0.230 E/Each	0.23	2.45	9/ 3/96
I- 223 CORD 15,250V 6'	1.00	4.370 E/Each	4.37	0.230 E/Each	0.23	2.45	9/ 3/96
I- 1762 KICK PLATES	78.00	5.538 C/Pcs.	4.32	1.090 C/Pcs.	0.85	9.05	9/ 3/96
I- 56 BOX PLASTIC OCT/BAR	2.00	196.000 C/Pcs.	3.92	18.000 C/Pcs.	0.36	3.83	9/ 2/96
I- 755 GROUND CLAMP 1/2" DB	2.00	190.000 C/Pcs.	3.80	12.000 C/Pcs.	0.24	2.56	9/ 3/96
I- 2393 WIRE #8 CU 600V	22.00	159.545 M/Feet	3.51	5.455 M/Feet	0.12	1.28	9/ 3/96
I- 1760 ROMEX STAPLE SMALL	312.00	1.093 C/Pcs.	3.41	0.904 C/Pcs.	2.82	30.03	9/ 6/96
I- 1900 SEAL CONN STR 1/2"	2.00	169.000 C/Pcs.	3.38	5.000 C/Pcs.	0.10	1.07	9/ 3/96
I- 2421 WIRE BARE #6 COPPER	15.00	222.000 M/Feet	3.33	6.667 M/Feet	0.10	1.07	9/ 3/96
I- 2422 WIRE BARE #4 COPPER	10.00	331.000 M/Feet	3.31	8.000 M/Feet	0.08	0.85	9/ 3/96
I- 1270 PLATE PLAS 3G SWITCH	3.00	108.000 C/Pcs.	3.24	5.333 C/Pcs.	0.16	1.70	9/ 3/96
I- 195 BUSHING PLASTIC 2"	2.00	152.000 C/Pcs.	3.04	11.500 C/Pcs.	0.23	2.45	9/ 2/96
I- 651 FLEX ST CONN 1/2"	5.00	60.000 C/Pcs.	3.00	5.400 C/Pcs.	0.27	2.88	9/ 4/96
I- 355 EMT COUP SS 1/2"	11.00	26.182 C/Pcs.	2.88	1.818 C/Pcs.	0.20	2.13	9/ 2/96
I- 1271 PLATE PLAS 4G SWITCH	2.00	141.000 C	2.82	6.500 C	0.13	1.38	9/ 2/96
I- 318 EMT CONN SS 1 1/4"	2.00	118.000 C/Pcs.	2.36	6.500 C/Pcs.	0.13	1.38	9/ 2/96
I- 1920 SEAL CONN 90 1/2"	1.00	224.000 C/Pcs.	2.24	8.000 C/Pcs.	0.08	0.85	9/ 3/96
I- 1286 PLATE IND GFCI	2.00	108.000 C	2.16	5.500 C	0.11	1.17	9/ 3/96

MIKE HOLT ENTERPRISESVersion [091889]

[8008812580]

Report #5

```
                         Material Cost Analysis - Report No. 5
                                  MIKE HOLT ENTERPRISES                                  Page:  3
                         Based on ELECTRICAL WIRING - Residential, Ray Mullin
Residential Bid          000000002                       3,744   sqft
                                                                  Labor      Labor
Id Code Description       Quantity     Cost/Unit   Cost Ext    Labor/Unit  Hour Ext   Cost Ext   Last Updt
I- 1350 PVC 1/2" SCHEDULE 40    15.00    14.200 C/Feet   2.13   1.733 C/Feet   0.26     2.77     9/ 3/96
I-  193 BUSHING PLAST 1 1/4"     2.00    97.000 C/Pcs.   1.94   8.000 C/Pcs.   0.16     1.70     9/ 2/96
I-  757 GROUND CLAMP 1/2-1"      1.00   179.000 C/Pcs.   1.79  21.000 C/Pcs.   0.21     2.24     9/ 3/96
I-  671 FLEX 90 CONN 3/4"        1.00   168.000 C/Pcs.   1.68   7.000 C/Pcs.   0.07     0.75     9/ 3/96
I-  670 FLEX 90 CONN 1/2"        2.00    83.000 C/Pcs.   1.66   7.000 C/Pcs.   0.14     1.49     9/ 3/96
I- 1425 PVC MALE ADAPT 2"        2.00    78.500 C/Pcs.   1.57  11.500 C/Pcs.   0.23     2.45     7/14/96
I- 2392 WIRE #10 CU 600V        18.00    85.556 M/Feet   1.54   5.000 M/Feet   0.09     0.96     9/ 3/96
I-  915 LOCKNUTS STEEL 2"        2.00    75.500 C/Pcs.   1.51   5.500 C/Pcs.   0.11     1.17     9/ 3/96
I- 1410 PVC LB 1/2"              1.00   151.000 C/Pcs.   1.51   9.000 C/Pcs.   0.09     0.96     9/ 3/96
I-  317 EMT CONN SS 1"           2.00    70.000 C/Pcs.   1.40   4.500 C/Pcs.   0.09     0.96     9/ 2/96
I-   26 BOX MET 2*4 REGULAR      2.00    64.000 C/Pcs.   1.28  16.000 C/Pcs.   0.32     3.41     9/ 2/96
I-  418 EMT STRAPS 1H 1 1/4"     6.00    20.833 C/Pcs.   1.25   2.500 C/Pcs.   0.15     1.60     9/ 3/96
I- 1266 PLATE PLAS 30/50 AMP     1.00   108.000 C/Pcs.   1.08   5.000 C/Pcs.   0.05     0.53     9/ 3/96
I- 1279 PLATE IND 4X4 SWITCH     1.00   108.000 C/Pcs.   1.08   5.000 C/Pcs.   0.05     0.53     9/ 4/96
I- 1280 PLATE IND 4X4 DB SW      1.00   108.000 C/Pcs.   1.08   5.000 C/Pcs.   0.05     0.53     9/ 3/96
I- 1282 PLATE IND 4X4 RECEPT     1.00   108.000 C/Pcs.   1.08   5.000 C/Pcs.   0.05     0.53     9/ 3/96
I-  653 FLEX ST CONN 1"          1.00   104.000 C/Pcs.   1.04   8.000 C/Pcs.   0.08     0.85     9/ 3/96
I-  357 EMT COUP SS 1"           1.50    68.000 C/Pcs.   1.02   4.667 C/Pcs.   0.07     0.75     9/ 2/96
I- 1276 PLATE IND 2X4 SWITCH     2.00    51.000 C/Pcs.   1.02   4.000 C/Pcs.   0.08     0.85     9/ 3/96
I- 1682 RIGID STR 1H 1"          2.00    43.500 C/Pcs.   0.87   2.500 C/Pcs.   0.05     0.53     9/ 3/96
I-  415 EMT STRAPS 1H 1/2"      12.00     6.500 C/Pcs.   0.78   2.417 C/Pcs.   0.29     3.09     9/ 3/96
I- 1752 ROMEX CONN 3/4"          3.00    26.000 C/Pcs.   0.78   2.000 C/Pcs.   0.06     0.64     9/ 6/96
I- 1375 PVC COUPLINGS 2"         1.00    74.000 C/Pcs.   0.74   5.000 C/Pcs.   0.05     0.53     9/ 3/96
I- 1680 RIGID STR 1H 1/2"        4.00    17.250 C/Pcs.   0.69   2.250 C/Pcs.   0.09     0.96     9/ 3/96
I- 2443 WIRE NUTS GREY           5.00    11.000 C/Pcs.   0.55   1.000 C/Pcs.   0.05     0.53     9/ 6/96
I- 1262 PLATE PLAS 1G SGL RE     1.00    51.000 C/Pcs.   0.51   3.000 C/Pcs.   0.03     0.32     9/ 3/96
I- 1277 PLATE IND 2X4 SNG RE     1.00    51.000 C/Pcs.   0.51   4.000 C/Pcs.   0.04     0.43     9/ 3/96
I-  417 EMT STRAPS 1H 1"         2.00    18.500 C/Pcs.   0.37   2.500 C/Pcs.   0.05     0.53     9/ 3/96
I- 1685 RIGID STR 2H 2"          1.00    31.000 C/Pcs.   0.31   3.000 C/Pcs.   0.03     0.32     9/ 3/96
I- 1420 PVC MALE ADAPT 1/2"      1.00    24.000 C/Pcs.   0.24   5.000 C/Pcs.   0.05     0.53     9/ 3/96
I-  910 LOCKNUTS STEEL 1/2"      1.00    16.000 C/Pcs.   0.16   4.000 C/Pcs.   0.04     0.43     9/ 3/96
I- 1761 ROMEX STAPLE LARGE      14.00     1.071 C/Pcs.   0.15   0.929 C/Pcs.   0.13     1.38     9/ 6/96
I-  100 BREAKER 15AMP 1 POLE    15.00           E/Each          0.090 E/Each   1.35    14.38     1/ 3/91
I-  101 BREAKER 20AMP 1 POLE    10.00           E               0.090 E        0.90     9.59     2/21/96
I-  103 BREAKER 20AMP 2 POLE     1.00           E/Each          0.140 E/Each   0.14     1.49     2/21/96
I-  104 BREAKER 20AMP 2 POLE     1.00           E               0.140 E        0.14     1.49     2/21/96
I-  105 BREAKER 30AMP 2 POLE     2.00           E               0.140 E        0.28     2.98     2/21/96
I-  106 BREAKER 40AMP 2 POLE     2.00           E               0.140 E        0.28     2.98     2/21/96
I-  109 BREAKER 70AMP 2 POLE     2.00           E               0.180 E        0.36     3.83     2/21/96
I-  251 DISC. 60 AMP 1PHASE      1.00           E               0.900 E        0.90     9.59     2/21/96
I-  252 DISC. 70 AMP 1PHASE      1.00           E               0.900 E        0.90     9.59     2/21/96
I-  473 FAN AND LIGHT            3.00           E/Each          0.677 E/Each   2.03    21.62     9/ 3/96
I-  474 FAN, EXHAUST             4.00           E/Each          0.900 E/Each   3.60    38.34     1/ 3/91
I-  476 FAN, HOOD                1.00           E/Each          0.450 E/Each   0.45     4.79     1/ 3/91
I-  478 RECESS FIXTURE          16.00           E/Each          0.900 E/Each  14.40   153.36    11/ 7/96
I-  480 PULL CHAIN               3.00           E/Each          0.227 E/Each   0.68     7.24     1/ 3/91
I-  491 OUTSIDE FIXTURE          6.00           E/              0.225 E/       1.35    14.38     2/19/96


                              MIKE HOLT ENTERPRISESVersion [091889]
                                                                             [8008812580]
```

Report #5

```
                    Material Cost Analysis - Report No. 5
                            MIKE HOLT ENTERPRISES                                    Page:  4
              Based on ELECTRICAL WIRING - Residential, Ray Mullin
Residential Bid        000000002                   3,744   sqft
                                                              Labor      Labor
Id Code Description          Quantity  Cost/Unit  Cost Ext  Labor/Unit  Hour Ext  Cost Ext   Last Updt
I-  492 INSIDE FIXTURE          9.00 _____ E _____    0.226 E      2.03    21.62    9/ 3/96
I-  500 SMOKE DETECTOR          4.00 _____ E/ _____    0.225 E/     0.90     9.59    2/19/96
I-  501 BATH HEAT/FAN/LIGHT     2.00 _____ E/ _____    0.900 E/     1.80    19.17    2/19/96
I-  502 KEYLESS FIXTURE         6.00 _____ E/ _____    0.225 E/     1.35    14.38    2/22/96
I-  505 POST LIGHT              1.00 _____ E/ _____    1.800 E/     1.80    19.17    9/ 3/96
I-  576 FLUO STR 1-8' 120V      1.00 _____ E/Each _____   0.680 E/Each 0.68     7.24    1/ 3/91
I-  579 FLUO WRAP 2-2' 120V     3.00 _____ E/Each _____   0.677 E/Each 2.03    21.62    1/ 3/91
I-  583 FLUO LAYIN 2X4' 120V    6.00 _____ E/Each _____   0.675 E/Each 4.05    43.13   11/ 7/96
I-  620 FIX TRACK 4'            1.00 _____ E _____    0.680 E      0.68     7.24    2/26/96
I-  621 FIX TRACK 8'            1.00 _____ E/Each _____   0.900 E/Each 0.90     9.59    1/ 3/91
I-  941 CHIME                   1.00 _____ E/Each _____   0.900 E/Each 0.90     9.59    9/ 3/96
I-  946 THERMOSTAT              1.00 _____ E _____    0.450 E      0.45     4.79    9/ 3/96
I- 1005 METER/MAIN 200 AMP      1.00 _____ E/Each _____   1.800 E/Each 1.80    19.17   10/ 7/96
I- 1190 PANEL 100 AMP 1PHASE    1.00 _____ E/Each _____   0.900 E/Each 0.90     9.59    1/ 3/91
I- 1191 PANEL 200 AMP 1PHASE    1.00 _____ E/Each _____   0.900 E/Each 0.90     9.59    2/22/96
I- 1521 MULTIOUTLET ASSEMBLY    1.00 _____ E _____    0.900 E      0.90     9.59    9/ 3/96
I- 2450 WIRE TERM # 12-6        3.00 _____ E/Each _____   0.137 E/Each 0.41     4.37    1/ 3/91

                                                 $  3,096.41              173.67  1,849.66
     NOTE: These Figures do NOT Include Tax.        ==========            ======  ========
```

Report #6

```
                           Work Phase Analysis - Report No. 6
                                  MIKE HOLT ENTERPRISES                                    Page:   1
                       Based on ELECTRICAL WIRING - Residential, Ray Mullin
Residential Bid        000000002                      3,744   sqft
                                        0: PRECONSTRUCTION COST          Labor      Labor
Id Code Description         Quantity    Cost/Unit    Cost Ext  Labor/Unit  Hour Ext  Cost Ext   Last Updt
D-   1 Permit                   1.00    375.000 E     375.00           E                         11/ 6/96

                                              $       375.00
    NOTE: These Figures do NOT Include Tax.         ============      ==========  ============

                                        1: SLAB/UNDERGROUND              Labor      Labor
Id Code Description         Quantity    Cost/Unit    Cost Ext  Labor/Unit  Hour Ext  Cost Ext   Last Updt
I- 1350 PVC 1/2" SCHEDULE 40   15.00    14.200 C/Feet    2.13  1.733 C/Feet   0.26      2.77    9/ 3/96
I- 1355 PVC 2" SCHEDULE 40     10.00    53.500 C/Feet    5.35  4.100 C/Feet   0.41      4.37    9/ 3/96
I- 1375 PVC COUPLINGS 2"        1.00    74.000 C/Pcs.    0.74  5.000 C/Pcs.   0.05      0.53    9/ 3/96
I- 2290 UF CABLE 14/2 COPPER   40.00   145.250 M/Feet    5.81  4.500 M/Feet   0.18      1.92    9/ 3/96

                                              $        14.03                  0.90      9.59
    NOTE: These Figures do NOT Include Tax.         ============      ==========  ============

                                        3: BOXES AND PIPE ROUGH          Labor      Labor
Id Code Description         Quantity    Cost/Unit    Cost Ext  Labor/Unit  Hour Ext  Cost Ext   Last Updt
I-   26 BOX MET 2*4 REGULAR     2.00    64.000 C/Pcs.    1.28  16.000 C/Pcs.  0.32      3.41    9/ 2/96
I-   31 BOX MET 4*4 REGULAR    13.00    64.154 C/Pcs.    8.34  16.077 C/Pcs.  2.09     22.26    9/ 2/96
I-   50 BOX PLASTIC 1 GANG     90.00   105.978 C/Pcs.   95.38   9.000 C/Pcs.  8.10     86.27    9/ 2/96
I-   51 BOX PLASTIC 2 GANG      9.00   182.333 C/Pcs.   16.41  10.778 C/Pcs.  0.97     10.33    9/ 2/96
I-   52 BOX PLASTIC 3 GANG      3.00   190.000 C/Pcs.    5.70  14.333 C/Pcs.  0.43      4.58    9/ 2/96
I-   53 BOX PLASTIC 4 GANG      2.00   246.000 C/Pcs.    4.92  18.000 C/Pcs.  0.36      3.83    9/ 2/96
I-   55 BOX PLASTIC OCTAGON    24.00    97.167 C/Pcs.   23.32   9.000 C/Pcs.  2.16     23.00    9/ 2/96
I-   56 BOX PLASTIC OCT/BAR     2.00   196.000 C/Pcs.    3.92  18.000 C/Pcs.  0.36      3.83    9/ 2/96
I-   59 BOX PLASTIC FAN         3.00   436.000 C        13.08  16.333 C       0.49      5.22    9/ 3/96
I-  193 BUSHING PLAST 1 1/4"    2.00    97.000 C/Pcs.    1.94   8.000 C/Pcs.  0.16      1.70    9/ 2/96
I-  195 BUSHING PLASTIC 2"      2.00   152.000 C/Pcs.    3.04  11.500 C/Pcs.  0.23      2.45    9/ 2/96
I-  251 DISC. 60 AMP 1PHASE     1.00            E              0.900 E        0.90      9.59    2/21/96
I-  252 DISC. 70 AMP 1PHASE     1.00            E              0.900 E        0.90      9.59    2/21/96
I-  305 EMT 1/2"              170.00    14.200 C/Feet   24.14  2.041 C/Feet   3.47     36.96    9/ 2/96
I-  307 EMT 1"                 15.00    33.867 C/Feet    5.08  3.333 C/Feet   0.50      5.33    9/ 3/96
I-  308 EMT 1 1/4"             50.00    54.620 C/Feet   27.31  3.600 C/Feet   1.80     19.17    9/ 2/96
I-  315 EMT CONN SS 1/2"       22.00    23.000 C/Pcs.    5.06  1.909 C/Pcs.   0.42      4.47    9/ 2/96
I-  317 EMT CONN SS 1"          2.00    70.000 C/Pcs.    1.40  4.500 C/Pcs.   0.09      0.96    9/ 2/96
I-  318 EMT CONN SS 1 1/4"      2.00   118.000 C/Pcs.    2.36  6.500 C/Pcs.   0.13      1.38    9/ 2/96
I-  355 EMT COUP SS 1/2"       11.00    26.182 C/Pcs.    2.88  1.818 C/Pcs.   0.20      2.13    9/ 2/96
I-  357 EMT COUP SS 1"          1.50    68.000 C/Pcs.    1.02  4.667 C/Pcs.   0.07      0.75    9/ 2/96
I-  358 EMT COUP SS 1 1/4"      9.00   108.222 C/Pcs.    9.74  6.333 C/Pcs.   0.57      6.07    9/ 2/96
I-  398 EMT ELLS 90'S 1 1/4"    3.00   339.667 C/Pcs.   10.19 13.667 C/Pcs.   0.41      4.37    9/ 2/96
I-  415 EMT STRAPS 1H 1/2"     12.00     6.500 C/Pcs.    0.78  2.417 C/Pcs.   0.29      3.09    9/ 3/96
I-  417 EMT STRAPS 1H 1"        2.00    18.500 C/Pcs.    0.37  2.500 C/Pcs.   0.05      0.53    9/ 3/96
I-  418 EMT STRAPS 1H 1 1/4"    6.00    20.833 C/Pcs.    1.25  2.500 C/Pcs.   0.15      1.60    9/ 3/96
I-  501 BATH HEAT/FAN/LIGHT     2.00            E/             0.900 E/       1.80     19.17    2/19/96
I-  502 KEYLESS FIXTURE         6.00            E/             0.225 E/       1.35     14.38    2/22/96
I-  620 FIX TRACK 4'            1.00            E              0.680 E        0.68      7.24    2/26/96
I-  640 FLEX STEEL 1/2"        21.00    33.857 C/Feet    7.11  3.000 C/Feet   0.63      6.71    9/ 3/96
```

MIKE HOLT ENTERPRISESVersion [091889]

[8008812580]

Report #6

```
                    Work Phase Analysis - Report No. 6
                           MIKE HOLT ENTERPRISES                              Page:  2
                    Based on ELECTRICAL WIRING - Residential, Ray Mullin
Residential Bid        000000002                   3,744   sqft
                            3: BOXES AND PIPE ROUGH          Labor      Labor
Id Code Description       Quantity    Cost/Unit   Cost Ext   Labor/Unit  Hour Ext  Cost Ext  Last Updt
I-  642 FLEX STEEL 1"        6.00    89.500 C/Feet    5.37   3.667 C/Feet    0.22     2.34    9/ 3/96
I-  651 FLEX ST CONN 1/2"    5.00    60.000 C/Pcs.    3.00   5.400 C/Pcs.    0.27     2.88    9/ 4/96
I-  653 FLEX ST CONN 1"      1.00   104.000 C/Pcs.    1.04   8.000 C/Pcs.    0.08     0.85    9/ 3/96
I-  670 FLEX 90 CONN 1/2"    2.00    83.000 C/Pcs.    1.66   7.000 C/Pcs.    0.14     1.49    9/ 3/96
I-  671 FLEX 90 CONN 3/4"    1.00   168.000 C/Pcs.    1.68   7.000 C/Pcs.    0.07     0.75    9/ 3/96
I-  672 FLEX 90 CONN 1"      1.00   541.000 C/Pcs.    5.41   9.000 C/Pcs.    0.09     0.96    9/ 4/96
I-  755 GROUND CLAMP 1/2" DB 2.00   190.000 C/Pcs.    3.80  12.000 C/Pcs.    0.24     2.56    9/ 3/96
I-  757 GROUND CLAMP 1/2-1"  1.00   179.000 C/Pcs.    1.79  21.000 C/Pcs.    0.21     2.24    9/ 3/96
I-  762 GROUND ROD 1/2X8' CU 2.00 1,092.500 C/Pcs.   21.85  45.000 C/Pcs.    0.90     9.59    9/ 3/96
I-  910 LOCKNUTS STEEL 1/2"  1.00    16.000 C/Pcs.    0.16   4.000 C/Pcs.    0.04     0.43    9/ 3/96
I-  915 LOCKNUTS STEEL 2"    2.00    75.500 C/Pcs.    1.51   5.500 C/Pcs.    0.11     1.17    9/ 3/96
I- 1005 METER/MAIN 200 AMP   1.00           E/Each           1.800 E/Each    1.80    19.17   10/ 7/96
I- 1041 GREEN GROMMENTS 1" 284.00    54.754 M/Pcs.   15.55   0.915 M/Pcs.    0.26     2.77    9/12/96
I- 1190 PANEL 100 AMP 1PHASE 1.00           E/Each           0.900 E/Each    0.90     9.59    1/ 3/91
I- 1191 PANEL 200 AMP 1PHASE 1.00           E/Each           0.900 E/Each    0.90     9.59    2/22/96
I- 1276 PLATE IND 2X4 SWITCH 2.00    51.000 C/Pcs.    1.02   4.000 C/Pcs.    0.08     0.85    9/ 3/96
I- 1277 PLATE IND 2X4 SNG RE 1.00    51.000 C/Pcs.    0.51   4.000 C/Pcs.    0.04     0.43    9/ 3/96
I- 1279 PLATE IND 4X4 SWITCH 1.00   108.000 C/Pcs.    1.08   5.000 C/Pcs.    0.05     0.53    9/ 4/96
I- 1280 PLATE IND 4X4 DB SW  1.00   108.000 C/Pcs.    1.08   5.000 C/Pcs.    0.05     0.53    9/ 3/96
I- 1282 PLATE IND 4X4 RECEPT 1.00   108.000 C/Pcs.    1.08   5.000 C/Pcs.    0.05     0.53    9/ 3/96
I- 1285 PLATE IND 4X4 BLANK  5.00   108.000 C        5.40    5.200 C        0.26     2.77    9/ 3/96
I- 1286 PLATE IND GFCI       2.00   108.000 C        2.16    5.500 C        0.11     1.17    9/ 3/96
I- 1410 PVC LB 1/2"          1.00   151.000 C/Pcs.    1.51   9.000 C/Pcs.    0.09     0.96    9/ 3/96
I- 1415 PVC LB 2"            2.00   509.000 C/Pcs.   10.18  24.500 C/Pcs.    0.49     5.22    9/ 3/96
I- 1420 PVC MALE ADAPT 1/2"  1.00    24.000 C/Pcs.    0.24   5.000 C/Pcs.    0.05     0.53    9/ 3/96
I- 1425 PVC MALE ADAPT 2"    2.00    78.500 C/Pcs.    1.57  11.500 C/Pcs.    0.23     2.45    7/14/96
I- 1521 MULTIOUTLET ASSEMBLY 1.00           E                0.900 E        0.90     9.59    9/ 3/96
I- 1595 RIGID 2"            10.00   249.000 C/Feet   24.90   6.300 C/Feet    0.63     6.71    9/ 3/96
I- 1680 RIGID STR 1H 1/2"    4.00    17.250 C/Pcs.    0.69   2.250 C/Pcs.    0.09     0.96    9/ 3/96
I- 1682 RIGID STR 1H 1"      2.00    43.500 C/Pcs.    0.87   2.500 C/Pcs.    0.05     0.53    9/ 3/96
I- 1685 RIGID STR 2H 2"      1.00    31.000 C/Pcs.    0.31   3.000 C/Pcs.    0.03     0.32    9/ 3/96
I- 1711 RING 1 GANG         24.00    42.667 C       10.24    4.250 C        1.02    10.86    9/ 3/96
I- 2360 WEATHERHEADS 2"      1.00     7.650 E/Each    7.65   0.360 E/Each    0.36     3.83    9/ 3/96
Q-    1 Switchgear           1.00   425.000 E      425.00           E                 0.00   11/ 6/96
D-    2 Temporary            1.00   330.000 E      330.00           E                 0.00   11/ 6/96

                                                  $ 1,164.33               40.89    435.52
     NOTE: These Figures do NOT Include Tax.      ============             =====   ========

                            4: WIRE/TERMINATIONS             Labor      Labor
Id Code Description       Quantity    Cost/Unit   Cost Ext   Labor/Unit  Hour Ext  Cost Ext  Last Updt
I- 1740 ROMEX 14/2 COPPER  1,616.00  117.970 M/Feet  190.64   9.004 M/Feet   14.55   154.96    9/ 3/96
I- 1741 ROMEX 14/3 COPPER  1,002.00  203.164 M/Feet  203.57  10.798 M/Feet   10.82   115.23    9/ 3/96
I- 1742 ROMEX 12/2 COPPER    434.00  161.659 M/Feet   70.16  10.806 M/Feet    4.69    49.95    9/ 3/96
I- 1743 ROMEX 12/3 COPPER     42.00  291.429 M/Feet   12.24  14.286 M/Feet    0.60     6.39    9/ 3/96
I- 1745 ROMEX 10/3 COPPER     50.00  433.800 M/Feet   21.69  14.400 M/Feet    0.72     7.67    9/ 3/96
I- 1746 ROMEX 8/2 COPPER      10.00  656.000 M/Feet    6.56  14.000 M/Feet    0.14     1.49    9/ 3/96
```

MIKE HOLT ENTERPRISES Version [091889]

Report #6

```
                    Work Phase Analysis - Report No. 6
                              MIKE HOLT ENTERPRISES                              Page:  3
                    Based on ELECTRICAL WIRING - Residential, Ray Mullin
Residential Bid          000000002                3,744   sqft
                                        4: WIRE/TERMINATIONS         Labor    Labor
Id Code Description       Quantity    Cost/Unit    Cost Ext   Labor/Unit  Hour Ext  Cost Ext   Last Updt
I- 1747 ROMEX 8/3 COPPER      20.00  895.500 M/Feet   17.91  18.000 M/Feet   0.36     3.83     9/ 4/96
I- 1751 ROMEX CONN 1/2"       48.00   17.417 C/Pcs.    8.36   0.938 C/Pcs.   0.45     4.79     9/ 6/96
I- 1752 ROMEX CONN 3/4"        3.00   26.000 C/Pcs.    0.78   2.000 C/Pcs.   0.06     0.64     9/ 6/96
I- 1760 ROMEX STAPLE SMALL   312.00    1.093 C/Pcs.    3.41   0.904 C/Pcs.   2.82    30.03     9/ 6/96
I- 1761 ROMEX STAPLE LARGE    14.00    1.071 C/Pcs.    0.15   0.929 C/Pcs.   0.13     1.38     9/ 6/96
I- 1762 KICK PLATES           78.00    5.538 C/Pcs.    4.32   1.090 C/Pcs.   0.85     9.05     9/ 3/96
I- 2228 TELEPHONE WIRE TWIST 288.00   98.333 M/Feet   28.32  10.764 M/Feet   3.10    33.02     9/ 3/96
I- 2390 WIRE #14 CU 600V     218.00   41.514 M/Feet    9.05   3.257 M/Feet   0.71     7.56     9/ 3/96
I- 2391 WIRE #12 CU 600V     111.00   52.432 M/Feet    5.82   3.964 M/Feet   0.44     4.69     9/ 3/96
I- 2392 WIRE #10 CU 600V      18.00   85.556 M/Feet    1.54   5.000 M/Feet   0.09     0.96     9/ 3/96
I- 2393 WIRE #8 CU 600V       22.00  159.545 M/Feet    3.51   5.455 M/Feet   0.12     1.28     9/ 3/96
I- 2394 WIRE #6 CU 600V       78.00  273.077 M/Feet   21.30   6.282 M/Feet   0.49     5.22     9/ 3/96
I- 2395 WIRE #4 CU 600V       31.00  382.258 M/Feet   11.85   6.774 M/Feet   0.21     2.24     9/ 3/96
I- 2396 WIRE #3 CU 600V      150.00  409.733 M/Feet   61.46   6.800 M/Feet   1.02    10.86     9/ 3/96
I- 2398 WIRE #1 CU 600V       30.00  646.000 M/FEET   19.38   7.333 M/FEET   0.22     2.34     9/ 3/96
I- 2405 WIRE 2/0 CU 600V      60.00  891.333 M/Feet   53.48  13.667 M/Feet   0.82     8.73     9/ 3/96
I- 2421 WIRE BARE #6 COPPER   15.00  222.000 M/Feet    3.33   6.667 M/Feet   0.10     1.07     9/ 3/96
I- 2422 WIRE BARE #4 COPPER   10.00  331.000 M/Feet    3.31   8.000 M/Feet   0.08     0.85     9/ 3/96
I- 2425 THERMO WIRE 18/2      70.00   76.429 M/Feet    5.35  11.714 M/Feet   0.82     8.73     9/ 3/96
I- 2426 THERMO WIRE 18/5     100.00  131.000 M/Feet   13.10  11.800 M/Feet   1.18    12.57     9/ 3/96
I- 2441 WIRE NUTS YELLOW     139.00    6.597 C/Pcs.    9.17   0.935 C/Pcs.   1.30    13.85     9/ 3/96
I- 2442 WIRE NUTS RED        285.00    8.709 C/Pcs.   24.82   0.916 C/Pcs.   2.61    27.80     9/ 6/96
I- 2443 WIRE NUTS GREY         5.00   11.000 C/Pcs.    0.55   1.000 C/Pcs.   0.05     0.53     9/ 6/96
I- 2445 WIRE NUTS GREEN       64.00   10.844 C/Pcs.    6.94   1.125 C/Pcs.   0.72     7.67     9/ 6/96
I- 2450 WIRE TERM # 12-6       3.00          E/Each           0.137 E/Each   0.41     4.37     1/ 3/91
                                              ──────────────────                   ─────────────────
                                         $    822.07                                50.68    539.75
         NOTE: These Figures do NOT Include Tax.      ==========                    ======= =========

                                        6: FIXTURE/FINAL/TRIM          Labor    Labor
Id Code Description       Quantity    Cost/Unit    Cost Ext   Labor/Unit  Hour Ext  Cost Ext   Last Updt
I-  100 BREAKER 15AMP 1 POLE  15.00          E/Each            0.090 E/Each  1.35    14.38     1/ 3/91
I-  101 BREAKER 20AMP 1 POLE  10.00          E                 0.090 E       0.90     9.59     2/21/96
I-  103 BREAKER 20AMP 2 POLE   1.00          E/Each            0.140 E/Each  0.14     1.49     2/21/96
I-  104 BREAKER 20AMP 2 POLE   1.00          E                 0.140 E       0.14     1.49     2/21/96
I-  105 BREAKER 30AMP 2 POLE   2.00          E                 0.140 E       0.28     2.98     2/21/96
I-  106 BREAKER 40AMP 2 POLE   2.00          E                 0.140 E       0.28     2.98     2/21/96
I-  109 BREAKER 70AMP 2 POLE   2.00          E                 0.180 E       0.36     3.83     2/21/96
I-  222 CORD 15.250V 3'        1.00    4.370 E/Each    4.37    0.230 E/Each  0.23     2.45     9/ 2/96
I-  223 CORD 15,250V 6'        1.00    4.370 E/Each    4.37    0.230 E/Each  0.23     2.45     9/ 2/96
I-  228 CORD DRYER 30A 4WIRE   1.00   13.110 E/Each   13.11    0.270 E/Each  0.27     2.88     9/ 2/96
I-  235 DIM 600W 120V 1POLE    2.00    6.665 E/Each   13.33    0.225 E/Each  0.45     4.79     9/ 2/96
I-  473 FAN AND LIGHT          3.00          E/Each             0.677 E/Each 2.03    21.62     9/ 3/96
I-  474 FAN, EXHAUST           4.00          E/Each             0.900 E/Each 3.60    38.34     1/ 3/91
I-  476 FAN, HOOD              1.00          E/Each             0.450 E/Each 0.45     4.79     1/ 3/91
I-  478 RECESS FIXTURE        16.00          E/Each             0.900 E/Each 14.40  153.36    11/ 7/96
I-  480 PULL CHAIN             3.00          E/Each             0.227 E/Each 0.68     7.24     1/ 3/91


                               MIKE HOLT ENTERPRISESVersion [091889]
                                                                                      [8008812580]
```

Report #6

```
                       Work Phase Analysis - Report No. 6
                              MIKE HOLT ENTERPRISES                              Page:  4
                      Based on ELECTRICAL WIRING - Residential, Ray Mullin
Residential Bid          000000002            3,744    sqft
                                6: FIXTURE/FINAL/TRIM          Labor      Labor
Id Code Description       Quantity    Cost/Unit    Cost Ext   Labor/Unit   Hour Ext  Cost Ext   Last Updt
I-  491 OUTSIDE FIXTURE     6.00    _____ E/     _____    0.225 E/      1.35     14.38      2/19/96
I-  492 INSIDE FIXTURE      9.00    _____ E      _____    0.226 E       2.03     21.62      9/ 3/96
I-  500 SMOKE DETECTOR      4.00    _____ E/     _____    0.225 E/      0.90      9.59      2/19/96
I-  505 POST LIGHT          1.00    _____ E/     _____    1.800 E/      1.80     19.17      9/ 3/96
I-  576 FLUO STR 1-8' 120V  1.00    _____ E/Each _____    0.680 E/Each  0.68      7.24      1/ 3/91
I-  579 FLUO WRAP 2-2' 120V 3.00    _____ E/Each _____    0.677 E/Each  2.03     21.62      1/ 3/91
I-  583 FLUO LAYIN 2X4' 120V 6.00   _____ E/Each _____    0.675 E/Each  4.05     43.13     11/ 7/96
I-  621 FIX TRACK 8'        1.00    _____ E/Each _____    0.900 E/Each  0.90      9.59      1/ 3/91
I-  941 CHIME               1.00    _____ E/Each _____    0.900 E/Each  0.90      9.59      9/ 3/96
I-  946 THERMOSTAT          1.00    _____ E      _____    0.450 E       0.45      4.79      9/ 3/96
I-  948 PUSH BUTTON         3.00       3.277 E/Each    9.83   0.227 E/Each  0.68      7.24      9/ 3/96
I- 1215 PLATE DECORA 1 GANG 8.00      64.500 C/Pcs.    5.16   2.375 C/Pcs.  0.19      2.02      9/ 3/96
I- 1261 PLATE PLAS 1G SWITCH 22.00    51.318 C/Pcs.   11.29   2.318 C/Pcs.  0.51      5.43      9/ 3/96
I- 1262 PLATE PLAS 1G SGL RE 1.00     51.000 C/Pcs.    0.51   3.000 C/Pcs.  0.03      0.32      9/ 3/96
I- 1263 PLATE PLAS 1G DPX RE 46.00    51.304 C/Pcs.   23.60   2.370 C/Pcs.  1.09     11.61      9/ 3/96
I- 1265 PLATE PLAS 2G SWITCH 8.00     88.500 C/Pcs.    7.08   3.625 C/Pcs.  0.29      3.09      9/ 3/96
I- 1266 PLATE PLAS 30/50 AMP 1.00    108.000 C/Pcs.    1.08   5.000 C/Pcs.  0.05      0.53      9/ 3/96
I- 1270 PLATE PLAS 3G SWITCH 3.00    108.000 C/Pcs.    3.24   5.333 C/Pcs.  0.16      1.70      9/ 3/96
I- 1271 PLATE PLAS 4G SWITCH 2.00    141.000 C        2.82    6.500 C       0.13      1.38      9/ 3/96
I- 1303 PLATE WP 1G DPX REC  5.00    425.000 C/Pcs.   21.25   3.600 C/Pcs.  0.18      1.92      9/ 3/96
I- 1490 REC DUPLEX 15A 125V 47.00     56.830 C/Pcs.   26.71  16.170 C/Pcs.  7.60     80.94      9/ 3/96
I- 1492 REC GFCI 15A 125V   15.00     13.109 E/Each  196.63   0.270 E/Each  4.05     43.13      9/ 3/96
I- 1495 REC SINGLE 15A 125V  1.00      5.460 E/Each    5.46   0.180 E/Each  0.18      1.92      9/ 3/96
I- 1496 REC SINGLE 20, 125V  1.00      5.460 E/Each    5.46   0.230 E/Each  0.23      2.45      9/ 3/96
I- 1498 REC DRYER 30A 4 WIRE 1.00      7.650 E/Each    7.65   0.230 E/Each  0.23      2.45      9/ 3/96
I- 1504 CLOCK OUTLET & PLATE 1.00    468.000 C/Pcs.    4.68  41.000 C/Pcs.  0.41      4.37      9/ 3/96
I- 1890 SEALTIGHT 1/2"     12.00      51.333 C/Feet    6.16   3.000 C/Feet  0.36      3.83      9/ 3/96
I- 1900 SEAL CONN STR 1/2"  2.00     169.000 C/Pcs.    3.38   5.000 C/Pcs.  0.10      1.07      9/ 3/96
I- 1920 SEAL CONN 90 1/2"   1.00     224.000 C/Pcs.    2.24   8.000 C/Pcs.  0.08      0.85      9/ 3/96
I- 2091 SW 15,125V 1POLE   31.00      62.258 C/Pcs.   19.30  18.000 C/Pcs.  5.58     59.43      9/ 3/96
I- 2092 SW 15,125V 3WAY    22.00     138.727 C/Pcs.   30.52  22.500 C/Pcs.  4.95     52.72      9/ 3/96
I- 2093 SW 15,125V 4WAY     2.00     785.500 C/Pcs.   15.71  25.000 C/Pcs.  0.50      5.33      9/ 3/96
I- 2097 SW 20,125V 2POLE    1.00   1,311.000 C/Pcs.   13.11  29.000 C/Pcs.  0.29      3.09      9/ 3/96
I- 2104 SW 15 AMP PILOT 125V 1.00    554.000 C/Pcs.    5.54  23.000 C/Pcs.  0.23      2.45      9/ 3/96
I- 2108 SW BATH/FAN/LIGHT   2.00      18.570 E        37.14   0.450 E       0.90      9.59      9/ 3/96
I- 2109 ATTIC FAN SWITCH    1.00      16.380 E        16.38   0.450 E       0.45      4.79      4/22/96
I- 2110 SW 15,125V 1POLE FAN 3.00  1,016.000 C        30.48  22.667 C       0.68      7.24      9/ 3/96
I- 2222 TV RECEPTACLE      12.00       3.278 E/Each   39.33   0.226 E/Each  2.71     28.86      9/ 3/96
I- 2223 WIRE COAX TV CABLE 504.00    196.627 M/FEET   99.10  10.794 M/FEET  5.44     57.94      9/ 3/96
I- 2229 PHONE JACK INDOOR   8.00       3.278 E/Each   26.22   0.226 E/Each  1.81     19.28      9/ 3/96
I- 2230 PHONE JACK WP.      1.00       8.740 E/Each    8.74   0.230 E/Each  0.23      2.45      4/22/96

                                            $      720.98                  81.20   864.80
       NOTE: These Figures do NOT Include Tax.       ============              ====== ========
```

MIKE HOLT ENTERPRISES Version [091889]

[8008812580]

Report #7

```
                    Job Proposal - Report No. 7
                         MIKE HOLT ENTERPRISES                        Page:    1
                   Based on ELECTRICAL WIRING - Residential, Ray Mullin
           Job #: 000000002   Type:      Square footage:      3,744

Tran I.D.   Description                          Quantity    Cost Each     Extension
   0  B   0 ==> START OF BID <==                   1.00      ******.**    *******.**
   1  B   1 Computer Estimate                      1.00      ******.**    *******.**
   2  C   1 Fixtures                               1.00      ******.**    *******.**
   3  A 600 Post Light                             1.00          39.72         39.72
   4  A 602 Outside Wall Light                     6.00           8.81         52.85
   5  A 610 Attic Light                            6.00           8.81         52.85
   6  A 614 Flourscent Workbench                   1.00          10.19         10.19
   7  A 615 Flourscent Lay-In                      6.00          15.79         94.75
   8  A 616 Flourscent Strip                       1.00          13.74         13.74
   9  A 617 Flourscent Wrap Fix.                   2.00          16.95         33.90
  10  A 617 Flourscent Wrap Fix.                   1.00          16.96         16.96
  11  A 618 Paddle Fan                             3.00          23.62         70.86
  12  A 619 Pull Chain Fixture                     3.00           5.62         16.87
  13  A 621 Recessed Light                        16.00          21.25        340.01
  14  A 624 Track Light 4'                         1.00          19.75         19.75
  15  A 625 Track Light 8'                         1.00          23.60         23.60
  16  A 626 Wall Fixture                           3.00           8.84         26.52
  17  A 627 Ceiling Fixture                        6.00           8.81         52.85
  18  A 629 Smoke Detector                         4.00           8.97         35.89
  19  A 630 Exhaust Fan                            4.00          17.63         70.53
  20  A 728 Range Hood Fan                         1.00           8.17          8.17
                                                 -----------------------------------
      C   1 Fixtures                               1.00      ******.**        980.01

                                                 ===================================
      C   1 Fixtures                               1.00      ******.**        980.01
                                                 ===================================

  21  C   2 Switches                               1.00      ******.**    *******.**
  22   2091 SW 15,125V 1POLE                      26.00           4.16        108.13
  23   2092 SW 15,125V 3WAY                       22.00           6.25        137.46
  24   2093 SW 15,125V 4WAY                        2.00          17.86         35.71
  25    235 DIM 600W 120V 1POLE                    2.00          15.34         30.67
  26   2110 SW 15,125V 1POLE FAN                   3.00          21.36         64.08
  27   2104 SW 15 AMP PILOT 125V                   1.00          13.41         13.41
  28  A 637 1G Switch Assembly                    21.00           6.38        133.96
  29  A 638 2G Switch Assembly                     8.00           8.88         71.05
  30  A 639 3G Switch Assembly                     3.00          10.29         30.88
  31  A 640 4G Switch Assembly                     2.00          12.54         25.07
                                                 -----------------------------------
      C   2 Switches                               1.00      ******.**        650.43

                                                 ===================================
      C   2 Switches                               1.00      ******.**        650.43
                                                 ===================================

  32  C   3 Receptacles                            1.00      ******.**    *******.**
  33  A 645 GFCI Weather Proof                     5.00          40.25        201.27
  34  A 646 Duplex Receptacle                      5.00          10.15         50.76
  35  A 646 Duplex Receptacle                      9.00          10.13         91.15
  36  A 647 Switched Receptacle                   21.00          10.13        212.83
  37  A 648 GFCI Recept 15A Ckt                    4.00          33.82        135.28
  38  A 649 Clock Receptacle                       1.00          20.08         20.08

            MIKE HOLT ENTERPRISESVersion [091889]
                                                                      [8008812580]
```

Report #7

```
                   Job Proposal - Report No. 7
                      MIKE HOLT ENTERPRISES                     Page:  2
               Based on ELECTRICAL WIRING - Residential, Ray Mullin
            Job #: 000000002  Type:    Square footage:    3,744

Tran I.D.  Description                   Quantity    Cost Each    Extension
 39 A 650  Garage Door Recept.               1.00       21.04        21.04
 40 A 651  GFCI Kitchen Recept.              2.00       36.04        72.07
 41 A 652  GFCI Bath Receptacle              2.00       36.04        72.07
 42 A 653  Kitchen Receptacle                7.00       12.39        86.73
 43 A 654  Bath Receptacle                   1.00       12.43        12.43
                                         ------------------------------------
     C   3 Receptacles                       1.00     ******.**      975.74


                                         ====================================
     C   3 Receptacles                       1.00     ******.**      975.74
                                         ====================================
 44 C   4  EMT Devices                       1.00     ******.**    ******.**
 45 A 657  2G Switch (EMT)                   1.00       26.54        26.54
 46 A 658  GFCI Receptacle EMT               2.00       36.17        72.34
 47 A 659  Duplex Recept. EMT                1.00       12.72        12.72
 48 A 660  Multioutlet Assembly              1.00       22.91        22.91
                                         ------------------------------------
     C   4 EMT Devices                       1.00     ******.**      134.52


                                         ====================================
     C   4 EMT Devices                       1.00     ******.**      134.52
                                         ====================================
 49 C   5  Branch Circuit                    1.00     ******.**    ******.**
 50 1740   ROMEX 14/2 COPPER               908.00        0.36        324.64
 51 1741   ROMEX 14/3 COPPER               982.00        0.54        525.48
 52 1742   ROMEX 12/2 COPPER               134.00        0.46         62.16
 53 2290   UF CABLE 14/2 COPPER             40.00        0.33         13.10
                                         ------------------------------------
     C   5 Branch Circuit                    1.00     ******.**      925.37


                                         ====================================
     C   5 Branch Circuit                    1.00     ******.**      925.37
                                         ====================================
 54 C   6  Special Systems                   1.00     ******.**    ******.**
 55 A 664  Chime W/Push Buttons              1.00       75.22         75.22
 56 A 666  TV Outlet                        11.00       32.94        362.30
 57 A 667  Phone Outlet                      7.00       22.27        155.89
 58 A 668  Phone Weather Proof               1.00       31.70         31.70
 59 A 669  Phone EMT                         1.00       32.72         32.72
 60 A 670  TV Outlet EMT                     1.00       43.38         43.38
 61 A 672  Thermostat                        1.00       47.93         47.93
 62 A 675  Home Run 15 Amp Ckt.             12.00       13.08        157.01
 63 A 676  Home Run 20 Amp Ckt.              5.00       17.76         88.82
 64 A 678  Home Run EMT 20 Amp               2.00       22.02         44.03
                                         ------------------------------------
     C   6 Special Systems                   1.00     ******.**     1039.01


                                         ====================================

              MIKE HOLT ENTERPRISESVersion [091889]
                                                             [8008812580]
```

Report #7

```
                    Job Proposal - Report No. 7
                         MIKE HOLT ENTERPRISES                    Page:  3
                 Based on ELECTRICAL WIRING - Residential, Ray Mullin
              Job #: 000000002   Type:    Square footage:    3,744

  Tran I.D.   Description                  Quantity     Cost Each     Extension
      C    6  Special Systems                  1.00     ******.**       1039.01
                                           ======================================

   65 C    7  Separate Circuits                1.00     ******.**     *******.**
   66 A  722  Heat 70 Ampere EMT               1.00         68.52         68.52
   67 A  251  FEEDER #4 CU 1PH EMT            15.00          2.13         31.96
   68 A  735  Water Pump                       1.00         19.88         19.88
   69 A    1  1/2 EMT 2#14 THHN               60.00          0.94         56.50
   70 A  724  Hydromassage Bathtub             1.00         23.42         23.42
   71   1740  ROMEX 14/2 COPPER               51.00          0.36         18.24
   72 A  713  Bath Heat/Vent/Light             1.00         62.23         62.23
   73   1742  ROMEX 12/2 COPPER               50.00          0.46         23.18
   74 A  717  Freezer EMT                      1.00         21.83         21.83
   75 A    1  1/2 EMT 2#14 THHN               30.00          0.94         28.25
   76 A  711  Air Condition 40 Amp             1.00         51.65         51.65
   77   1746  ROMEX 8/2 COPPER                10.00          1.38         13.77
   78 A  734  Water Heater 20 Amp              1.00         51.89         51.89
   79 A   11  1/2 EMT 2#10 THHN                5.00          1.17          5.84
   80 A  712  Attic Fan & Switch               1.00         43.03         43.03
   81   1740  ROMEX 14/2 COPPER               45.00          0.36         16.09
   82 A  713  Bath Heat/Vent/Light             1.00         62.23         62.23
   83   1742  ROMEX 12/2 COPPER               40.00          0.46         18.55
   84 A  754  Panel "B" 100 Ampere EMT         1.00         96.91         96.91
   85    109  BREAKER 70AMP 2 POLE             1.00          3.09          3.09
   86 A  755  Feeder Panel "B" 100 Amp        50.00          4.66        233.05
   87 A  716  Dryer                            1.00         54.94         54.94
   88   1745  ROMEX 10/3 COPPER               35.00          0.99         34.80
   89 A  715  Dishwasher/Disposal              1.00         46.24         46.24
   90   1741  ROMEX 14/3 COPPER               20.00          0.54         10.70
   91 A  714  Cook Top 40 Ampere               1.00         14.73         14.73
   92   1747  ROMEX 8/3 COPPER                20.00          1.85         37.02
   93 A  725  Oven 30 Ampere                   1.00         19.02         19.02
   94   1745  ROMEX 10/3 COPPER               15.00          1.00         14.93
   95 A  733  Washing Machine                  1.00         23.48         23.48
   96   1742  ROMEX 12/2 COPPER               40.00          0.46         18.55
   97 A  750  Meter/Main 200 Amp               1.00        269.45        269.45
   98 A  751  Panel Back/Back 200A             1.00        151.63        151.63
   99 Q    1  Switchgear                       1.00        732.15        732.15
  100 D    1  Permit                           1.00        603.75        603.75
  101 D    2  Temporary                        1.00        531.30        531.30
                                           ---------------------------------------
      C    7  Separate Circuits                1.00     ******.**       3512.80


                                           ======================================
      C    7  Separate Circuits                1.00     ******.**       3512.80
                                           ======================================


                                           ======================================
      B    1  Computer Estimate                1.00     ******.**       8217.87
                                           ======================================


                     MIKE HOLT ENTERPRISESVersion [091889]
                                                                  [8008812580]
```

Report #7

```
                    Job Proposal - Report No. 7
                         MIKE HOLT ENTERPRISES                    Page:    4
               Based on ELECTRICAL WIRING - Residential, Ray Mullin
              Job #: 000000002  Type:    Square footage:     3,744

Tran I.D. Description                         Quantity   Cost Each    Extension

*****************************************    ===================================
***          Sub-Total            ***            1.00    ******.**      8217.87
*****************************************    ===================================

*****************************************                              **********
***           BONDING             ***                                       0.00
*****************************************                              **********

*****************************************                              **********
***          INSURANCE            ***                                       0.00
*****************************************                              **********

*****************************************                              **********
***          CONTINGENT           ***                                       0.00
*****************************************                              **********

*****************************************                              **********
***           PERMIT              ***                                       0.00
*****************************************                              **********

*****************************************                              **********
***          Bid Total            ***                                    8217.87
*****************************************                              **********
```

MIKE HOLT ENTERPRISESVersion [091889]

[8008812580]

Report #8

```
                    Draw Schedule - Report No. 8
                         MIKE HOLT ENTERPRISES                              Page: 1
               Based on ELECTRICAL WIRING - Residential, Ray Mullin
                        Job Name: Residential Bid
           Job Number: 000000002  Type:    Square footage:   3,744
           Comment: Based on ELECTRICAL WIRING - Residential, Ray Mullin
```

0 PRECONSTRUCTION COST

		Total Hours	Cost Extension	Labor Extension	Sales Tax	Overhead	Profit	INSURANCE	Total	% of Bid	Cost / Sq. Ft
Material	:		0.00		0.00	0.00	0.00	0.00	0.00	0.00	.00
Labor	:	0.00		0.00		0.00	0.00	0.00	0.00	0.00	.00
Direct Costs	:	0.00	375.00	0.00	0	150.00	78.75	0.00	603.75	7.35	.16
* PRECONSTRUCTION COST	:	0.00	375.00	0.00	0.00	150.00	78.75	0.00	603.75	7.35	.16

1 SLAB/UNDERGROUND

		Total Hours	Cost Extension	Labor Extension	Sales Tax	Overhead	Profit	INSURANCE	Total	% of Bid	Cost / Sq. Ft
Material	:		14.04		0.98	6.01	3.15	0.00	24.18	0.29	.00
Labor	:	0.90		9.54		3.81	2.00	0.00	15.35	0.19	.00
* SLAB/UNDERGROUND	:	0.90	14.04	9.54	0.98	9.82	5.16	0.00	39.53	0.48	.01

3 BOXES AND PIPE ROUGH

		Total Hours	Cost Extension	Labor Extension	Sales Tax	Overhead	Profit	INSURANCE	Total	% of Bid	Cost / Sq. Ft
Material	:		409.35		28.65	175.20	91.98	0.00	705.19	8.58	.18
Labor	:	40.74		433.85		173.54	91.11	0.00	698.49	8.50	.18
Direct Costs	:	0.00	330.00	0.00	0	132.00	69.30	0.00	531.30	6.47	.14
Quotes	:	0.00	425.00	0.00	29.75	181.90	95.50	0.00	732.15	8.91	.19
* BOXES AND PIPE ROUGH	:	40.74	1,164.35	433.85	58.40	662.64	347.89	0.00	2,667.13	32.46	.71

4 WIRE/TERMINATIONS

		Total Hours	Cost Extension	Labor Extension	Sales Tax	Overhead	Profit	INSURANCE	Total	% of Bid	Cost / Sq. Ft
Material	:		822.23		57.56	351.91	184.75	0.00	1,416.45	17.24	.37
Labor	:	50.16		534.16		213.67	112.17	0.00	860.00	10.47	.22
* WIRE/TERMINATIONS	:	50.16	822.23	534.16	57.56	565.58	296.93	0.00	2,276.45	27.70	.60

MIKE HOLT ENTERPRISESVersion [091889]

[8008812580]

Report #8

Labor Analysis - Report No. 3
MIKE HOLT ENTERPRISES
Based on ELECTRICAL WIRING - Residential, Ray Mullin

Page: 2

Residential Bid   000000002   3,744 sqft

Id Code Description	Quantity	Cost/Unit	Cost Ext	Labor/Unit	Labor Hour Ext	Labor Cost Ext	Last Updt
I- 1762 KICK PLATES	78.00	5.538 C/Pcs.	4.32	1.090 C/Pcs.	0.85	9.05	9/ 3/96
I- 2405 WIRE 2/0 CU 600V	60.00	891.333 M/Feet	53.48	13.667 M/Feet	0.82	8.73	9/ 3/96
I- 2425 THERMO WIRE 18/2	70.00	76.429 M/Feet	5.35	11.714 M/Feet	0.82	8.73	9/ 3/96
I- 1745 ROMEX 10/3 COPPER	50.00	433.800 M/Feet	21.69	14.400 M/Feet	0.72	7.67	9/ 3/96
I- 2445 WIRE NUTS GREEN	64.00	10.844 C/Pcs.	6.94	1.125 C/Pcs.	0.72	7.67	9/ 6/96
I- 2390 WIRE #14 CU 600V	218.00	41.514 M/Feet	9.05	3.257 M/Feet	0.71	7.56	9/ 3/96
I- 480 PULL CHAIN	3.00	E/Each		0.227 E/Each	0.68	7.24	1/ 3/91
I- 576 FLUO STR 1-8' 120V	1.00	E/Each		0.680 E/Each	0.68	7.24	1/ 3/91
I- 620 FIX TRACK 4'	1.00	E		0.680 E	0.68	7.24	2/26/96
I- 948 PUSH BUTTON	3.00	3.277 E/Each	9.83	0.227 E/Each	0.68	7.24	9/ 3/96
I- 2110 SW 15,125V 1POLE FAN	3.00	1,016.000 C	30.48	22.667 C	0.68	7.24	9/ 3/96
I- 640 FLEX STEEL 1/2"	21.00	33.857 C/Feet	7.11	3.000 C/Feet	0.63	6.71	9/ 3/96
I- 1595 RIGID 2"	10.00	249.000 C/Feet	24.90	6.300 C/Feet	0.63	6.71	9/ 3/96
I- 1743 ROMEX 12/3 COPPER	42.00	291.429 M/Feet	12.24	14.286 M/Feet	0.60	6.39	9/ 3/96
I- 358 EMT COUP SS 1 1/4"	9.00	108.222 C/Pcs.	9.74	6.333 C/Pcs.	0.57	6.07	9/ 2/96
I- 1261 PLATE PLAS 1G SWITCH	22.00	51.318 C/Pcs.	11.29	2.318 C/Pcs.	0.51	5.43	9/ 3/96
I- 307 EMT 1"	15.00	33.867 C/Feet	5.08	3.333 C/Feet	0.50	5.33	9/ 3/96
I- 2093 SW 15,125V 4WAY	2.00	785.500 C/Pcs.	15.71	25.000 C/Pcs.	0.50	5.33	9/ 3/96
I- 59 BOX PLASTIC FAN	3.00	436.000 C	13.08	16.333 C	0.49	5.22	9/ 3/96
I- 1415 PVC LB 2"	2.00	509.000 C/Pcs.	10.18	24.500 C/Pcs.	0.49	5.22	9/ 3/96
I- 2394 WIRE #6 CU 600V	78.00	273.077 M/Feet	21.30	6.282 M/Feet	0.49	5.22	9/ 3/96
T- 235 DIM 600W 120V 1POLE	2.00	6.665 E/Each	13.33	0.225 E/Each	0.45	4.79	9/ 2/96
I- 476 FAN, HOOD	1.00	E/Each		0.450 E/Each	0.45	4.79	1/ 3/91
I- 946 THERMOSTAT	1.00	E		0.450 E	0.45	4.79	9/ 3/96
I- 1751 ROMEX CONN 1/2"	48.00	17.417 C/Pcs.	8.36	0.938 C/Pcs.	0.45	4.79	9/ 6/96
I- 2109 ATTIC FAN SWITCH	1.00	16.380 E	16.38	0.450 E	0.45	4.79	4/22/96
I- 2391 WIRE #12 CU 600V	111.00	52.432 M/Feet	5.82	3.964 M/Feet	0.44	4.69	9/ 3/96
I- 52 BOX PLASTIC 3 GANG	3.00	190.000 C/Pcs.	5.70	14.333 C/Pcs.	0.43	4.58	9/ 2/96
I- 315 EMT CONN SS 1/2"	22.00	23.000 C/Pcs.	5.06	1.909 C/Pcs.	0.42	4.47	9/ 2/96
I- 398 EMT ELLS 90'S 1 1/4"	3.00	339.667 C/Pcs.	10.19	13.667 C/Pcs.	0.41	4.37	9/ 2/96
I- 1355 PVC 2" SCHEDULE 40	10.00	53.500 C/Feet	5.35	4.100 C/Feet	0.41	4.37	9/ 3/96
I- 1504 CLOCK OUTLET & PLATE	1.00	468.000 C/Pcs.	4.68	41.000 C/Pcs.	0.41	4.37	9/ 3/96
I- 2450 WIRE TERM # 12-6	3.00	E/Each		0.137 E/Each	0.41	4.37	1/ 3/91
I- 53 BOX PLASTIC 4 GANG	2.00	246.000 C/Pcs.	4.92	18.000 C/Pcs.	0.36	3.83	9/ 2/96
I- 56 BOX PLASTIC OCT/BAR	2.00	196.000 C/Pcs.	3.92	18.000 C/Pcs.	0.36	3.83	9/ 2/96
I- 109 BREAKER 70AMP 2 POLE	2.00	E		0.180 E	0.36	3.83	2/21/96
I- 1747 ROMEX 8/3 COPPER	20.00	895.500 M/Feet	17.91	18.000 M/Feet	0.36	3.83	9/ 4/96
I- 1890 SEALTIGHT 1/2"	12.00	51.333 C/Feet	6.16	3.000 C/Feet	0.36	3.83	9/ 3/96
I- 2360 WEATHERHEADS 2"	1.00	7.650 E/Each	7.65	0.360 E/Each	0.36	3.83	9/ 3/96
I- 26 BOX MET 2*4 REGULAR	2.00	64.000 C/Pcs.	1.28	16.000 C/Pcs.	0.32	3.41	9/ 2/96
I- 415 EMT STRAPS 1H 1/2"	12.00	6.500 C/Pcs.	0.78	2.417 C/Pcs.	0.29	3.09	9/ 3/96
I- 1265 PLATE PLAS 2G SWITCH	8.00	88.500 C/Pcs.	7.08	3.625 C/Pcs.	0.29	3.09	9/ 3/96
I- 2097 SW 20,125V 2POLE	1.00	1,311.000 C/Pcs.	13.11	29.000 C/Pcs.	0.29	3.09	9/ 3/96
I- 105 BREAKER 30AMP 2 POLE	2.00	E		0.140 E	0.28	2.98	2/21/96
I- 106 BREAKER 40AMP 2 POLE	2.00	E		0.140 E	0.28	2.98	2/21/96
I- 228 CORD DRYER 30A 4WIRE	1.00	13.110 E/Each	13.11	0.270 E/Each	0.27	2.88	9/ 2/96
I- 651 FLEX ST CONN 1/2"	5.00	60.000 C/Pcs.	3.00	5.400 C/Pcs.	0.27	2.88	9/ 4/96

MIKE HOLT ENTERPRISES Version [091889]

6.14 THE PROPOSAL

PROPOSAL Page 1 of 2

Submitted By:
Mike Holt Enterprises, Inc.
Attention: Mike Holt
7310 West McNab Road #201
Tamarac, Florida 33071-5821
1-888-NEC® Code, Fax 1-954-720-7944

Date: June 15, xxxx

Submitted To:
U.S. Builders, Inc.
1 Lake View
Clermont, Florida 33000
1-800-555-1212, Fax 1-444-777-1499

Job: Custom Home to be built in the Cypress Glen Subdivision of Coral Springs, Florida.

Scope of Work: Complete electrical installation including the installation of lighting fixtures.

Includes: One temporary electrical service pole, 100 amperes 120/240 volt single phase with two duplex 15 ampere 125 volt receptacles and one 30 ampere 240 volt single phase receptacle.

Plans and Specifications: Proposal is based on the Blueprints E–1, and E–2, dated January 1, xxxx.

Price: $8,171.00

Price shall remain in effect until June 1, xxxx. Any work required under this agreement after this date is not covered and will be considered an extra and charged and paid accordingly. All change orders shall become part of this agreement.

Payment Schedule: Slab Inspection (20%) = $1,650
Rough Inspection (55%) = $4,500
Final Inspection (25%) = $2,021

Payments not received within 30 days of invoice due date shall be considered past due and will accrue an additional interest charge at 1.5% per month of the unpaid balance until paid in full. No work shall be performed (including warranty) if any amount is past due (including change orders). In addition, no release of lien shall be signed unless all payments are paid in full.

Acceptance of proposal:

This proposal may be withdrawn if not accepted within fifteen (15) days from date of submission. When signed by both parties, this instrument, including the conditions on the front and reverse side constitutes a legal and binding contract.

Mike Holt Enterprises, Inc. _____   Date: _____

U.S. Builders, Inc. _____   Date: _____

Change Orders: Any deviation or alteration from this proposal will be executed only on receipt of written orders of same, and will become an extra charge. Said charges shall in no way affect or make void the proposal. Charges for extras will be based on a labor rate of Forty-five ($45.00) Dollars per hour. This includes labor, labor benefits, supervision, overhead, warranty, and profit. Material shall be charged at contractor's list price. The Electrical Contractor must receive written authorization by any of the individuals listed below prior to commencement of the work.

NO WORK SHALL COMMENCE UNTIL WRITTEN AUTHORIZATION IS
RECEIVED BY THIS ELECTRICAL CONTRACTOR.

Individuals with authorization to sign written change orders shall be:

Name: _____ Title: _____

Name: _____ Title: _____

ELECTRICAL CONTRACTOR SHALL NOT BE LIABLE: For failure to perform if prevented by strikes, or other labor disputes, accidents, acts of God, governmental or municipal regulation or interference, shortages of labor or materials, delays in transportation, non-availability of the same from manufacturer or supplier, or other causes beyond Electrical Contractor's control. In no event shall the Electrical Contractor be liable for special or consequential damages whatsoever or however caused.

EXCLUSIONS: This proposal does not include concrete, forming, painting, patching, trenching, core drilling, venting and sealing of roof penetrations. All waste created by Electrical Contractor will be removed to a specific area on the construction site.

LIGHTING FIXTURES AND EQUIPMENT SUPPLIED BY OTHERS: (1) Price includes installation of lighting fixtures furnished by others, if lighting fixtures are on job at time of electrical trim out. (2) Price does not cover the warranty of lighting fixtures and equipment supplied by others. (3) Price does not cover the assembly of lighting fixtures and/or equipment supplied by others. (4) Price does not cover lighting fixtures weighing more than fifty (50) pounds. (5) Fluorescent lighting fixtures supplied by others shall be assembled, pre-whipped, and pre-lamped with in-line fuses. (6) Equipment supplied by others shall be installed by others except lighting fixtures according to conditions above. (7) Electrical Contractor shall not be responsible for Owner-provided lighting fixtures and equipment. Losses due to theft, damage, vandalism, etc. are not the responsibility of this Electrical Contractor. Electrical Contractor shall not receive nor store Owner-provided lighting fixtures or equipment.

MATERIALS AND EQUIPMENT: All material and equipment shall be as warranted by the manufacturer and will be installed in a manner consistent with standard practices at this time. It is agreed that title to all material required according to this proposal will remain the property of this Electrical Contractor until paid in full. It is understood that this Electrical Contractor shall have the authorization to enter upon owner/contractor property for the purpose of repossessing material and equipment whether or not installed without liability to owner/contractor for trespass or any other reason.

NATIONAL AND LOCAL CODES: Electrical installation shall meet the National Electrical Code. Errors in design by the architect and/or engineer are not the responsibility of the Electrical Contractor. Any additional outlets, wiring, lighting fixtures, equipment, etc. not indicated on blueprints and specifications that are required by others (i.e., Inspectors) shall not be part of this proposal.

NON-COMPETE CLAUSE: Owner and all authorized representatives of owner/contractor are not to contract or employ any contractor employees for a period of one (1) year from the completion of any electrical work performed by this Contractor with said Owner/Agent within an area of fifty (50) miles radius from this job site.

OWNER/CONTRACTOR DEFAULTS: Owner/contractor will be in default if (1) any payment called for under this proposal and all authorized change orders becomes past due; (2) any written agreement made by the owner/contractor is not promptly performed; or (3) any conditions warranted by the owner/contractor prove to be untrue; (4) failure of owner/contractor to comply with any of the conditions of this proposal.

Electrical Contractor's remedies in the event of owner/contractor defaults, Electrical Contractor may do any or all of the following: (1) Suspend the work and remove its material/equipment from the premises; (2) remove any Electrical Contractor-supplied material/equipment, whether or not it has been installed and whether or not is has been placed in operation. In this regard, owner/contractor agrees that Electrical Contractor may enter upon owner/contractor property for the purpose of repossessing such equipment without liability to owner/contractor for trespass or any other reason; (3) retain all money paid hereunder, regardless of the stage of completion of the work and bring any appropriate action in court to enforce its rights. The owner/contractor agrees to pay all costs and expenses, attorney's fees, court costs, collection fees (including fees incurred in connection with appeals) incurred by Electrical Contractor in enforcing its rights under this proposal.

PERFORMANCE: Contractor agrees that where a written construction schedule is provided with the signing of this proposal and fails to comply with said schedule, Contractor shall pay all overtime costs necessary to complete construction in a timely manner. If a written construction schedule is not provided with the signing of this proposal, Electrical Contractor shall not pay for any overtime to complete project and any overtime required shall be considered an extra and authorization shall be required according to Changes Orders referred to above. Reasonable time shall be given to Electrical Contractor to complete each phase of the electrical job.

WARRANTY: (1) Warranties apply exclusively to the electrical installation of the material, lighting fixtures, equipment, and other items supplied by the Electrical Contractor, (2) Warranty does not apply to material, lighting fixtures, equipment and other items supplied by others. (3) Warranty does not apply to extensions or additions to the original installation if made by others. (4) Warranty shall commence from the final electrical inspection date for a maximum period of one (1) year. Warranty Does Not Apply If Any Payments According To This Proposal Become Past Due Including Change Orders.

Chapter 6

Review Questions

Practice Estimate

The following questions are in reference to the residential blueprints marked RE-1, RE-2, R-3, and RE-4 located in the *Estimating Workbook*. Before answering any of the following questions be sure to review these blueprints completely.

1. Determine the count for the following symbols as shown on Blueprint RE-1:

COUNT OF SYMBOLS	
Description	Quantity
Fixtures	
Outside Wall Light	
Attic Light	
Fluorescent Wrap	
Paddle Fan	
Recessed Light	
Wall Fixture	
Ceiling Fixture	
Smoke Detector	
Exhaust Fan	
Cooktop Fan	
Switches	
Single Pole Switch	
Three Way Switch	
Fan Switch	
Dimmer	
1 Gang Switch Box	
2 Gang Switch Box	
Receptacle	
Weatherproof GFCI	
Garage GFCI	
Garage Door	
Convenience	
Bath GFCI	
Bath GFCI Protected	
Kitchen GFCI	
Low Voltage	
TV Outlets	
Telephone Outlets	

179

2. Determine the length of the wiring for the TV and telephones according to the symbols listed in blueprint RE–2.

| SPECIAL SYSTEMS ||
Description	Length
TV Wire	
Telephone Wire	

3. Determine the total length of 14/2, 14/3, and 12/2 NM cable required (not including homeruns) according to the circuits listed in blueprint RE–3.

| MEASURE – CIRCUIT ||
Description	Length
NM Cable 14/2	
NM Cable 14/3	
NM Cable 12/2	

4. Determine the length of home run wiring according to the wiring diagrams listed in blueprint RE–4.

| HOME RUN LENGTHS |||
Circuit #	Description	Length
A 1-3	Air Handler/Heat Strip	
A 2-4	Air Condition Compressor	
A 5-7	Cook Top	
A 6-8	Oven	
A 9-11	Water Heater	
A 10-12	Dryer	
A 13	Kitchen Receptacle	
A 14-16	Dishwasher/Disposal	
A 15	Kitchen Receptacle	
A 17	Kitchen Receptacle	
A 18	Washer	
A 19	1/2 Bath	
A 20	Master Bath Receptacle	
A 21	Garage	
A 22	1/2 Bath/Hall Lighting	
A 23	Living Room	
A 24	Entry	
A 25	Guest Room	
A 26	Front Bedroom	
A 27	Hall/Bath	
A 28	Master Bedroom	
A 29	Refrigerator	

5. Determine the bill-of-material for the lighting fixtures.

BILL-OF-MATERIAL, LIGHTING FIXTURES		Box		Labor
	Quantity	Round	Fan	Total Hrs.
Outside Wall Light				1.50
Keyless (Attic)				0.25
Fluorescent Fixture				2.25
Paddle Fan/Light				3.75
Recessed Light				4.00
Wall Fixture (Bath)				0.75
Ceiling Fixture				1.00
Smoke Detector				0.50
Exhaust Fans				3.00
Range Exhaust Fan				0.50
Total		19	5	17.50

6. Determine the bill-of-material for the switches.

BILL-OF-MATERIAL, SWITCHES		Box				Switch			
	Qnty.	1G	2G	1P	3W	Dim	Fan	1G	2G
1 Pole									
3 Way									
Dimmer									
Fan Switch									
Box									
1 Gang Plastic Box									
2 Gang Plastic Box									
Total		12	7	16	6	1	3	12	7

7. Determine the bill-of-material for the receptacles.

BILL-OF-MATERIAL, RECEPTACLES		Box	Receptacle		Plate		
	Quantity	1 Gang	Std	GFCI	Std	GFCI	WP
15 Ampere Circuit Wiring							
Weatherproof							
Garage GFCI							
Grage Door							
Convenience							
20 Ampere Circuit Wiring							
Bath GFCI							
Bath GFCI Protected							
Kitchen GFCI							
Kitchen GFCI Protected							
Total		40	33	7	33	5	2

8. Determine the bill-of-material for the TV and telephone.

			Plates		Cable	
	Qnty.	Ring	TV	Phone	CATV	Phone
Television Outlet						
Phone Outlet						
Other						
Total	13	13	6	7	332	208

BILL-OF-MATERIAL, SPECIAL SYSTEMS (table title)

9. Determine the NM cable requirements for the lighting, switches, and receptacles, including home runs.

Circuit #		14/2	14/3	12/2
Branch Circuit Outlet Wiring		765	202	85
Branch Circuit Homeruns				
A 13	Kitchen Receptacles			
A 15	Kitchen Receptacles			
A 17	Kitchen Receptacles			
A 19	½ Bath/Hall			
A 20	Bath Receptacles			
A 21	Garage Circuit			
A 22	½ Bath And Hall Circuit			
A 23	Living Room			
A 24	Entry			
A 25	Guest Bedroom			
A 26	Front Bedroom			
A 27	Hall/Bath			
A 28	Master Bedroom			
Total		1,116	202	192

10. Complete the pricing/laboring sheet and determine the total material cost and labor-hours for the following:

PRICING/LABORING WORKSHEET (1 OF 4)							
	Quantity	Cost	U	Extension	Labor	U	Extension
Fixtures							
Round Plastic Box	19						
Paddle Fan Box	5						
Fixture Labor	1						
Switches							
1 Gang Switch Box	12						
2 Gang Switch Box	7						
1 Pole Switch	16						
3 Way Switch	6						
Dimmer Switch	1						
Fan Switch	3						
1 Gang Switch Plate	12						
2 Gang Switch Plate	7						
Receptacles							
Box — 1 Gang	40						
Receptacle	33						
Receptacle GFCI	7						
Plate – Receptacle	33						
Plate – GFCI (Decorator)	5						
Plate – WP Receptacle	2						
Special Systems							
Ring – 1 Gang	13						
TV Outlet	6						
Phone Outlet	7						
TV Cable Wire	332						
Phone Cable Wire	208						
Branch Circuit Wiring And Homerun Wiring							
NM Cable 14/2	1,116						
NM Cable 14/3	202						
NM Cable 12/2	192						
Total				$597.51			63.17

11. Price, labor, and total the following material requirements for the special circuits.

PRICING/LABORING WORKSHEET (2 OF 4)							
Description	Quantity	Cost	U	Extension	Labor	U	Extension
Electric Heat (A1-3)							
Breaker – 60 Ampere (2 Pole)	1						
Disconnect – 60 Ampere	1						
NM Cable 6/2	25						
3/4" Flex	6						
Flex – Straight Connector	1						
Flex – 90 Connector	1						
No. 6 Wire	15						
No. 10 Wire	7						
Air Conditioning Equipment (A2-4)							
Breaker – 30 Ampere (2 Pole)	1						
Disconnect – 30 Ampere	1						
NM Cable 12/2	25						
NM Cable Connector	1						
1/2" Liquidtight	6						
Liquidtight – Straight Connector	1						
Liquidtight – 90° Connector	1						
No. 12 Wire	20						
Cooktop (A5-7)							
Breaker — 40 Ampere (2 Pole)	1						
Box – Metal 4" × 4"	1						
Cover – 4" × 4" Raised	1						
NM Cable 8/3	14						
No. 8 Wire	14						
No. 10 Wire	7						
Oven (A6-8)							
Breaker — 30 Ampere (2 Pole)	1						
Box – Metal 4" × 4"	1						
Cover – 4" × 4" Metal	1						
1/2" Flex	6						
1/2" Flex – Straight Connector	2						
NM Cable 10/3	28						
NM Cable Connector	1						
No. 10 Wire	20						
Water Heater (A9 + 11)							
Breaker – 30 ampere (2 Pole)	1						
Box – Metal 4" × 4"	1						
Cover – 4" × 4" Metal	1						
NM Cable 10/2	22						
1/2" Flex	6						
1/2" Flex – Straight Connector	1						
1/2" Flex – 90° Connector	1						
Total				$340.19			13.54

| \multicolumn{8}{c}{PRICING/LABORING WORKSHEET (3 OF 4)} |
Description	Quantity	Cost	U	Extension	Labor	U	Extension
Dryer (A10 + 12)							
Breaker — 30 Ampere (2 Pole)	1						
Box – Plastic 2 Gang	1						
Receptacle – 30 Ampere	1						
Plate (Plastic) – 30 Ampere	1						
NM Cable 10/3	20						
NM Cable Connector	1						
Dryer Cord 4 Wire	1						
Dishwasher/Disposal (A14-16)							
Breaker – 20 Amp (2 Pole)	1						
Box – Plastic 1 Gang	2						
Receptacle – 15 Ampere	1						
Plate (Plastic) – Receptacle	1						
Switch – Single Pole	1						
Plate (Plastic) – Switch	1						
NM Cable 12/3	25						
NM Cable Connector	2						
Cord – 3'	1						
Cord – 6'	1						
Washing Machine (A18)							
Breaker – 20 Ampere (1 Pole)	1						
Box – Plastic 1 Gang	1						
Receptacle — Single 20 Amp	1						
Plate (Plastic) – Receptacle	1						
NM Cable 12/2	23						
NM Cable Connector	1						
Refrigerator (A29)							
Breaker – 20 Ampere (1 Pole)	1						
Box – Plastic 1 Gang	1						
Receptacle — Single 20 Amp	1						
Plate (Plastic) – Receptacle	1						
NM Cable 12/2	5						
NM Cable Connector	1						
Total				$66.07			4.18

PRICING/LABORING WORKSHEET (4 OF 4)							
Description	Quantity	Cost	U	Extension	Labor	U	Extension
Meter							
Meter/Main	1						
2" Rigid Conduit	10						
2" Rigid Conduit – Strap	2						
2" Weather-Head	1						
No. 2/0 Wire	30						
No. 1 Wire	15						
Grounding							
Ground Rod – ½" Copper	2						
Ground Clamp – Direct Burial	2						
No. 6 Copper Wire (Bare)	15						
½" PVC	10						
½" PVC Male Adapter	1						
½" Locknut	1						
Bonding							
No. 4 Wire	10						
Ground Clamp – Water Pipe	1						
Panel Feed							
Panel – 200 Ampere	1						
2" PVC	10						
2" PVC – LB	1						
2" PVC – 90°	1						
2" PVC – Coupling	2						
2" PVC – Male Adapter	2						
2" Locknut	2						
2" Bushing	2						
No. 2/0 Wire	30						
No. 1 Wire	15						
Total				$147.98			8.92

12. Determine the total adjusted labor hours for the following:

TOTAL ADJUSTED LABOR HOURS	
Labor-Hours – Pricing/Laboring Worksheets	89.81
Labor-Unit Adjustment, + 8%	
Total Adjusted Hours	
Efficiency Adjustment, –10%	
Total Adjusted Labor-Hours (Rounded)	87.00

13. Determine the total labor cost based on question twelve total adjusted labor hours.

LABOR COST CALCULATION
Labor Cost = Labor-Hours × Labor Rate
Labor Cost = 87 hours × $10.65
Labor Cost =

14. Determine the total material cost for the following:

MATERIAL COST CALCULATION	
Material Cost – Pricing Worksheets	$1,152.00
Miscellaneous Materials, + 10%	
Waste And Theft, + 5%	
Small Tools, + 3%	
Quotes	$00.00
Total Taxable Materials (Rounded)	$1,659.00
Sales Tax, + 7%	
Total Material Cost (Rounded)	

15. Determine the estimated prime cost based on questions twelve through fourteen.

ESTIMATED PRIME COST	
Labor Cost – (87 Hours)	$922.00
Material Cost Including Quotes And Tax	
Other Direct Cost – Permit And Temporary	$500.00
Estimated Prime Cost (Rounded)	

16. Determine the overhead for the following, not to exceed cost or $13.00 per man hour.

OVERHEAD CALCULATION	
Estimated Prime Cost	$3,197.00
Overhead 87 hours × $13.00 Per hr. =	
Break Even (Rounded)	$4,328.00

17. Determine the bid price based on questions twelve through sixteen.

TOTAL BID PRICE	
Estimated Break Even Cost	$4,328.00
Profit, + 15%	
Other Cost: None	$0.00
Total Bid Price (Rounded)	

18. Determine the percentage distribution for the following:

Cost Distribution
Labor $922/$4,977   _____%
Material $1,775/$4,977   _____%
Direct Cost $500/$4,977   _____%
Overhead $1,131/$4,977   _____%
Profit $649/$4,977   _____%

19. Determine the cost per square foot for the residence, based on question seventeen.

Cost per sq. ft. = $4,977/2,148 sq. ft.

Cost per sq. ft. = _____ per square foot

20. Complete the following labor analysis.

Labor Analysis

Total Hours = _____ Hours

Total 8 Hr. Days (2 men)

(_____ hours/16 hours) = _____ Days

Total 5 Day Weeks

(_____ days/5 days) = _____ Weeks

Chapter 7

Estimating Commercial Wiring

OBJECTIVE

After reading this chapter, you should
- be able to manually estimate commercial wiring and understand how commercial estimating is accomplished by the use of a computer.

INTRODUCTION

This chapter explains in detail the manual and computer assisted estimating process for commercial electrical wiring. If you understand the electrical wiring and material requirements for blueprints EC–1, EC–2, EC–3 (Figures 7–1, 7–2, and 7–3), and the specification details listed in Section 7.01, you should be able to manually estimate this job in about twelve hours. If you use a computer you should be able to estimate this project in less than three hours.

Note. The estimate in this chapter is based on blueprints extracted from *Electrical Wiring – Commercial* (with some modifications). A complete set of blueprints at ¼ inch scale are contained in *Electrical Wiring – Commercial*.

If you don't understand the blueprints or the specification of this chapter, be sure to order *Electrical Wiring – Commercial* published by Delmar Publishers, authored by Ray Mullin and Bob Smith.

PART A – PLANS AND SPECIFICATIONS

This part contains the blueprint and specification details required to understand the estimate contained in Parts B and C of this chapter. Take a few moments at this time to quickly review the blueprints, graphic illustrations and electrical specifications in this part.

1. Figure 7–1 Blueprint EC–1 Basement Electrical
2. Figure 7–2 Blueprint EC–2 First Floor Electrical, page 191
3. Figure 7–3 Blueprint EC–3 Second Floor Electrical, page 192
4. Figures 7–4, 7–5, and 7–6 (pages 193-196) contain the graphics needed to understand the general wiring requirements for this commercial building.

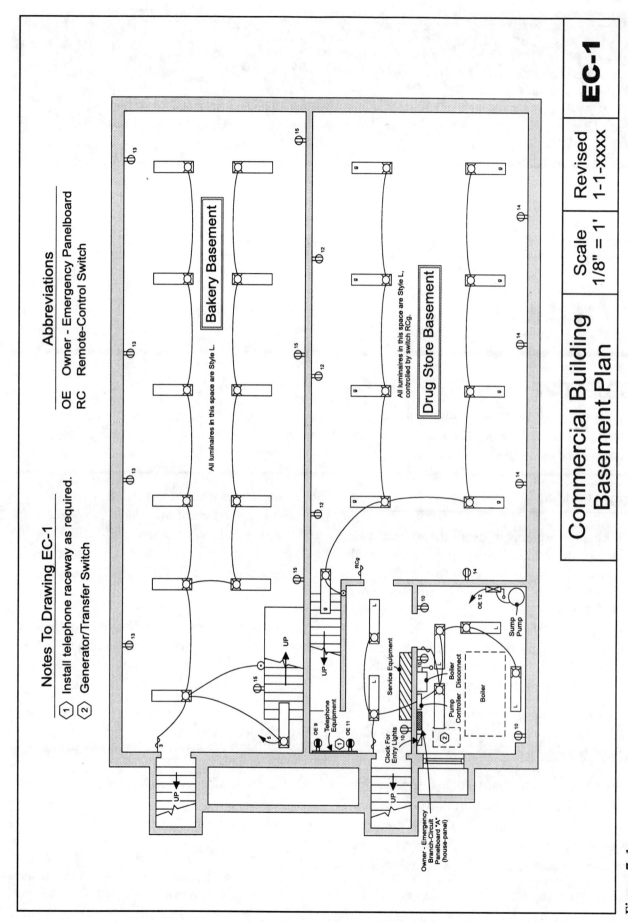

Figure 7–1
Commercial Building Basement Plan C–1

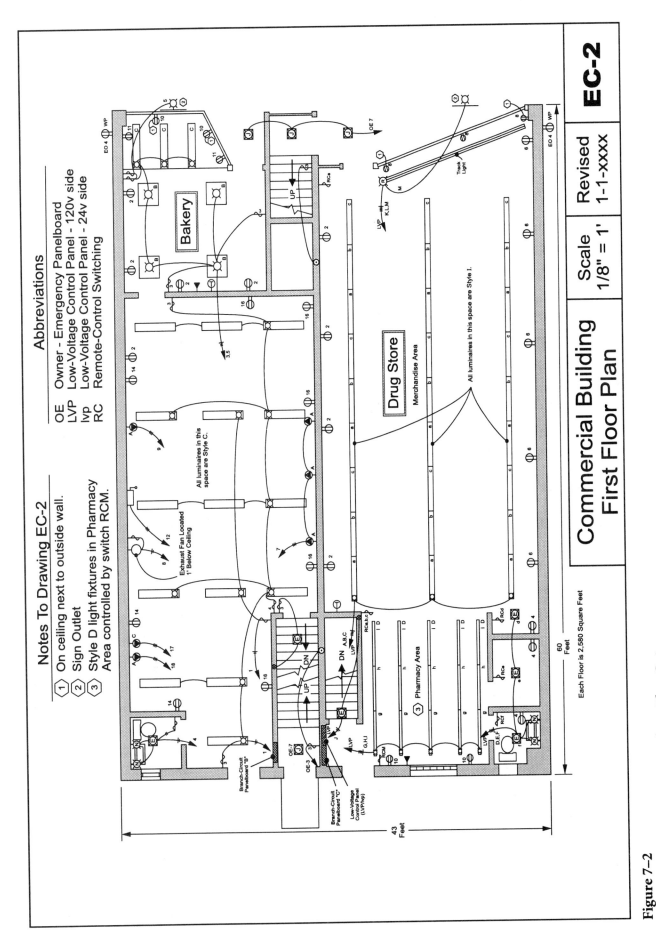

Figure 7–2
Commercial Building First Floor Plan C–2

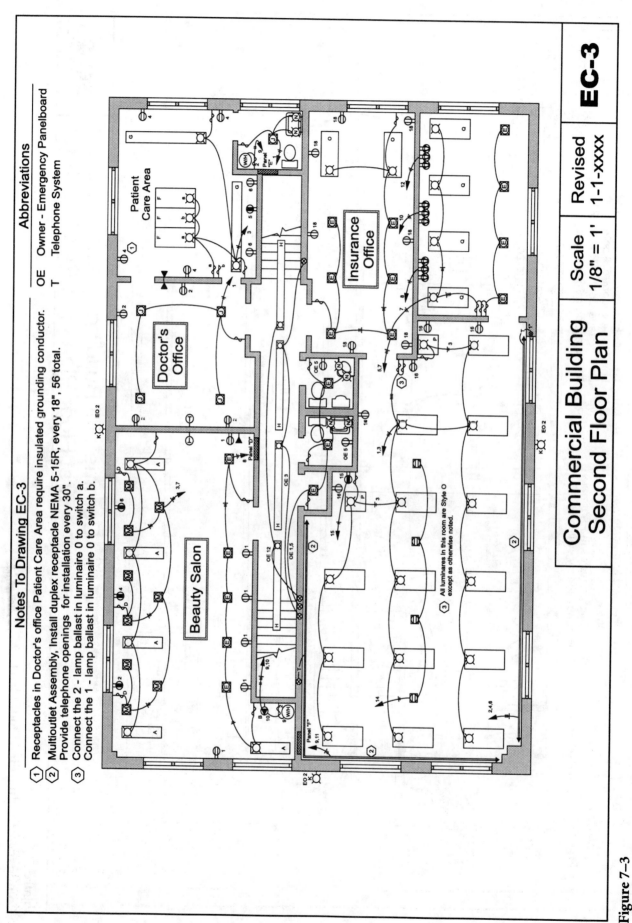

Figure 7-3
Commercial Building Second Floor Plan C-3

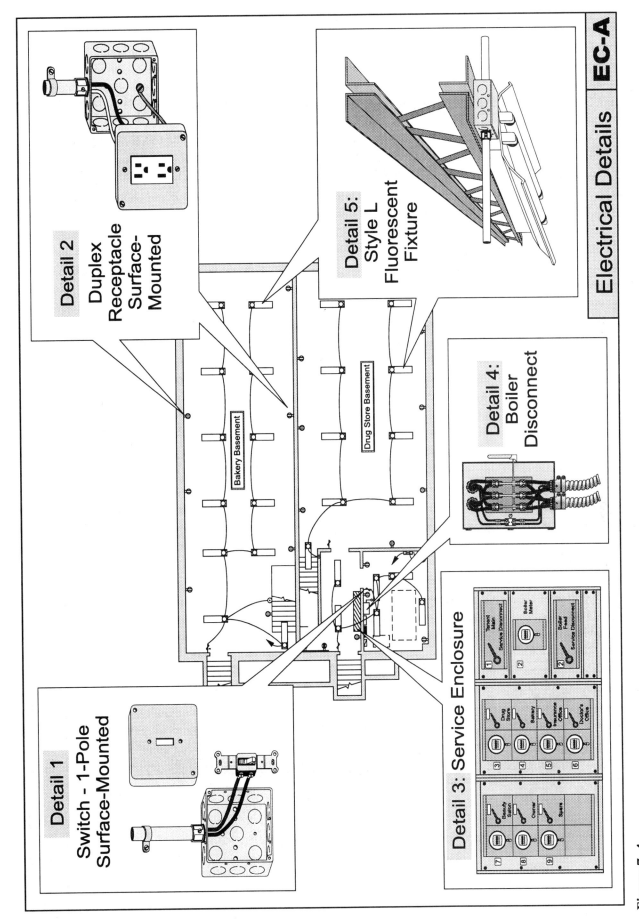

Figure 7-4
Electrical Details

194 Chapter 7 – Estimating Commercial Wiring

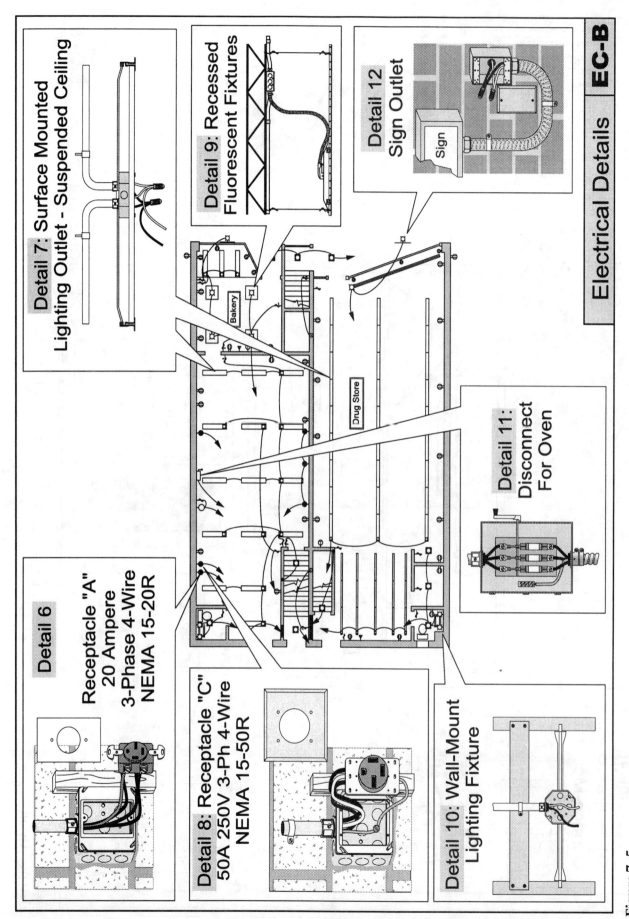

Figure 7–5
Electrical Details

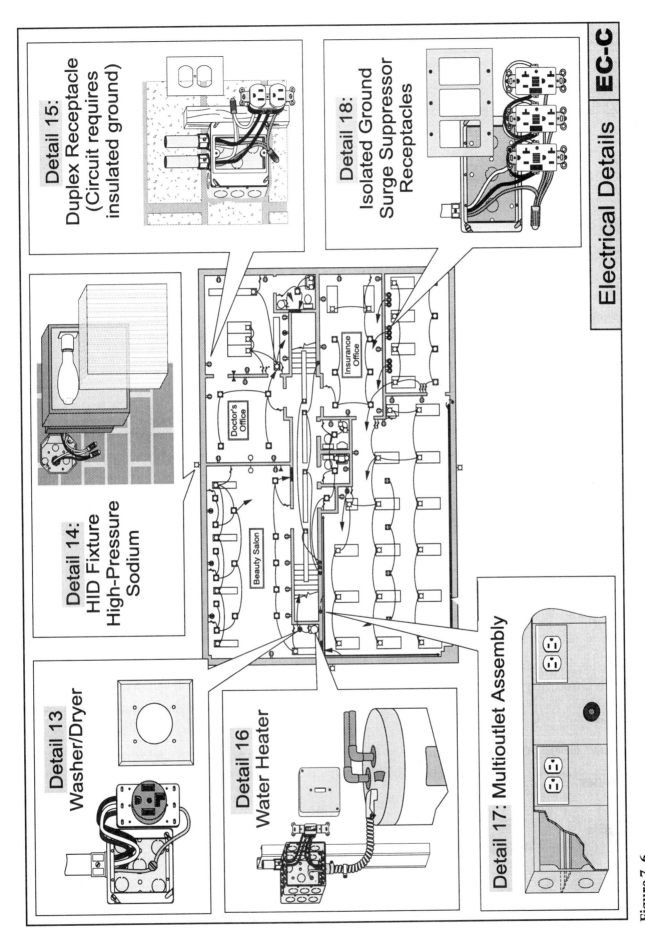

Figure 7-6
Electrical Details

7.01 ELECTRICAL SPECIFICATIONS

Electrical Specifications for

American Professional Plaza

Bid Due Date: July 15, xxxx

Job Location: 100 Main Street

City: West Palm Beach, Florida

Job Owner: Linda Robertson

Phone: 1-444-432-4523

Section 16 – Electrical Specification Table of Contents

16.01 – General Provisions

16.02 – Boxes

16.03 – Conductors

16.04 – Devices – Switches And Receptacles

16.05 – Electrical Equipment

16.06 – Grounding

16.07 – Lighting

16.08 – Low-Voltage Remote-Control Switching

16.09 – Motor-Generator

16.10 – Panelboards

16.11 – Raceways

16.12 – Safety Switches, Circuit Breakers, And Fuses

16.13 – Service Equipment

16.14 – Supports

16.15 – Telephone System

16.01 – General Provisions

The general clauses and conditions and the supplementary general conditions are hereby made a part of the Electrical Specifications.

Scope

The Electrical Contractor shall furnish all materials, equipment, labor, supervision and all related items necessary to complete the work as indicated on the drawing and/or specified herein. The Electrical Contractor shall secure all necessary permits and licenses required and in accepting the contract agrees to have all equipment [1] in working order at the completion of the project.

Author's Comment. See the footnotes at the end of the specifications for the reasons of the underlines.

As-Built Drawings And Records

Maintain a complete set of electrical prints for indicating all changes. Use a colored pencil or pen to make changes at the time of execution. Deliver the as-builts [2] to the Owner's Representative upon completion. The As-Builts will be checked each month for compliance prior to release of any progress payments.

Drawings

The drawings are schematics showing relative locations and connections and shall not be scaled for exact locations. Unless specific dimensions are shown, the structural, architectural and site conditions shall govern the exact locations. Should any difficulty occur in the running of devices of connections at the points shown, provided the necessary minor deviations therefrom as approved without additional cost. See Figure 7–7 for blueprint symbol details.

Electrical Shock Protection

Provide ground-fault protection for all temporary wiring according to the National Electrical Code, Section 305–6.

Materials

All materials used shall be new, listed by a Nationally Recognized Testing Laboratory (NRTL) and shall meet the requirements of the drawings and specifications. Damaged materials shall not be installed. See the National Electrical Code, Section 110–12(c).

Material Protection

Store and protect all materials from injury and weather prior to installation. Materials shall not be stored directly on the ground or floor and shall be kept as clean and dry as possible and free from damage or deteriorating elements and must comply with the National Electrical Code, Section 110–11.

Related Work

See Figure 7–8 (page 200) architectural transverse section for additional details.

Workmanship

All electrical work shall be in accordance with the National Electrical Code, shall be executed in a neat and workmanlike manner in accordance with the National Electrical Institute Standards [3] (NEIS), and shall present a neat and symmetrical appearance when completed. See National Electrical Code, Section 110–12.

Shop Drawings

The contractor shall submit for approval descriptive literature for all equipment installed as part of this contract.

16.02 – Boxes

Boxes And Accessories

Minimum size outlet box shall be 4" square by 1fi" deep, unless approved or indicated otherwise. No "handy" boxes shall be permitted.

16.03 – Conductors

Conductors

1. *Copper:* Conductors shall be copper, 600 volt, types THHN, THHW, or XHHW. Aluminum conductors shall not be permitted.

Splices And Terminations

1. *Connections:* All terminations and connections shall be to listed devices. See National Electrical Code, Section 110–14.

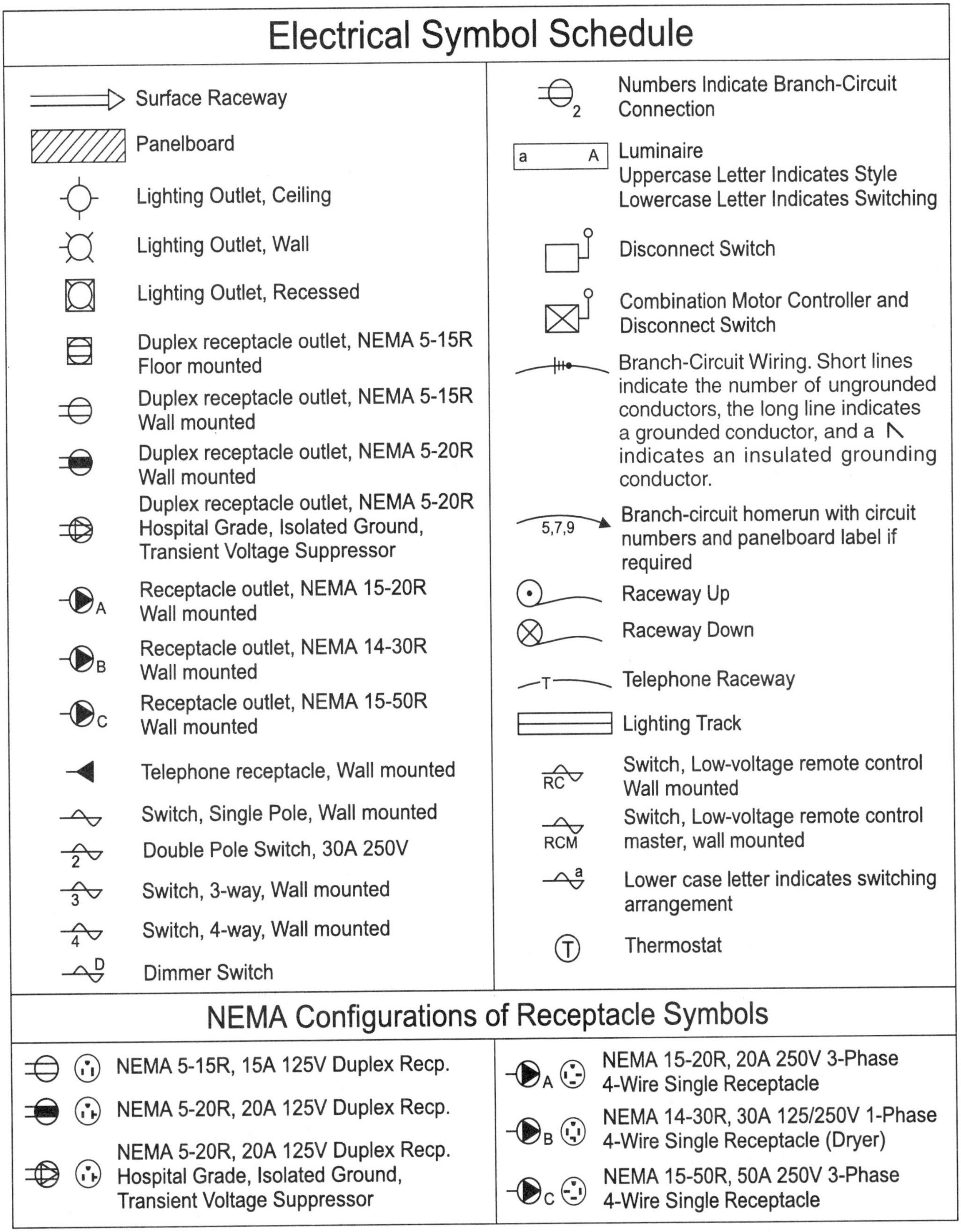

Electrical Symbol Schedule

- Surface Raceway
- Panelboard
- Lighting Outlet, Ceiling
- Lighting Outlet, Wall
- Lighting Outlet, Recessed
- Duplex receptacle outlet, NEMA 5-15R Floor mounted
- Duplex receptacle outlet, NEMA 5-15R Wall mounted
- Duplex receptacle outlet, NEMA 5-20R Wall mounted
- Duplex receptacle outlet, NEMA 5-20R Hospital Grade, Isolated Ground, Transient Voltage Suppressor
- Receptacle outlet, NEMA 15-20R Wall mounted
- Receptacle outlet, NEMA 14-30R Wall mounted
- Receptacle outlet, NEMA 15-50R Wall mounted
- Telephone receptacle, Wall mounted
- Switch, Single Pole, Wall mounted
- Double Pole Switch, 30A 250V
- Switch, 3-way, Wall mounted
- Switch, 4-way, Wall mounted
- Dimmer Switch
- Numbers Indicate Branch-Circuit Connection
- Luminaire — Uppercase Letter Indicates Style, Lowercase Letter Indicates Switching
- Disconnect Switch
- Combination Motor Controller and Disconnect Switch
- Branch-Circuit Wiring. Short lines indicate the number of ungrounded conductors, the long line indicates a grounded conductor, and a ∧ indicates an insulated grounding conductor.
- Branch-circuit homerun with circuit numbers and panelboard label if required
- Raceway Up
- Raceway Down
- Telephone Raceway
- Lighting Track
- Switch, Low-voltage remote control Wall mounted
- Switch, Low-voltage remote control master, wall mounted
- Lower case letter indicates switching arrangement
- Thermostat

NEMA Configurations of Receptacle Symbols

- NEMA 5-15R, 15A 125V Duplex Recp.
- NEMA 5-20R, 20A 125V Duplex Recp.
- NEMA 5-20R, 20A 125V Duplex Recp. Hospital Grade, Isolated Ground, Transient Voltage Suppressor
- NEMA 15-20R, 20A 250V 3-Phase 4-Wire Single Receptacle
- NEMA 14-30R, 30A 125/250V 1-Phase 4-Wire Single Receptacle (Dryer)
- NEMA 15-50R, 50A 250V 3-Phase 4-Wire Single Receptacle

Figure 7–7
Blueprint Symbol Details

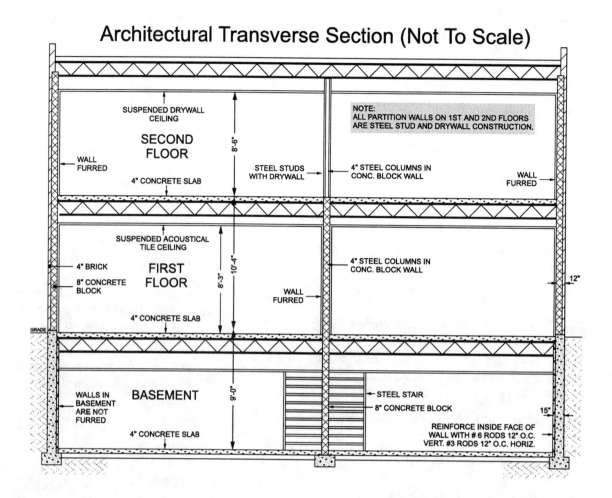

**Figure 7–8
Architectural Transverse Section**

Execution

1. *Bonding:* Bonding conductors size No. 10 and smaller shall have a green covering and shall be sized according to the National Electrical Code, Section 250–95, unless otherwise indicated.

2. *Color Code:* Conductors shall be identified as follows: 277/480 volts, brown, orange, yellow and gray; and 120/208 volts, black, red, blue and white.

3. *Conductor Installation:* Installation of conductors shall be made only in completed raceway systems and all conductors in any raceway shall be pulled in together. See the National Electrical Code, Section 300–18 for additional requirements.

4. *Pull Wire Lubricants:* Wire pulling compounds or lubricants shall be listed.

5. *Splices And Terminations:*

(a) *Lugs:* Use solderless terminal lugs on all stranded conductors. Use listed solderless connectors for all splices. Keep splices to a minimum and all splices and terminations must comply with the National Electrical Code, Section 110–14.

(b) *Neutrals:* Splice all neutrals prior to connection to wiring devices where required by the National Electrical Code, Section 300–13(b).

16.04 – Devices – Switches And Receptacles

Receptacles

The Electrical Contractor shall furnish all receptacles and install as indicated on the blueprints. Provide GFCI receptacles as required by the National Electrical Code in Sections 210–8 and 210–63. Receptacles for patient care areas in the Doctor's office must be installed in accordance with the National Electrical Code, Section 517–13(b).

Switches

The Electrical Contractor shall furnish all switches and install as indicated on the blueprints.

1. *Gang Switches:* Gang switches where feasible.
2. *Grouping:* Group the switches in locations indicated on blueprints, unless otherwise noted.

Mounting

1. *Ceramic Tile Walls:* Adjust outlet heights in ceramic tile walls to be entirely in or entirely out of the tile.
2. *Height To Center Of Box:* Mount receptacles and telephones 12" above finish floor. Mount switches and dimmers at 48" above finished floor, and weatherproof receptacles at 18" above finish grade.

Plates

1. *Plates:* Install plastic plates and covers on all outlets.

16.05 – Electrical Equipment

Air-Conditioning Equipment

Air-conditioning equipment will be furnished and installed by the heating contractor in the following areas:

Note. Bakery is not air conditioned.

1. *Drugstore:* A split system packaged unit will be installed with a rooftop compressor-condenser and a remote evaporator located in the basement. Voltage: 208 volts, three-phase, 3-wire, 60 hertz. Minimum circuit ampacity: 31.65 amperes (No. 8 AWG). Maximum overcurrent protection: 50 amperes, time-delay fuse located in service disconnect on roof. Circuit C–16 from Panel "C."
2. *Beauty Salon:* A single package unit will be installed on the roof. Voltage: 208 volts, three-phase, 3-wire, 60 hertz. Minimum circuit ampacity: 20.92 amperes (No. 10 AWG). Maximum overcurrent protection: 35 amperes, time-delay fuse located in service disconnect on roof. Circuit D–5 of Panel "D."
3. *Doctor's Office:* A single package unit will be installed on the roof. Voltage: 208 volts, single-phase, 2-wire, 60 hertz. Minimum circuit ampacity: 24.7 amperes (No. 10 AWG). Maximum overcurrent protection: 40 amperes, time-delay fuse located in service disconnect on roof. Circuit E–7 of Panel "E."
4. *Insurance Office:* A single packaged unit will be installed on the roof. Voltage: 208 volts, three-phase, 3-wire, 60 hertz. Minimum circuit ampacity: 31.65 amperes (No. 8 AWG). Maximum overcurrent protection: 50 amperes, time-delay fuse located in service disconnect on roof. Circuit F–13 from Panel "F."

Air-Conditioning Wiring And Disconnects

The Electrical Contractor shall furnish raceway and conductors between the panelboards located in each occupancy to the A/C disconnects on the roof as indicated on the plans, Figure 7–9.

Air Conditioning Service Receptacle

The Electrical Contractor shall install a weatherproof GFCI protected receptacle outlet for each A/C unit located on the roof in accordance with National Electrical Code Sections 210–8(b)(2) and 210–63. These receptacle outlets are to be mounted directly on the A/C unit. See Figure 7–9 for details.

Boiler Equipment

The heating contractor will furnish and install a 200-kW electric hot water heating boiler completely equipped with safety, operating, and sequencing controls for 208-volt, three-phase electric power. The Electrical Contractor shall furnish and install two parallel three inch raceways containing: three- 500 kcmil and 1- No. 1/0 AWG each, from the service equipment to the boiler disconnect. The boiler disconnect shall be supplied by the heating contractor. All conductor terminations and connections from the boiler disconnect and the boiler will be made by the heating contractor.

1. *Hot Water Circulating Pumps:* The heating contractor will furnish and install five circulating pumps and the pump control panel. <u>All wiring to the motors and the connection of motors shall also be completed by the heating contractor</u> [4]. Pumps will be 1/6 horsepower, 120 volts, single phase and overload protection will be provided by a manual motor starter. Electrical Contractor shall supply the five overcurrent protection de-

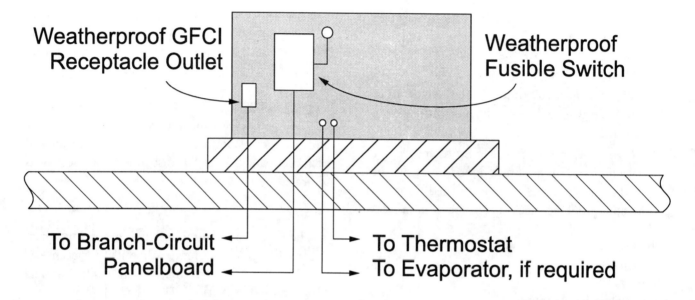

Figure 7–9
Detail Of Typical Roof Type Cooling Unit System

vices for the pump motors in Panel "A" and install the five 2-wire circuits to the pump controller panel located adjacent to Panel "A." See Figure 7–14, page 206.

2. *Hot Water Circulating Pump Controls:* The heating contractor will furnish and install all wiring for the thermostats [4] located in each unit.

3. *Hot Water Circulating Pump Emergency Power:* The Electrical Contractor shall furnish and install the circuit breakers, raceways, and conductors between the Owner-Emergency panelboard "A" and the circulating pump control panel located adjacent to panelboard "A." All electrical connections in the circulating pump control panel will be made by the heating contractor. See Figure 7–14 for details.

4. *Thermostat Controls:* The heating and cooling will be controlled by a combination thermostat located in the proper area as shown on the electrical plans. The heating contractor will furnish and install all equipment and wiring associated with the thermostats located in each unit, including all raceways, conductors and cables.

Sump Pump Motor

The plumbing contractor shall furnish and install a fi horsepower, 120 volt electric motor-driven fully automatic sump pump with float switch. Overload protection will be provided by a manual motor starter supplied by the plumbing contractor and all electrical power wiring to the manual starter and motor shall be made by the Electrical Contractor, Figure 7–10.

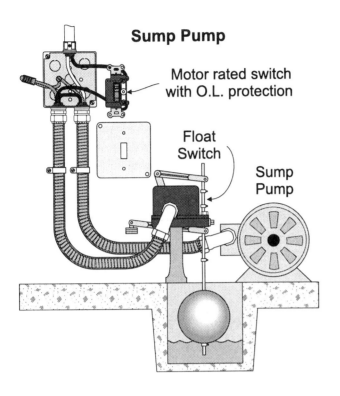

**Figure 7–10
Sump Pump**

16.06 – Grounding

Installation

1. *Bonding Metal Parts:* Bond all metal parts with suitable lugs or clamps in accordance with National Electrical Code Sections 250–112, 114, 115 and 117.
2. *Flexible Conduit:* Provide a bond wire in all flexible metal conduits and connect to the boxes at each end in accordance with National Electrical Code Sections 250–70(f) and 250–79(e).
3. *Protection:* Use rigid nonmetallic conduit for sleeving the grounding electrode conductor. Where sleeves are subject to extreme injury, use rigid metal conduit and bond both ends in accordance with National Electrical Code Sections 250–71(a)(3), 250–72, and 250–92.
4. *Receptacles:* Grounding contacts of receptacles shall be grounded in accordance with the National Electrical Code, Section 250–74. The grounding contacts for receptacle in patient care area shall be connected to an insulated grounding conductor, not smaller than No. 12 AWG in accordance with the National Electrical Code, Section 517–13 . The ground contacts of the receptacle shall be mounted down, except in patient care areas, where the contacts shall be mounted up.

Ground Rods

Grounding accessories shall be listed and the supplemental grounding electrode shall be a $3/4" \times 10'$ copper ground rod installed in accordance with National Electrode Code Sections 250–81(a), 250–83 and 250–84.

16.07 – Lighting

Fixtures, General Requirements

Furnish all materials, equipment, labor, supervision and related items necessary to complete the work as indicated on the drawings and/or specifications. This shall include but is not be limited to: lighting fixtures, lamps, trim, ballast, and accessories. See Figure 7–11, page 205, for Luminaire Lamp Schedule.

Note. Site lighting fixtures are not included within the scope of this contract and shall be provided by others.

Installation

1. *Accessories:* Provide all necessary hangers and mounting accessories for a complete installation.
2. *Alignment:* Lighting fixture bottoms, edges, and ends of rows shall be even. Rows shall be straight, aligned and equally spaced in distinct areas.
3. *Clean:* Clean all fixtures of debris and fingerprints and adjust trim to fit surfaces snugly.
4. *Coordinate:* Coordinate lighting fixture locations with air grilles, pipes and ductwork.
5. *Finish:* Except as indicated or specified otherwise, the metal parts of light fixtures shall be of corrosion resistant metal or shall be suitably finished to resist corrosion; metal portions of fixtures which will be visible after installation shall have an unblemished finish.
6. *Incandescent Fixtures:* Incandescent lamps shall have a rating of 125–130 volts unless specified otherwise and shall have an intermediate base unless specified otherwise.

7. *Lenses:* Lens frames shall be supported to avoid sagging, and shall be readily removable or suitably hinged and latched. Removable frames shall have adequate means of retention for use when servicing.

8. *Location:* Locate the fluorescent fixture in the electrical equipment rooms to best illuminate the equipment installed. Use chains or rods to support fixtures below ducts and pipes as required. Install fixtures after pipes and ducts are in.

9. *Test:* Test all fixtures, switches, and controls for operation. Replace all lamps if their estimated operating period is less than 80% of rated lamp life prior to final acceptance.

10. *Time Clock:* A time clock shall be installed adjacent to Panel "A" to control the lighting in the front and rear entries. The clock will be supplied from circuit No. OE–7. The clock shall be 120 volts, one circuit, with astronomic control and a spring-wound carry-over mechanism. See Figure 7–14.

Fluorescent Lamps And Accessories

1. *Fuse Holder:* Each ballast shall be individually fused in the primary side of each ungrounded conductor, factory installed.

2. *Lamps:* Fluorescent lamps shall be hot-cathode, rapid start, cool white type unless indicated or specified otherwise. The fluorescent lamps shall be suitable for operation with the ballasts to which they are connected. Fluorescent lamps shall be 35 watt and be rated a minimum of 3050 lumens, initial.

3. *Lenses:* Plastic lenses shall be made of heat-resistant methyl-methacrylate.

4. *Listed:* Fluorescent ballasts shall be listed and CBM certified, low energy type, Class P and have a sound rating of A.

5. *Warranty:* Ballasts shall be warranted for 3 years. See Figure 7–11, Luminaire Lamp Schedule.

16.08 – Low-Voltage Remote-Control Switching

A low-voltage, remote-control switch system shall be installed in the drugstore adjacent to Panel "C" by the Electrical Contractor as shown on the plans and detailed herein. Low voltage panel, remote control switches and control cable shall be supplied by the owner. All components shall be constructed to operate on 24-volt control power. The transformer shall be a 120/24-volt, energy-limiting type for use on a Class II signal system installed in accordance with the requirements of Article 725 of the National Electrical Code.

1. *Cabinet:* A metal cabinet matching the panelboard cabinets shall be installed adjacent to Panel "C" for the installation of relays and other components. A barrier will separate the control section from the power wiring.

2. *Relays:* A 24-volt AC, split-coil design relay rated to control 20 amperes of tungsten or fluorescent lamp loads shall be provided.

3. *Switches:* Remote control switches shall be complete with wall plates and mounting brackets. They shall be normally open, single-pole, double-throw, momentary contact switches with on-off identification.

4. *Wire:* Each switch shall be wired with three-conductor cable, color coded, No. 20 AWG wire, installed in 1/2" raceway.

5. *Switching Schedule:* Connections at the remote control panel and all switches shall be made to accomplish the lighting control as shown in the following switching schedule. See Figures 7–12 and 7–13 for remote control switch details.

16.09 – Motor-Generator

Motor-Generator

The motor-generator plant shall be capable of delivering 12 kVA at 208Y/120 volts, three phase. Motor generator shall come complete with 12-volt batteries and a battery charger. The generator shall be mounted on anti-vibration mounts with all necessary accessories, including mufflers, exhaust piping, fuel tanks, fuel lines, and remote derangement annunciator.

1. *Installed and Supplied By:* The motor generator shall be installed and supplied by the motor-generator contractor.

2. *Wiring Responsibilities:* All power and control wiring from the motor-generator to the transfer switch shall be the responsibility of the motor-generator contractor.

Transfer Switch

The automatic load transfer switch shall be capable of handling 12 kVA at 208Y/120 volts, three phase.

Luminaire - Lamp Schedule

Style	Nominal Size	Lamp	Ballast	Mounting	Symbol
A	18 inches by 4 feet	Two F40/Spec 30/RS	Energy Saving	Surface	
B	2 by 2 feet	Two FB40/Spec 30/6	Energy Saving	Surface	
C	9 feet by 8 5/8 in.	Two F40/Spec 30/RS	Energy Saving	Surface	
D	Strip 4 feet long	One F48T12/CW	Standard	Surface	
E	9 inches diameter	Two 26W Quad T4	Compatible	Recessed	
F	2 by 4 feet	Four F023/35K	Matching Electronic	Recessed	
G	1 by 8 feet	Four F023/35K	Matching Electronic	Recessed	
H	9 inches by 8 feet	Two F40/Spec 41/RS	Energy Saving	Surface	
I	9 inches by 4 feet	Two F40/Spec 41/RS	Energy Saving	Surface	
J	8 inches diameter	One 150W, A21	NA	Recessed	
K	16 inches square	One 70W, HPS	Standard	Surface	
L	13 inches by 4 feet	Two F40/Spec 41/RS	Energy Saving	Surface or hung	
M	7 inches diameter	12V 50W NFL	Transformer	Recessed	
N	20 inches long	One 60W, 120V	NA	Surface	
O	4 feet by 20 inches	Three F40/Spec 41/RS	Two Ballast	Recessed	
P	2 feet by 20 inches	Two FB40/Spec 41/RS	One Ballast	Recessed	
Q	4 feet by 20 inches	Three F40/Spec 41/RS	Two Ballast	Recessed	
R	15 feet by 3 inches	Eight R40 75W A19	NA	Surface	

Figure 7–11
Luminaire Lamp Schedule

LOW VOLTAGE RELAYS AND CIRCUITS

Relay	Area Controlled	Wire	Panel C #
LVP – A	Drug Store Lighting	No. 12	1
LVP – B	Drug Store Lighting	No. 12	3
LVP – C	Drug Store Lighting	No. 12	3
LVP – D	Makeup Area Lighting	No. 12	5
LVP – E	Storage Area Lighting	No. 12	5
LVP – F	Toilet Lighting	No. 12	5
LVP – G	Pharmacy Lighting	No. 12	7
LVP – H	Pharmacy Lighting	No. 12	7
LVP – I	Pharmacy Lighting	No. 12	7
LVP – J	Stairways/Basement Lighting	No. 12	9
LVP – K	Show Window Lighting	No. 12	11
LVP – L	Show Window Lighting	No. 12	13
LVP – M	Sign	No. 12	15

Figure 7–12
Low Voltage Relays and Circuits

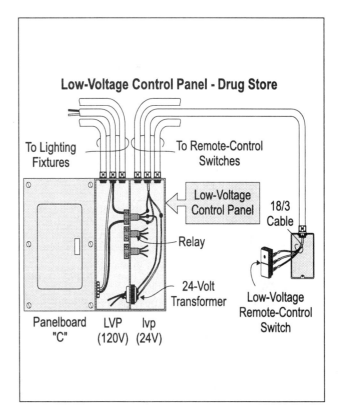

Figure 7–13
Low Voltage Control Panel – Drug Store

The transfer switch control will sense a loss of power on any phase and signal the motor-generator to start. When emergency power reaches 80% of voltage and frequency, the transfer switch will automatically transfer to the generator source. When the normal power has been restored for a minimum of five minutes, the transfer switch will reconnect the load to the regular power and shut off the motor-generator. Transfer switch shall be in a NEMA 1 enclosure.

1. *Installed and Supplied By:* The transfer switch shall be supplied by the motor-generator contractor and installed by the Electrical Contractor.
2. *Electrical Contractor:* The Electrical Contractor shall furnish and install the 100 ampere three-phase Owner-Emergency Panelboard identified as Panel "A." The Electrical Contractor shall supply a 3/4" raceway with four No. 6 THHN conductors to the transfer switch from the service equipment for normal power and a 3/4" raceway with four No. 6 THHN conductors from the load terminals of the transfer switch to Panel "A." All other wiring and control requirements shall be the responsibility of the motor-generator contractor. See Figure 7–14 for motor generator and transfer switch details.

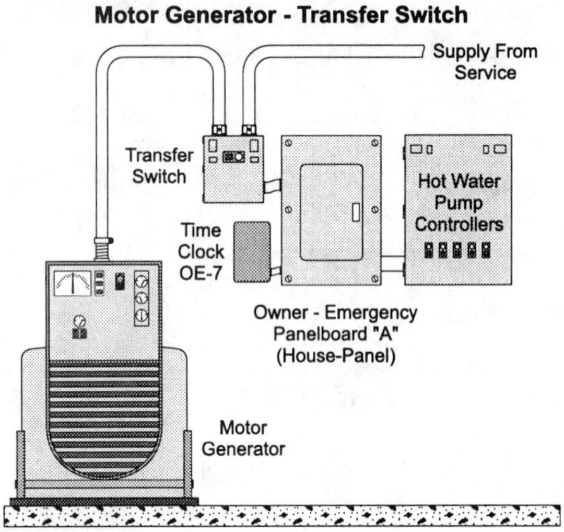

Figure 7–14
Motor Generator And Transfer Switch Details

16.10 – Panelboards

The Electrical Contractor shall furnish panelboards as shown on the plans and detailed in the panelboard schedules.

1. *Adjacent Poles:* Adjacent poles of single pole devices shall be of opposite polarity with split-phase bussing.
2. *Cabinet:* Panelboard cabinets shall be of galvanized sheet steel and colored pearl gray.
3. *Circuit Numbered:* Circuits shall be numbered serially from top to bottom with odd numbers on the left.
4. *Directory:* Provide a laser printer quality (minimum 300 dpi) circuit directory with a transparent protective cover on the inside of the panelboard cover.
5. *Finish:* Panelboard cabinet fronts shall be finished to resist corrosion with not less than one priming coat and one pearl gray finishing coat.
6. *Fronts:* Panelboard cabinet fronts must be suitable for either flush or surface installation and shall be equipped with a keyed lock and a directory card holder.
7. *Grounding:* All panelboards shall have an equipment grounding bus bonded to the cabinet.
8. *Keys:* Provide keys, each of which will operate all the panelboard cabinet locks.
9. *Type:* Panelboards shall be listed and be of load center type construction, factory assembled and tested.

16.11 – Raceways
General
1. *Bends:* The contractor shall use factory bends and sweeps wherever possible. Where field sweeps, offsets and returns are necessary, they shall be done in a neat and workmanlike manner.
2. *Branch Circuits:* Electrical metallic tubing may be used for branch circuit wiring in areas above grade and within the building.
3. *Clean:* Clean raceway systems by pulling a rat brush and mandrel throughout after installation.
4. *Condensation:* Raceways shall be installed in such manner as to ensure against the collection of trapped condensation, and runs of raceway shall be arranged to be devoid of such traps. Underground raceways shall be sloped away from building.
5. *Concealed:* Conceal all raceways in walls, above ceilings, in or under slabs, or in furring, except in mechanical and electrical rooms.
6. *Coordinate:* Run raceway and install outlets carefully and coordinate with other trades to avoid pipings, ducts and other equipment. The exact location of outlets or risers for mechanical equipment shall be determined on the job.

PANEL A – HOUSE PANEL (100 Ampere MLO – 208/120 Volt, 3-Phase, 4-Wire)					
Circuit Number	Purpose	Ampere	Phase	Wire
1	2nd Floor Area Lighting	20 Ampere	1 Pole	No. 12
2	Exterior Lighting	20 Ampere	1 Pole	No. 12
3	2nd Floor Corridor Lights	20 Ampere	1 Pole	No. 12
4	Exterior Receptacles	20 Ampere	1 Pole	No. 12
5	2nd Floor Area Receptacles	20 Ampere	1 Pole	No. 12
6	Spare			
7	Entry Lights/Time Clock	20 Ampere	1 Pole	No. 12
8	Utility Receptacle	20 Ampere	1 Pole	No. 12
9	Telephone Receptacle	20 Ampere	1 Pole	No. 12
10	Utility Receptacle GFCI	20 Ampere	1 Pole	No. 12
11	Telephone Receptacle	20 Ampere	1 Pole	No. 12
12	2nd Floor Corridor Lights	20 Ampere	1 Pole	No. 12
13	Drug Store Hot Water Pump	15 Ampere	1 Pole	No. 12
14	Bakery Hot Water Pump	15 Ampere	1 Pole	No. 12
15	Insurance Hot Water Pump	15 Ampere	1 Pole	No. 12
16	Beauty Salon Hot Water Pump	15 Ampere	1 Pole	No. 12
17	Doctor's Hot Water Pump	15 Ampere	1 Pole	No. 12
18	Sump Pump	20 Ampere	1 Pole	No. 12
Terminate Circuits 13 Through 18 To Adjacent Water Circulating Pump Controller					

PANEL B – BAKERY (225 Ampere MLO – 208Y/120 Volt, 3-Phase, 4-Wire)					
Circuit Number	Purpose	Ampere	Phase	Wire
1	Bake Area Lighting	20 Ampere	1 Pole	No. 12
2	Sales Area Receptacles	20 Ampere	1 Pole	No. 12
3	Sales/Show Room Lighting	20 Ampere	1 Pole	No. 12
4	Toilet Lighting	20 Ampere	1 Pole	No. 12
5	Sign Outlet	20 Ampere	1 Pole	No. 12
6	Basement Lighting	20 Ampere	1 Pole	No. 12
7	Mixer/Divider Receptacle	20 Ampere	3 Pole	No. 12
8	Exhaust Fan Motor	20 Ampere	1 Pole	No. 12
9	Doughnut Machine Receptacle	20 Ampere	3 Pole	No. 12
10	Show Window Receptacles	20 Ampere	1 Pole	No. 12
11	Show Window Receptacles	20 Ampere	1 Pole	No. 12
12	Oven	60 Ampere	3 Pole	No. 6
13	Basement Receptacles (North)	20 Ampere	1 Pole	No. 12
14	Bake Area Receptacles (North)	20 Ampere	1 Pole	No. 12
15	Basement Receptacles (South)	20 Ampere	1 Pole	No. 12
16	Bake Area Receptacles (South)	20 Ampere	1 Pole	No. 12
17	Dishwasher	40 Ampere	3 Pole	No. 8
18	Disposal	20 Ampere	3 Pole	No. 12

PANEL C – DRUG STORE (100 Ampere MLO – 208Y/120 Volt, 3-Phase, 4-Wire)				
Circuit Number	Purpose	Ampere	Phase	Wire
1	Merchandise Area Lighting	20 Ampere	1 Pole	No. 12
2	Merchandise Receptacles	20 Ampere	1 Pole	No. 12
3	Merchandise Area Lighting	20 Ampere	1 Pole	No. 12
4	Storage/Toilet Receptacles	20 Ampere	1 Pole	No. 12
5	Storage/Toilet Lighting	20 Ampere	1 Pole	No. 12
6	Merchandise Receptacles	20 Ampere	1 Pole	No. 12
7	Pharmacy Lighting	20 Ampere	1 Pole	No. 12
8	Show Window Receptacles	20 Ampere	1 Pole	No. 12
9	Basement Lighting	20 Ampere	1 Pole	No. 12
10	Pharmacy Receptacles	20 Ampere	1 Pole	No. 12
11	Show Window Lighting	20 Ampere	1 Pole	No. 12
12	Basement Receptacles	20 Ampere	1 Pole	No. 12
13	Show Window Lighting	20 Ampere	1 Pole	No. 12
14	Basement Receptacles	20 Ampere	1 Pole	No. 12
15	Sign	20 Ampere	1 Pole	No. 12
16	Cooling System	50 Ampere	3 Pole	No. 8
17	Roof Receptacle	20 Ampere	1 Pole	No. 12

Circuits 1, 3, 5, 7, 9, 11, 13, and 15 terminate in adjacent remote control panel

PANEL D – BEAUTY SALON (100 Ampere MLO – 208Y/120 Volt, 3-Phase, 4-Wire)				
Circuit Number	Purpose	Ampere	Phase	Wire
1	General Receptacles	20 Ampere	1 Pole	No. 12
2	Station A Receptacles	20 Ampere	1 Pole	No. 12
3	Station Spots	20 Ampere	1 Pole	No. 12
4	Station B Receptacles	20 Ampere	1 Pole	No. 12
5	Cooling System	35 Ampere	3 Pole	No. 10
6	Station C Receptacles	20 Ampere	1 Pole	No. 12
7	Station Fluorescence	20 Ampere	1 Pole	No. 12
8	Waiting Area Lighting	20 Ampere	1 Pole	No. 12
9	Water Heater/Washer & Dryer	30 Ampere	2 Pole	No. 10
10	Washer and Dryer Combination	30 Ampere	2 Pole	No. 10
11	Roof Receptacle	20 Ampere	1 Pole	No. 12

7. *Expansion Fittings:* Provide expansion fittings in raceway runs crossing expansion joints in the structure. Refer to Architectural drawings for expansion joint locations.

8. *Exposed:* Install exposed raceways parallel with or at right angles to the building lines.

9. *Feeders:* Feeders shall be installed in either rigid metal conduit or rigid nonmetallic conduit.

10. *Fire Stopping:* Provide listed <u>fire stopping materials</u> [5] at all chases to prevent drafts in accordance with the National Electrical Code, Section 300–21.

11. *Flexible Conduit:* Use flexible conduit for all

PANEL E – DOCTOR'S OFFICE (225 Ampere MLO – 208Y/120 Volt, 3-Phase, 4-Wire)				
Circuit Number	Purpose	Ampere	Phase	Wire
1	Waiting Room Lights	20 Ampere	1 Pole	No. 12
2	Waiting Room Receptacles	20 Ampere	1 Pole	No. 12
3	Examining Room Lights	20 Ampere	1 Pole	No. 12
4	Examining Room Receptacles	20 Ampere	1 Pole	No. 12
5	Laboratory Equipment Receptacles	20 Ampere	1 Pole	No. 12
6	Laboratory Table Receptacles	20 Ampere	1 Pole	No. 12
7	Cooling System	40 Ampere	2 Pole	No. 10
8	Roof Receptacle	20 Ampere	1 Pole	No. 12
9	Water Heater	30 Ampere	2 Pole	No. 10

PANEL F – INSURANCE OFFICE (225 Ampere MLO – 208Y/120 Volt, 3-Phase, 4-Wire)				
Circuit Number	Purpose	Ampere	Phase	Wire
1	Staff Office Lighting	20 Ampere	1 Pole	No. 12
2	Multioutlet Assembly (South)	20 Ampere	1 Pole	No. 12
3	Staff Office Lighting	20 Ampere	1 Pole	No. 12
4	Multioutlet Assembly (South)	20 Ampere	1 Pole	No. 12
5	Reception Area Lighting	20 Ampere	1 Pole	No. 12
6	Multioutlet Assembly (South)	20 Ampere	1 Pole	No. 12
7	Computer Room Receptacles	20 Ampere	1 Pole	No. 12
8	Computer Room Receptacle	20 Ampere	1 Pole	No. 12
9	Multioutlet Assembly (East)	20 Ampere	1 Pole	No. 12
10	Computer Room Receptacle	20 Ampere	1 Pole	No. 12
11	Multioutlet Assembly (East)	20 Ampere	1 Pole	No. 12
12	Computer Room Receptacle	20 Ampere	1 Pole	No. 12
13	Cooling System	50 Ampere	3 Pole	No. 8
14	Staff Office Floor Receptacle	20 Ampere	1 Pole	No. 12
15	Copy Machine Receptacle	20 Ampere	1 Pole	No. 12
16	Staff Room Receptacles	20 Ampere	1 Pole	No. 12
17	Roof Receptacle	20 Ampere	1 Pole	No. 12
18	Reception Room Receptacles	20 Ampere	1 Pole	No. 12

connections to vibrating equipment and for short connections to control devices, recessed fixtures, and similar items.

12. *Liquidtight Flexible Conduit:* Use liquidtight flexible conduit connection to all equipment in damp locations including mechanical rooms.

13. *Minimum Size:* All raceway shall be minimum 1/2" trade size, including flexible metal conduit.

14. *Plugged:* Raceways shall be plugged during construction to prevent dirt and water from entering. Raceways intended as spares shall be per-

manently capped with a device manufactured for the purpose.
15. *Route:* Route feeders, home runs and raceways where indicated, except minor deviations as approved.
16. *Slab:* Raceways in the slab shall be at least 6 inches from parallel runs of other raceways, pipes or reinforcing steel and shall be located so as not to affect the structural strength of the slab.
17. *Support:* Raceways shall not be supported from pipes, mechanical equipment or hangers of other trades. Raceways within suspended ceilings must be supported in accordance with the National Electrical Code, Section 300–11(a).
18. *Underground:* All underground conduit on the exterior of the building shall be two feet below finish grade and all trenches shall be backfilled and compacted in accordance with the National Electrical Code, Section 300–5.

Raceway Types Permitted

1. Electrical Metallic Tubing shall be galvanized.
2. Flexible Metal Steel Conduit shall be galvanized and listed.
3. Liquidtight Flexible Metal Conduit shall be listed.
4. Rigid Metal Conduit shall be hot dipped galvanized, and aluminum conduit is not permitted in the ground or in slabs. Threadless fittings shall not be permitted.
5. Polyvinyl chloride conduit and fittings shall be provided with end bells, spacers, plugs, and couplings.
6. Wireways shall be listed and of galvanized steel with removable covers.

16.12 – Safety Switches, Circuit Breakers, And Fuses

Circuit Breakers, Molded Case

Molded case circuit breakers shall be installed in branch-circuit panelboards as indicated on the panelboard schedules and shall be listed for the panel in which they are installed.

1. *Trip:* Each circuit breaker shall have a trip unit for each pole with elements providing inverse time delay under overload conditions and instantaneous magnetic trip for short circuit protection, unless indicated as non-automatic. The breaker operating mechanism shall be trip-free and multi-pole shall operate a common trip bar to open all elements.
2. *Type:* Circuit breakers shall be of the ampere rating, voltage rating, number of poles and class of interrupting capacity to interrupt 10,000 RMS symmetrical amperes. Lugs and terminals shall be listed for copper-aluminum.

Fuses

1. *Coordination:* All fuses shall be so selected to assure positive selective coordination.
2. *Spare:* Spare fuses [6] shall be provided in the number of 20% of each size and type installed, but in no case shall less than three spares of a specific size and type be supplied. These spare fuses shall be delivered to the owner at the time of acceptance of the project, and shall be placed in a spare fuse cabinet supplied by the owner.
3. *Type:* Fuses 601 amperes and larger shall have an interrupting rating of 200,000 symmetrical RMS amperes. They shall provide time-delay of not less than 4 seconds at 500% of their ampere rating. They shall be current limiting and of silver-sand Class L construction.

Fuses 600 amperes and less shall have an interrupting rating of 200,000 symmetrical RMS amperes. They shall be of the rejection Class RK1 type, having a time delay of not less than 10 seconds at 500% of their ampere rating.

Safety Switches/Circuit Breaker Disconnects

1. *Circuit Breakers/Switch:* Circuit breaker disconnects may be used in lieu of safety switches providing they comply with the safety switch requirements and are applied within their ratings.
2. *Enclosures:* Enclosures for safety switches shall be NEMA 1, general purpose, except that weatherproof switches shall be NEMA 3R, unless marked NEMA 4. Provide hubs as required for NEMA 3R enclosures with suitable gaskets and bonding means.
3. *Type:* Safety switches shall be heavy duty type H, fusible or non-fusible, with the poles, ampere,

voltage, and horsepower ratings indicated and shall have solid neutrals. Lugs shall be listed for copper-aluminum conductors.

4. *Labels:* Provide permanent labels on all fusible equipment [7] indicating the type and size replacement fuse required and equipment being serviced.

5. *Mounting:* Mount all switches and disconnects as close as possible to equipment served on backboards or channel hangers with proper hardware. Mount switches and disconnects so that they are readily accessible, between 4' and 5' above the finish floor.

6. *Working Space:* Maintain working space as required by the National Electrical Code in Sections 110–16 and 384–4.

16.13 – Service Equipment

Service Conductors

The Electrical Contractor will furnish and install the service entrance conductors as indicated in Figure 7–15. Transformers and primary service shall be provided by the electric utility.

Service Equipment

The Electrical Contractor shall furnish and install service-entrance equipment, as shown in the plans and detailed herein. See Figure 7–15 for service equipment details.

The service equipment will consist of nine switches, and the metering equipment will be fabricated in three type NEMA 1 sections. A continuous neutral buss shall be furnished for the length of the equipment and shall be isolated except for the main bonding jumper connection to the grounding buss.

1. *Interrupting Rating:* The switchboard shall be braced for 50,000 RMS symmetrical amperes.

2. *Metering:* The metering will be located in one section and shall consist of seven meters. Five meters shall be for the occupants of the building and two meters shall serve the owner's equipment.

3. *Switches:* The bolted pressure switches shall be knife-type switches constructed with a mechanism that automatically applies a high pressure to the blade when the switch is closed. Switches shall be rated to interrupt 200,000 symmetrical RMS amperes when used with current-limiting fuses having an equal rating. The quick make-quick break switches shall be constructed with a device that assists the operator in opening or

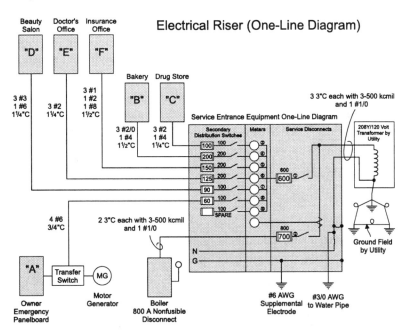

Figure 7–15
Service Equipment Details

closing the switch to minimize arcing. They shall have rejection type R fuse clips for Class R fuses. The switches shall be as follows:

1. Bolted pressure switch, three-pole, 600 ampere, w/three 600-ampere fuses.
2. Bolted pressure switch, three-pole, 800 ampere, w/three 700-ampere fuses.
3. Quick make-quick break, three-pole, 100 ampere, w/three 100-ampere fuses.
4. Quick make-quick break, three-pole, 200 ampere, w/three 200-ampere fuses.
5. Quick make-quick break, three-pole, 200 ampere, w/three 150-ampere fuses.
6. Quick make-quick break, three-pole, 200 ampere, w/two 125-ampere fuses.
7. Quick make-quick break, three-pole, 100 ampere, w/three 90-ampere fuses.
8. Quick make-quick break, three-pole, 100 ampere, w/three 60-ampere fuses.
9. Quick make-quick break, three-pole, 100 ampere, spare.

16.14 – Supports

Selection Of Products

1. *Atmosphere:* Use devices with corrosion resistant characteristics for the atmospheric conditions in which they are installed.
2. *Loading:* The weight of hangers or supports and of enclosed materials is part of the load. Supports shall be suitable for shear, straight pull, vibration, impact or external loads as applicable.
3. *Standard Hardware:* Make job fabricated hangers or supports from standard structural shapes and hardware.
4. *Type:* Devices, including anchors, fasteners, hangers and supports, shall be of a type designed or fabricated for the purpose, and shall adequately and safely secure the material and equipment and present a neat appearance.
5. *US Threads:* All bolts, screws, nuts and other threaded devices shall have US standard threads and heads as appropriate.

Fastening

Fasten all materials and equipment with approved devices.

1. *Wood:* Fasten to wood with screws, except nails may be used on wood partitions for outlet boxes and raceways up to 1" diameter.
2. *Masonry:* Fasten to masonry with threaded metal inserts, metal expansion screws, toggle bolts, power-actuated fasteners, or other approved means.
3. *Steel:* Fasten to metal joist with approved for the purpose clamps.
4. *Backboards:* Use wood backboards for surface mounting grouped electrical equipment. Paint the wall side of the wood backboards with an asphaltum coating when the walls are constructed of masonry.

Installation

1. *Design:* All parts of hanger and support assemblies, including all accessory hardware, shall be of types designed to be used together.
2. *Drill Holes:* Drill holes for devices in accordance with the manufacturer's recommendations, including diameter and depth.
3. *Removal:* Install equipment, including switches, controller, fixtures and transformers such that removal or replacement may be readily accomplished without damage to equipment or fasteners.
4. *Threads:* Internal and external threads of parts that are screwed or bolted together shall be of the same material, including coatings. All threads shall be fully engaged, using tools intended for the purpose.

16.15 – Telephone System

General

Electrical Contractor to provide raceway with pull wire, as indicated in plans. In addition, the Electrical Contractor shall provide appropriate multioutlet assembly enclosure in the insurance office to accommodate a telephone outlet every 30 inches. All telephone connections, plates and wiring to be installed by the telephone contractor. See Figure 7–16 for telephone details.

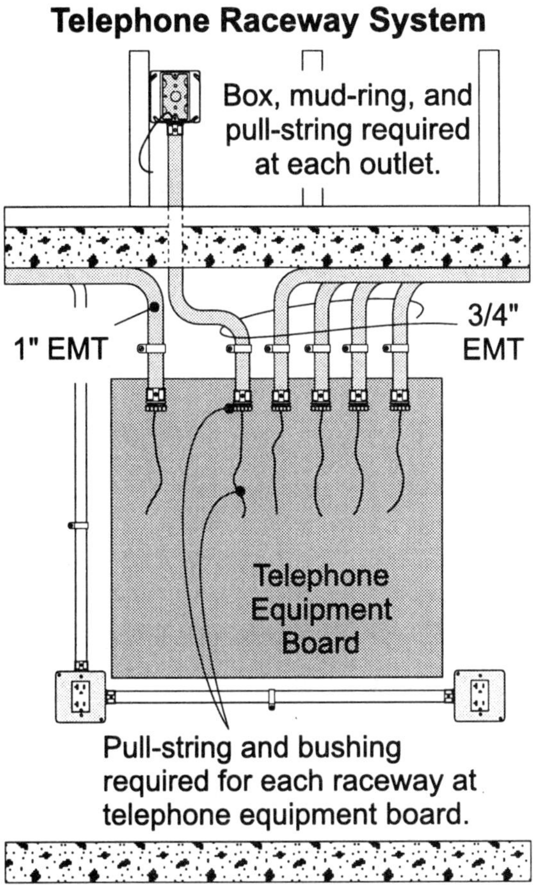

Telephone Raceway System

Box, mud-ring, and pull-string required at each outlet.

1" EMT

3/4" EMT

Telephone Equipment Board

Pull-string and bushing required for each raceway at telephone equipment board.

Figure 7–16
Telephone Raceway System

Specifications Footnotes

1. Make a note in the contract that the electrical contractor will only be responsible for equipment supplied by the electrical contractor and not other trades.
2. Include ten (10) man-hours in the estimate summary for as-builts.
3. Contact NECA at 1-301-657-3110 about the NEIS standards.
4. Heating contractor supplies all wiring and termination for hot water pump controls.
5. Add the labor and extra cost to the bid for firestops.
6. Be sure the supplier includes the extra fuses with the quote.
7. Include in the estimate the cost of labeling of fuse disconnects.

PART B – MANUAL ESTIMATE AND BID

7.02 PREPARING THE ESTIMATE

Create a job folder to contain the estimate notes and worksheets and then complete the Estimate Record Worksheet and hang it on the wall over your take-off bench. See Figure 7–17.

7.03 UNDERSTANDING SCOPE OF WORK

To get a good understanding on the scope of work, take a few minutes now and carefully read the specifications and all notes on the blueprints in Part A of this chapter.

Underline or highlight important and/or unusual items in the plans or specifications that impact the estimate. Once you have carefully reviewed the plans and specifications, complete the Specification Check List. See Figure 7–18.

7.04 ESTIMATE AND BID NOTES

Keep a note pad handy so you can keep track of any questions you have about the project. See Figure 7–19 for estimate and bid notes.

7.05 THE TAKE-OFF

The following take-off sequence is probably the most efficient method for manually estimating the wiring of this building.
1. Count lighting fixtures and develop quantities for lighting fixture quotes.
2. Count switches.
3. Count receptacles.
4. Count and measure special systems such as phone and low voltage switch control.
5. Measure general branch circuit wiring.
6. Measure individual branch circuits and home runs and develop quantities for gear quotes.
7. Count panels and measure feeder, service runs, service equipment.

Step 1 – Count Lighting Fixture Symbols

(See the *Estimating Workbook* for blueprints EC–4, EC–5, and EC–6.)

Prepare a take-off worksheet (Figure 7–21) for the lighting fixtures as listed on the lighting fixture schedule. See Figure 7–20.

Count the first lighting fixture symbol that catches your eye and continue counting that symbol until you have counted all similar symbols on EC–4. As you count each lighting fixture symbol color it yellow. Once you have counted all of the same lighting fixture symbols, write this number on the take-off worksheet. Now count the next lighting fixture symbol that catches your eye and continue this process for all fixtures on blueprint EC–4. Once you have completed counting all fixture symbols on EC–4, continue this process for blueprints EC–5 and EC–6.

See the lighting fixture count listed in Figure 7–21 for details.

Fixture Quote

Once you have counted all lighting fixture symbols for blueprints EC–4, EC–5, and EC–6, complete the Lighting Fixture Quote Worksheet. Fax this worksheet, Figure 7–22, with a copy of the lighting fixture schedule, Figure 7–11, to at least three suppliers for pricing.

Step 2 – Count Switch Symbols

(See the *Estimating Workbook* for blueprints EC–4, EC–5, and EC–6.)

As you count the single pole switch symbols, mark them with the color blue. Once you have counted all of the single pole switches, transfer the count to the take-off worksheet. Now count the three-way switch symbols, and continue this process for the four-way switch symbols, dimmers, and other switches on blueprint EC–4. Once you have completed counting all switch symbols on blueprint EC–4, continue this process for blueprints EC–5 and EC–6. The final step is to count the number of ganged switch boxes. See the switch count listed in Figure 7–23.

ESTIMATE RECORD WORKSHEET

Job Information Detail   Estimate Job Number: 0701 - xxxxxA

Job Name: American Professional   Bid Due Date: July 15, xxxx
Job Location: 100 Main Street
Job Owner: Linda Robertson   City: West Palm Beach, Florida
Fax: 1-444-432-2424   Phone: 1-444-432-4523

Plans And Specifications Detail
Date Of Plans: 1/1/xxxx   Blueprint Page Numbers: E-1 and E-2

Comments: _____
_____

Contractor Information Detail

Contractor: US Builders, Inc.   Contact: Mike Seiple
Address: 1 Lake View   City: Clermont, Florida
                                      E-mail: usbuilt@gate.net
Phone: 1-800-444-1234   Mobil: 1-444-777-1234
Beeper: 1-444-777-1498   FAX: 1-444-777-1499

Important Contacts

Architect: Summit Architectural Group   Phone: 1-444-444-1234
Engineer: Allied Engineering, Inc.   Phone: 1-444-444-1235
Telephone Utility: John Strous   Phone: 1-444-444-1236
Electric Utility: Susan Billings   Phone: 1-444-444-1237
Electrical Inspector: Mr. Johnson   Phone: 1-444-444-1238

Notes: _____
_____
_____
_____
_____

Figure 7–17
Estimate Record Sheet

Specification Check List

Labor-Unit Adjustment (See Section 3.09)

1. Building Conditions — + 5%
2. Change Orders — + 3%
3. Concealed/Exposed Wiring — + 3%
4. Construction Schedule — _____
5. Job Factors — _____
6. Labor Skill (Efficiency) — _____
7. Ladder and Scaffold — _____
8. Management — _____
9. Material — _____
10. Off Hours/Occupied — _____
11. Overtime — _____
12. Remodel (Old Work) — _____
13. Repetitive Factor — _____
14. Restrictive Conditions — _____
15. Shift Work — _____
16. Teamwork — _____
17. Temperature/Weather — _____

Labor Adjustment — + 11%

Additional Labor (See Section 5.01)

1. As-Builts — + 11 hours
2. Demolition — _____
3. Energized Parts — _____
4. Environmental Hazardous — _____
5. Excavation and Trenching — _____
6. Fire Seals — _____
7. Job Location — + 8 hours
8. Match-Up Of Existing — _____
9. Miscellaneous — + 5 hours
10. Mobilization (startup) — + 16 hours
11. Nonproductive Labor — _____
12. OSHA Compliance — _____
13. Plans and Specifications — _____
14. Public Safety — _____
15. Security Problems/Restrictions — _____
16. Site Conditions — _____
17. Shop Time — _____
18. Sub-Contract Supervision — + 10 hours
19. Temporary, Stand-By Power — + 10 hours

Total — 60 Hours

Direct Job Expenses (See Section 5.05)

1. As-Built Plans — $100
2. Bus. & Occupational Fees — $_____
3. Engineering Drawings — $_____
4. Equipment Rental — $_____
5. Field Office Expenses — $ 500
6. Fire Seals — $ 500
7. Guarantee — $_____
8. Insurance – Special — $_____
9. Miscellaneous, Labels — $ 100
10. Mobilization — $ 500
11. OSHA Compliance — $_____
12. Out of Town Expenses — $_____
13. Parking Fees — $_____
14. Permits/Inspection Fees — $ 2,500
15. Recycle Fees — $_____
16. Storage & Handling — $_____
17. Sub Contract: _____ — $_____
18. Supervision Cost — $_____
19. Temporary Wiring:
    Lighting — $ 400
    Power — $ 400
    Maintenance — $_____
20. Testing — $_____
21. Trash — $_____
22. Utility Cost — $_____

Total Direct Cost — **$5,000**

Other Final Costs (See Section 5.12)

1. Allowances — $_____
2. Back-Charges — $_____
3. Bond — $_____
4. Completion Penalty — $_____
5. Finance Cost — $_____
6. Gross Receipts And Net Profit Tax — $_____
7. Inspector Problems — $_____
8. Retainage — $_____

Total Other Cost

Other Considerations

1. Conductor Size - Minimum? — No. 12
2. Raceway Size - Minimum? — ½ inch
3. Control Wiring Responsibility? — Others
4. Cutting Responsibility? — N/A
5. Demolition Responsibility? — N/A
6. Excavation/Back Fill Responsibility? — N/A
7. Painting Responsibility? — N/A
8. Patching Responsibility? — N/A
9. Special Equipment? — Others
10. Specification Grade Devices or Fitting? — N/A

Figure 7–18
Specification Check List

Estimate And Bid Notes Page 1 of 1

Blueprint Questions or Comments:

1. Receptacles are not installed in every bathroom.
2. Outside receptacles are not GFCI protected.
3. Low voltage switches - The basement fixtures marked with lower case "g" on blueprint E-1, are connected to relay "J", but they are controlled by switch "g" located in the basement. Fixtures identified with lower case "g", "h" and "i" on blueprint E-2, are connected to relay "G", "H", and "I" and these fixtures are controlled by the remote control master switch by the rear entrance.

Specification Questions or Comments:

1. Be sure to include the cost of the permit in the estimate summary.
2. Make a note in the contract that the Electrical Contractor will only be responsible for equipment supplied by the Electrical Contractor and not other trades.
3. Include ten (10) man-hours in the estimate summary for as-built plans.
4. Contact NECA at 1-800-647-3156 about the NEIS™ standards.
5. The heating contractor supplies all wiring and termination for hot water pump controls.
6. The heating contractor to supply all equipment, wiring and termination of thermostats.
7. Add the labor and material cost for firestops.
8. Be sure the supplier includes the cost of extra fuses with the quote.
9. Include in the estimate, the cost of labeling the fuse disconnects.

Figure 7–19
Estimate And Bid Notes

Chapter 7 – Estimating Commercial Wiring

Fixture Schedule

Style	Description	Symbol
A	18" x 4' Surface Fluorescent	
B	2' x 2' Surface Fluorescent	
C	9' x 8 5/8" Enclosed Gasketed Fluorescent	
D	4' Fluorescent Strip	
E	9" Dia. Recessed Incandescent	
F	2' x 4' Recessed Fluorescent	
G	1' x 8' Recessed Fluorescent	
H	9" x 8' Surface Fluorescent	
I	9" x 4' Surface Fluorescent	
J	8" Dia. Recessed Incandescent	
K	16" Sq. WP Surface HPS	
L	13" x 4' Surface/Hung Fluorescent	
M	7" Dia. 12v Recessed Adjustable Spot	
N	20" Surface Incandescent	
O	4' x 20" Recessed Fluorescent	
P	2' x 20" Recessed Fluorescent	
Q	4' x 20" Recessed Fluorescent	
R	15' x 3" 2-ckt Track, w/ 8 Fixtures	

Figure 7–20
Lighting Fixture Schedule

LIGHTING FIXTURE COUNT Figure 7–21

Fixtures	EC-4	EC-5	EC-6	Total
A			5	5
B		4		4
C		18		18
D/*5		15		15
E		6	18	24
F			3	3
G			2	2
H			4	4
I/*3		27		27
J		4	5	9
K			3	3
L	27			27
M			9	9
N		4	6	10
O			16	16
P			2	2
Q			4	4
R		1		1

* Indicates Number Of Outlet Boxes
Total Lighting Outlet Boxes = 149

Step 3 – Count Receptacle Symbols

(See the *Estimating Workbook* for blueprints EC–4, EC–5, and EC–6.)

The counting of receptacle symbols is very quick. As you count the standard 15-ampere duplex receptacle symbols, mark them with the color green. Once you have counted all of the 15-ampere duplex receptacles symbols, write this number on the take-off worksheet. Now count the 20-ampere duplex receptacle symbols on blueprint page EC–4 and continue this process for all other receptacle symbols on blueprint EC–4. Continue this process for all receptacle symbols on blueprints EC–5 and EC–6. See the receptacle symbol count listed in Figure 7–24.

FIXTURE QUOTE WORKSHEET Figure 7–22

Type	Description	Qnty.	Unit	Extension
A	Surface	5		
B	Surface	4		
C	Surface	18		
D	Surface	15		
E	Recessed	24		
F	Recessed	3		
G	Surface	2		
H	Surface	4		
I	Surface	27		
J	Recessed	9		
K	Surface	3		
L	Surface	27		
M	Recessed	9		
N	Surface	10		
O	Recessed	16		
P	Recessed	2		
Q	Recessed	4		
R	Track Light	1		
	Total			

1. Include all associated costs such as delivery, freight, and any lighting fixture accessories listed on the blueprints or specifications.
2. Comply with Lighting Schedule, enclosed with this fax.

SWITCH COUNT Figure 7-23				
Switches	EC-4	EC-5	EC-6	Total
Single Pole Switch	2	4	21	27
Switch 3 Way	1	7		8
Switch 4 Way		1		1
Switch (Dimmer)			3	3
Switch 30A 2 Pole			2	2
Boxes				
Box With 1 Gang Ring	3	8	19	30
Box With 2 Gang Ring		2	2	4
Box With 3 Gang Ring			1	1
Total Number Of Switch Outlet Boxes = 35				

Author's Comment. According to the National Electrical Code, Section 210-8, GFCI protection is required for 120 volt convenience receptacles located in bathrooms and on roof tops of commercial occupancies, but GFCI protection is not required for receptacles located outside.

Step 4 – Special Systems

(See the *Estimating Workbook* for blueprints EC-4, EC-5, and EC-6.)

For this commercial building, special systems include the telephone and low voltage switching system. The first step is to list the components for each system on the take-off worksheet. Then as you count the device symbols, mark them with a red pencil to indicate that they have been counted, and transfer the count to the take-off worksheet. Once you have counted each of the device symbols, measure the raceway run lengths and transfer this value to the take-off worksheet. See the telephone and low voltage take-off listed in Figure 7-25.

Step 5 – Branch Circuit Wiring

(See the *Estimating Workbook* for blueprints EC-7, EC-8, and EC-9.)

Measure

Before you measure the branch circuit raceways, test the blueprint scale by using a ruler against a known distance. Highlight the three-wire circuits in blue, the four-wire circuits in purple and five-wire circuits in red.

RECEPTACLE COUNT Figure 7-24				
Receptacles	EC-4	EC-5	EC-6	Total
NEMA 5-15R	19	35	28	82
GFCI NEMA 5-15R		3	2	5
NEMA 5-20R	2		5	7
NEMA 5-15R Floor Outlets			4	4
NEMA 5-20R Hospital Grade			9	9
NEMA 15-20R		5		5
NEMA 14-30R			1	1
NEMA 15-50R		1		1
Total Number Of Receptacle Outlets = 114				

Author's Comment. Measuring the branch circuit wiring is rather tedious and tiring on the eyes. Use a magnifying glass. Your eyes will thank you.

Now measure the distance of the two-wire circuits, and write the distance next to the circuit. Continue doing this for the three-wire circuits, the four-wire circuits, etc.

Note. Draw a scaled line on the blueprint to represent the distance of the drops for the switches and receptacles. When you come across a drop for a switch or a receptacle, simply roll the distance from the pre-scaled line. See the circuit wiring measurements listed in Figure 7-26.

SPECIAL SYSTEMS TAKE-OFF Figure 7-25				
Description	EC-4	EC-5	EC-6	Totals
Telephone				
Outlet		2	3	5
3/4" EMT w/Pull Wire	166	40	12	218
1" EMT w/pull wire	60	11	3	74
Remote Control Switching				
RC Master Control Panel		1		1
RC – Switch	1	8		9
RC – Master Switch		1		1
1/2" EMT Raceway	12	186		198
3 Wire Low Voltage Cable	12	421		433

FIXTURES, SWITCHES, AND RECEPTACLES Figure 7-26

Drug Store Basement (EC-7)	Length
1/2" EMT w/2 – No. 12	400
1/2" EMT w/3 – No. 12	28
Bakery Basement (EC-7)	
1/2" EMT w/2 – No. 12	215
1/2" EMT w/3 – No. 12	44
1/2" EMT w/4 – No. 12	10
Drug Store (EC-8)	
1/2" EMT w/2 – No. 12	154
1/2" EMT w/3 – No. 12	78
1/2" EMT w/4 – No. 12	154
2-Wire No. 12 (No Pipe)	32
3-Wire No. 12 (No Pipe)	32
4-Wire No. 12 (No Pipe)	72
Bakery (EC-8)	
1/2" EMT w/2 – No. 12	428
1/2" EMT w/3 – No. 12	89
1/2" EMT w/4 – No. 12	95
1/2" EMT w/5 – No. 12	5
Insurance (EC-9)	
1/2" EMT w/2 – No. 12	321
1/2" EMT w/3 – No. 12	252
1/2" EMT w/4 – No. 12	51
Beauty and Doctor's (EC-9)	
1/2" EMT w/2 – No. 12	365
1/2" EMT w/3 – No. 12	58
1/2" EMT w/4 – No. 12	64

CIRCUIT WIRING MEASUREMENT – AVERAGE METHOD Figure 7-27

	No. Outlets
Lighting Fixtures	149 outlets
Switches	35 outlets
Receptacles	114 outlets
Total EMT	298 outlets
1/2" EMT Average	10 feet
Total 1/2" EMT	2,980 feet
Average Length No. 12 Wire Per Foot	3 feet
Total No. 12 Wire	8,940 feet

Average Run Length

The average run method is much simpler than measuring each circuit. You simply multiply the number of outlet boxes times the average run length. For this estimate, assume that past experience has indicated that the average run length per outlet is approximately ten feet of 1/2" EMT with three feet of No. 12 wire. See the average circuit wiring calculations listed in Figure 7-27.

Author's Comment. You must have historical experience to use an average run length per outlet. Measuring each circuit resulted in 2,821 feet of 1/2" EMT with 8,175 feet of No. 12 wire. The average method would have resulted in 2,980 feet of 1/2" EMT with 8,940 feet of No. 12 wire.

Step 6 – Separate Circuits And Home-Runs

(See the *Estimating Workbook* for blueprints EC-10, EC-11, and EC-12.) Taking-off the separate circuits and home runs should follow these steps:

1. Locate the termination symbol for each circuit in the blueprints and highlight them.
2. Draw a dotted line to represent the circuit wiring from the circuit termination symbol to the panelboard symbol.

   Note. Blueprint pages EC-10, EC-11, and EC-12 contain the dotted lines to represent this step.

3. Measure the distance of the drawn line and be sure to include the rise and drops at the termination. Make a note of the circuit length on the blueprints next to the circuit termination.
4. When you have measured all circuits, transfer the circuit lengths from the blueprints to the worksheet. See Figures 7-28 through 7-30.

Quotes

As you transfer the home run lengths to the take-off worksheet, determine if any items require special pricing; such as circuit breakers, disconnects, multioutlet assembly, time clock, etc. Transfer these items to the Gear Quote Worksheet for special pricing. See Figure 7-33.

Author's Comment. In an effort to save time, we will not determine every material item required for this project. That's okay since we will apply a miscellaneous material adjustment factor of 10% in the Summary Worksheet.

Step 7(a) – Service Equipment And Feeders

(See the *Estimating Workbook* for blueprints EC–10, EC–11, and EC–12.)

Step 7(b) – Service Equipment And Panelboard

Take-off the main equipment components that will provide you with enough detail to determine the bill-of-material for the service equipment and panelboards. See Figure 7–31 for service equipment and panelboard take-off details.

HOME RUN TAKE-OFF Figure 7-28

Panel A House Panel Circuits (EC-10)

Circuit Number	Purpose	Ampere	Wire	No. Wires	Length
1 and 5	2nd Floor Lighting	20 Ampere	No. 12	3 Wire	26 Feet
2 and 4	Exterior Lighting and Receptacle	20 Ampere	No. 12	2 Wire	9 Feet
3	2nd Floor Corridor Lights	20 Ampere	No. 12	2 Wire	17 Feet
6	Spare				
7	Entry Lights (Front And Back)	20 Ampere	No. 12	2 Wire	17 Feet
8	Utility Receptacle	20 Ampere	No. 12	2 Wire	10 Feet
9 and 11	Telephone/Utility Receptacle	20 Ampere	No. 12	3 Wire	19 Feet
10	Utility Receptacle GFCI	20 Ampere	No. 12	2 Wire	5 Feet
12	2nd Floor Corridor Lights	20 Ampere	No. 12	2 Wire	27 Feet
13	Drug Store Hot Water Pump	20 Ampere	No. 12	2 Wire	*
14	Baker Hot Water Pump	15 Ampere	No. 12	2 Wire	*
15	Insurance Hot Water Pump	15 Ampere	No. 12	2 Wire	*
16	Beauty Salon Hot Water Pump	15 Ampere	No. 12	2 Wire	*
17	Doctor's Hot Water Pump	15 Ampere	No. 12	2 Wire	*
18	Sump Pump	20 Ampere	No. 12	2 Wire	30 Feet

Circuits 13 through 17 terminate to adjacent water pump controller

Panel B Bakery Circuits (EC-10 And EC-11)

Circuit Number	Purpose	Ampere	Wire	No. Wires	Length
1	Bake Area Lighting	20 Ampere	No. 12	2 Wire	7 Feet
2, 10, 11	Sales Area Receptacles	20 Ampere	No. 12	5 Wire	62 Feet
3 and 5	Sales Area Lighting	20 Ampere	No. 12	3 Wire	53 Feet
4, 14, 16	Toilet Lighting, Wall Receptacles	20 Ampere	No. 12	5 Wire	10 Feet
6	Basement Lighting	20 Ampere	No. 12	2 Wire	8 Feet
7	Mixer/Divider Receptacle	20 Ampere	No. 12	3 Wire	37 Feet
8	Exhaust Fan Motor	20 Ampere	No. 12	2 Wire	37 Feet
9	Doughnut Machine Receptacle	20 Ampere	No. 12	3 Wire	55 Feet
12	Oven	60 Ampere	No. 6	3 Wire	47 Feet
13 and 15	Basement Receptacles	20 Ampere	No. 12	3 Wire	30 Feet
17	Dishwasher	40 Ampere	No. 8	3 Wire	36 Feet
18	Disposal	20 Ampere	No. 12	3 Wire	35 Feet

HOME RUN TAKE-OFF Figure 7-29

Panel C Drug Store Circuits (EC-10 And EC-11)

Circuit Number	Purpose	Ampere	Wire	No. Wires	Length
1	Merchandise Area Lighting	20 Ampere	No. 12	2 Wire	*
2	Merchandise Receptacles	20 Ampere	No. 12	2 Wire	31 Feet
3	Merchandise Area Lighting	20 Ampere	No. 12	2 Wire	*
4, 6, and 8	Toilet, Merchandise, Show Window Rec.	20 Ampere	No. 12	4 Wire	33 Feet
5	Storage/Toilet Lighting	20 Ampere	No. 12	2 Wire	*
6	Merchandise Receptacles	20 Ampere	No. 12	2 Wire	37 Feet
7	Pharmacy Lighting	20 Ampere	No. 12	2 Wire	*
8	Show Window Receptacles	20 Ampere	No. 12	2 Wire	37 Feet
9	Basement Lighting	20 Ampere	No. 12	2 Wire	*
10	Pharmacy Receptacles	20 Ampere	No. 12	2 Wire	20 Feet
11	Show Window Lighting	20 Ampere	No. 12	2 Wire	*
12 and 14	Basement Receptacles	20 Ampere	No. 12	2 Wire	41 Feet
13	Show Window Lighting	20 Ampere	No. 12	2 Wire	*
15	Sign	20 Ampere	No. 12	2 Wire	*
16	Cooling System	50 Ampere	No. 8	3 Wire	35 Feet
17	Roof Receptacle	20 Ampere	No. 12	2 Wire	35 Feet

* Circuits 1, 3, 5, 7, 9, 11, 13, and 15 terminate in adjacent remote control panel LVP

Panel D Beauty Salon Circuits (EC-12)

Circuit Number	Purpose	Ampere	Wire	No. Wires	Length
1	General Receptacles	20 Ampere	No. 12	2 Wire	15 Feet
2, 4, 6	Station A, B, C Receptacles	20 Ampere	No. 12	4 Wire	31 Feet
3 and 7	Station Spots/Fluorescence	20 Ampere	No. 12	3 Wire	19 Feet
5	Cooling System	35 Ampere	No. 10	3 Wire	36 Feet
8	Waiting Area Lighting	20 Ampere	No. 12	2 Wire	8 Feet
9 and 10	Water Heater/Washer & Dryer	30 Ampere	No. 10	5 Wire	35 Feet
11	Roof Receptacle	20 Ampere	No. 12	2 Wire	35 Feet

Panel E Doctor's Office Circuits (EC-12)

Circuit Number	Purpose	Ampere	Wire	No. Wires	Length
1	Waiting Room Lights	20 Ampere	No. 12	2 Wire	8 Feet
2 and 4	Waiting/Exam Receptacles	20 Ampere	No. 12	4 Wire	22 Feet
3	Exam Room Lights	20 Ampere	No. 12	2 Wire	16 Feet
5 and 6	Laboratory Receptacles	20 Ampere	No. 12	5 Wire	16 Feet
7	Cooling System	40 Ampere	No. 10	2 Wire	29 Feet
8	Roof Receptacle	20 Ampere	No. 12	2 Wire	28 Feet
9	Water Heater	30 Ampere	No. 10	2 Wire	2 Feet

HOME RUN TAKE-OFF Figure 7-30

Panel F Insurance Office Circuits (EC-12)

Circuit Number	Purpose	Ampere	Wire	No. Wires	Length
1 and 3	Staff Office Lighting	20 Ampere	No. 12	3 Wire	43 Feet
2, and 4	Multioutlet Assembly	20 Ampere	No. 12	3 Wire	34 Feet
5 and 7	Reception/Computer Lights	20 Ampere	No. 12	3 Wire	48 Feet
8	Computer Room Receptacle	20 Ampere	No. 12	3 Wire	66 Feet
9 and 11	Multioutlet Assembly	20 Ampere	No. 12	3 Wire	17 Feet
10	Computer Room Receptacle	20 Ampere	No. 12	3 Wire	71 Feet
12	Computer Room Receptacle	20 Ampere	No. 12	3 Wire	76 Feet
13	Cooling System	50 Ampere	No. 8	3 Wire	21 Feet
14	Staff Office Floor Receptacle	20 Ampere	No. 12	2 Wire	19 Feet
15	Copy Machine Receptacle	20 Ampere	No. 12	2 Wire	39 Feet
16 and 18	Staff/Reception Receptacles	20 Ampere	No. 12	2 Wire	39 Feet
17	Roof Receptacle	20 Ampere	No. 12	2 Wire	20 Feet

Quotes

As you transfer the service equipment and panelboards to the take-off worksheet, determine if any items require special pricing and transfer those items to the Gear Quote Worksheet, Figure 7-33.

Step 7(c) – Service And Feeder Conductors

To determine the service and feeder raceways and conductors, follow these steps:

1. Locate the symbols that indicate the service and feeder terminations. Highlight these symbols on the blueprint in a bright color and draw a dark line to represent the service and feeder runs.

SERVICE AND FEEDER EQUIPMENT Figure 7-31

Service Equipment and Panelboard Take-Off Worksheet

Item	Quantity
Service Switchboard (1600 Amperes)	1
Panelboards	
Panel A (100 Ampere MLO - 208/120 Volt, 3-Phase, 4 Wire)	1
Panel B (225 Ampere MLO - 208Y/120 Volt, 3-Phase, 4 Wire)	1
Panel C (100 Ampere MLO - 208Y/120 Volt, 3-Phase, 4 Wire)	1
Panel D (100 Ampere MLO - 208Y/120 Volt, 3-Phase, 4 Wire)	1
Panel E (225 Ampere MLO - 208Y/120 Volt, 3-Phase, 4 Wire)	1
Panel F (225 Ampere MLO - 208Y/120 Volt, 3-Phase, 4 Wire)	1

Note. Blueprint page EC-10, EC-11, and EC-12 contain the dark lines to represent this step.

2. Measure the lines that represent each feeder, and mark the length on the blueprints next to the feeder termination. Careful, be sure you verify the scale by using a ruler against a known distance.

3. Transfer the lengths to the take-off worksheet, Figure 7-32.

7.06 DETERMINING THE BILL-OF-MATERIAL

Since we are estimating this job manually, we'll need to determine the bill-of-material manually. The following spreadsheets determine the bill-of-material for the lighting fixtures, switches, receptacles, telephones, low voltage switching, branch circuits and home runs. See Figures 7-34 through 7-37.

7.07 PRICING, LABORING, EXTENSIONS, AND TOTALS

Transfer the bill-of-material from the spreadsheets, Figure 7-34 through 7-37 to the Price/Labor Worksheet, Figure 7-38 and 7-39. List the material items and their quantities for the remaining take-off items to the Price/Labor Worksheet as well, Figures 7-40 and 7-41. Now price and labor each item from the catalog in Appendix A, extend each line, and determine the total cost and total labor hours for each worksheet, see Figures 7-38 through 7-41.

SERVICE/FEEDER TAKE-OFF WORKSHEET Figure 7-32

From	To	Raceway	Wire	Length
Utility	Service Equipment	3 - 3" PVC	Three 500 kcmil and one No. 1/0	30 feet
Service Equipment	Panel A T/Switch	1 - 1 3/4" EMT	Four No. 6	10 feet
Service Equipment	Panel B	1 - 1 1/2" EMT	Three No. 2/0 And One No. 4	30 feet
Service Equipment	Panel C	1 - 1 1/4" EMT	Three No. 2 And One No. 4	20 feet
Service Equipment	Panel D	1 - 1 1/4" EMT	Three No. 3 And One No. 6	50 feet
Service Equipment	Panel E	1 - 1 1/4" EMT	Three No. 2 And One No. 4	70 feet
Service Equipment	Panel F	1 - 1 1/2" EMT	Three No. 1, One No. 2 And One No. 8	40 feet
Service Equipment	Boiler Disconnect	2 - 3" Nipple	Three 500 kcmil And One No. 1/0	2 feet

GEAR QUOTE WORKSHEET Figure 7-33

Circuit Breakers	Quantity					Cost	Unit	Extension
1 Pole 15 Ampere	5							
1 Pole 20 Ampere	73							
2 Pole 20-60 Ampere	4							
3 Pole 20-60 Ampere	8							
Disconnects								
60 Ampere, 3 Phase, NEMA 1	1							
60 Ampere, 3 Phase, NEMA 3R	4							
Miscellaneous Items								
Multiwire Assembly	50							
Time Clock, 120 Volts	1							
3/4" X 10' Copper Ground Rod	2							
NEMA 5-15R Floor Outlets	4							
NEMA 5-20R Hospital Grade	9							
NEMA 15-20R (Symbol A)	5							
NEMA 14-30R (Symbol B)	1							
NEMA 15-50R (Symbol C)	1							
Panelboards								
100 Ampere, 3-Phase, 4-Wire	3							
225 Ampere, 3-Phase, 4-Wire	3							
Service Switchboard (2,500 Pounds)								
Total								$7,500
Other Cost								$500
Total Quote Price								$8,000

Notes.
1. Price to include all taxes, freight and delivery charges.
2. The Electrical Supplier shall furnish all materials and equipment and all related items necessary to complete the work as indicated on the drawing and/or specifications. See attached specification details.

LIGHTING FIXTURE SPREADSHEET Figure 7-34						
Type	Number Fixtures	Outlets	Install Labor	Box 4" x 4"	Ring 1 Gang	Conn. 1/2"
Type A	5	5	3.75	5	5	10
Type B	4	4	3.00	4	4	8
Type C	18	18	13.50	18	18	36
Type D	15	5	11.25	5	5	10
Type E	24	24	18.00	24	24	48
Type F	3	3	2.25	3	3	6
Type G	2	2	1.50	2	2	4
Type H	4	4	3.00	4	4	8
Type I	27	3	20.25	3	3	6
Type J	9	9	6.75	9	9	18
Type K	3	3	2.25	3	3	6
Type L	27	27	20.25	27	27	54
Type M	9	9	6.75	9	9	18
Type N	10	10	7.50	10	10	20
Type O	16	16	12.00	16	16	32
Type P	2	2	1.50	2	2	4
Type Q	4	4	3.00	4	4	8
Type R	1	1	3.00	1	1	2
Total			139.50	149	149	298

Switch Spreadsheet														
	Qnty.	Switch					Plate			Box	Ring			Conn.
Item		1 Pole	3 Way	4 Way	Dimmer	2 Pole	1G	2G	3G	4 x 4	3G	1G	2G	3G	1/2"
Single Pole Switch	27	27													
Switch 3 Way	8		8												
Switch 4 Way	1			1											
Switch Dimmer	3				3										
Switch 2 Pole 30 Ampere	2					2	2			2		2			4
Box															
1 Gang Box With Ring	30						30			30		30			60
2 Gang Box With Ring	4							4		4			4		8
3 Gang Box With Ring	1								1		1			1	2
Total		27	8	1	3	2	32	4	1	36	1	32	4	1	74

RECEPTACLE SPREADSHEET Figure 7-35

	Qnty.	Receptacle				Plate			Box		Ring		Conn.
		5-15R	GFCI	5-20R	Special	1G	GFCI	3G	4" x 4"	3G	1G	3G	1/2"
NEMA 5-15R	82	82				82			82		82		164
GFCI NEMA 5-15R	5		5				3/2 wp		5		5		10
NEMA 5-20R	7			7		7			7		7		14
Hospital Grade	9				9			3	3			3	6
Floor Outlet	4	4											8
NEMA 15-20R	5				5	5			5		5		10
NEMA 14-30R	1				1				1		1		2
NEMA 15-50R	1				1				1		1		2
		86	5	7	16	94	3/2 wp	3	101	3	101	3	216

TELEPHONE SPREADSHEET

	Qnty.	EMT		Pull	Coupling		Box	Ring	Conn.	
		3/4"	1"	Wire	3/4"	1"	4" x 4"	1G	1/2"	1"
Telephone Outlet	5						5	5	10	2
Basement (EC-4)										
1/2" Emt w/Pull Wire	166	166		183	17					
1" EMT w/Pull Wire	60		60	66		6				
First Floor (EC-5)										
3/4" EMT w/Pull Wire	40	40		44	4					
1" EMT w/Pull Wire	11		11	15		1				
Second Floor (EC-6)										
3/4" EMT w/Pull Wire	12	12		15	1					
1" EMT w/Pull Wire	3		3	5		1				
		218	74	328	22	8	5	5	10	2

Remote Control Switch Spreadsheet

	Qnty.
RC Master Control Panel	1
RC – Switch	9
RC – Master Switch	1
1/2" EMT	198
Remote Control Cable	433

BRANCH CIRCUIT SPREADSHEET Figure 7-36				
	Quantity	EMT	Coupling	Wire (+ 10%)
		½"	½"	No. 12
Drug Store Basement (EC-7)				
½" EMT w/2 - No. 12	400	400	40	880
½" EMT w/3 - No. 12	28	28	3	92
Bakery Basement (EC-7)				
½" EMT w/2 - No. 12	215	215	22	473
½" EMT w/3 - No. 12	44	44	4	145
½" EMT w/4 - No. 12	10	10	1	44
Drug Store (EC-8)				
½" EMT w/2 - No. 12	154	154	15	339
½" EMT w/3 - No. 12	78	78	8	257
½" EMT w/4 - No. 12	154	154	15	678
2-Wire No. 12 (No Pipe)	32			70
3-Wire No. 12 (No Pipe)	32			106
4-Wire No. 12 (No Pipe)	72			317
Bakery (EC-8)				
½" EMT w/2 - No. 12	428	428	42	942
½" EMT w/3 - No. 12	89	89	9	294
½" EMT w/4 - No. 12	95	95	10	418
½" EMT w/5 - No. 12	15	15	2	83
Insurance Office (EC-9)				
½" EMT w/2 - No. 12	321	321	32	706
½" EMT w/3 - No. 12	252	252	25	832
½" EMT w/4 - No. 12	51	51	5	224
Beauty And Doctor's Office				
½" EMT w/2 - No. 12	365	365	37	803
½" EMT w/3 - No. 12	58	58	6	191
½" EMT w/4 - No. 12	64	64	6	282
Total		2,821	282	8,176

		HOME RUN SPREADSHEET Figure 7-37							
	Quantity	EMT		Coupling		Wire (+ 10%)			
		1/2"	3/4"	1/2"	3/4"	No. 12	No. 10	No. 8	No. 6
Basement Homeruns (EC-10)									
1/2" EMT w/2 No. 12	114	114		11		251			
1/2" EMT w/3 No. 12	125	125		13		413			
First Floor Home Runs (EC-11)									
1/2" EMT w/2 No. 12	136	136		14		399			
1/2" EMT w/3 No. 12	184	184		18		607			
1/2" EMT w/4 No. 12	33	33		3		145			
1/2" EMT w/5 No. 12	72	72		7		396			
3/4" EMT w/3 No. 8	71		71		7			234	
3/4" EMT w/3 No. 6	47		47		5				140
Second Floor Home Runs (EC-12)									
1/2" EMT w/2 No. 12	189	189		19		416			
1/2" EMT w/3 No. 12	379	379		38		1,251			
1/2" EMT w/4 No. 12	87	87		9		383			
1/2" EMT w/5 No. 12	16	16		2		88			
EMT w/2 No. 10	34	34		3			75		
1/2" EMT w/3 No. 10	36	36		4			119		
1/2" EMT w/5 No. 10	35	35		4			193		
3/4" EMT w/2 No. 8	21		21		2			46	
3/4" EMT w/3 No. 8	21		21		2			69	
Total		1,440	160	145	16	4,349	387	349	140

BILL-OF-MATERIAL (1 OF 4) Figure 7-38							
	Qnty.	Material Cost	Unit	Material Extension	Labor Hours	Unit	Labor Extension
Lighting Fixtures							
Lighting Fixture Labor	1	*	*	*	1	E	139.50
4" x 4" Box	149	$59.00	C	$87.91	18	C	26.82
1 Gang Ring	149	$39.00	C	$58.11	4.5	C	6.71
½" EMT Connector	298	$21.00	C	$62.58	2	C	5.96
Switches							
Single Pole Switch	27	$57.00	C	$15.39	20	C	5.40
3 Way Switch	8	$127.00	C	$10.16	25	C	2.00
4 Way Switch	1	$719.00	C	$7.19	28	C	0.28
Dimmer	3	$610.00	C	$18.30	25	C	0.75
2 Pole, 30 Amperes	2	$1,400.00	C	$28.00	36	C	0.72
Switch Plate 1 Gang	32	$47.00	C	$15.04	2.5	C	0.80
Switch Plate 2 Gang	4	$81.00	C	$3.24	4	C	0.16
Switch Plate 3 Gang	1	$99.00	C	$0.99	6	C	0.06
4" x 4" Box	36	$59.00	C	$21.24	18	C	6.48
3 Gang Box	1	$516.00	C	$5.16	20	C	0.20
1 Gang Ring	32	$39.00	C	$12.48	4.5	C	1.44
2 Gang Ring	4	$52.00	C	$2.08	5	C	0.20
3 Gang Ring	1	$100.00	C	$1.00	9	C	0.09
½" EMT Connector	74	$21.00	C	$15.54	2	C	1.48
Receptacles							
NEMA 5-15R (15A)	86	$52.00	C	$44.72	18	C	15.48
NEMA 5-15R GFCI (15A)	5	$12.00	E	$60.00	0.3	E	1.50
NEMA 5-20R (20A)	7	$180.00	C	$12.60	19	C	1.33
Special NEMA receptacles priced and labored on bill-of-material page 3 of 4							
Receptacle Plate 1 Gang	94	$44.18	C	$42.30	2.5	C	2.35
Receptacle Plate GFCI	3	$59.00	C	$1.77	2.5	C	0.08
Receptacle Plate 3 Gang	3	$99.00	C	$2.97	6	C	0.18
Receptacle Plate Weatherproof	2	$389.00	H	$7.78	4	H	0.08
4" x 4" Box	101	$59.00	C	$59.59	18	C	18.18
3 Gang Box	3	$516.00	C	$15.48	20	C	0.60
1 Gang Ring	101	$39.00	C	$39.39	4.5	C	4.55
3 Gang Ring	3	$100.00	C	$3.00	9	C	0.27
½" EMT Connector	216	$21.00	C	$45.36	2	C	4.32
Total				$701.25			247.97

BILL-OF-MATERIAL (2 OF 4) Figure 7-39							
	Qnty.	Material Cost	Unit	Material Extension	Labor Hours	Unit	Labor Extension
Telephone Raceways							
3/4" EMT	218	$20.00	C	$43.60	2.85	C	6.21
1" EMT	74	$31.00	C	$22.94	3.75	C	2.78
Pull Wire	328	$20.00	M	$6.56	6.50	M	2.13
3/4" EMT Coupling	22	$40.00	C	$8.80	3.00	C	0.66
1" EMT Coupling	8	$62.00	C	$4.96	5.00	C	0.40
4" X 4" Box	5	$59.00	C	$2.95	18.00	C	0.90
1 Gang Ring	5	$39.00	C	$1.95	4.50	C	0.23
3/4" EMT Connector	10	$41.00	C	$4.10	3.00	C	0.30
1" EMT Connector	2	$31.00	C	$0.62	3.75	C	0.08
Remote Control Wiring							
Master Control Panel	1	*	*	*	3.00	E	3.00
Remote Switch	9	*	*	*	1.00	E	9.00
Master Switch	1	*	*	*	0.50	E	0.50
1/2" EMT	198	$13.00	C	$25.74	2.25	C	4.46
Control Cable (3 Conductor)	433	*	*	*	13.00	M	5.63
Branch Circuit and Home Run Wiring							
1/2" EMT (2,821 + 1,440)	4,261	$13.00	C	$553.93	2.25	C	95.87
3/4" EMT	160	$20.00	C	$32.00	2.85	C	4.56
1/2" EMT Coupling (282 + 145)	427	$24.00	C	$102.48	2.00	C	8.54
3/4" EMT Coupling	16	$40.00	C	$6.40	3.00	C	0.48
No. 12 (8,176 + 4,349)	12,52.50	$48.00	M	$601.20	4.25	M	53.23
No. 10	387	$78.00	M	$30.19	5.10	M	1.97
No. 8	349	$146.00	M	$51.10	6.00	M	2.09
No. 6	140	$250.00	M	$35.00	7.00	M	0.98
Circuit Breakers							
1-Pole 15 Ampere	5	*	*	*	0.10	E	0.50
1-Pole 20 Ampere	73	*	*	*	0.10	E	7.30
2-Pole 20-60 Ampere	4	*	*	*	0.15	E	0.60
3-Pole 20-60 Ampere	8	*	*	*	0.20	E	1.60
Miscellaneous Items							
Multiwire Assembly	50	*	*	*	20.00	C	10.00
Time Clock, 120 Volts	1	$33.00	E	$33.00	1.00	E	1.00
1/2" X 10' Copper Ground Rod	2	$1,200.00	C	$24.00	60.00	C	1.20
Total				$1,591.36			226.20

BILL-OF-MATERIAL (3 OF 4) Figure 7-40							
	Qnty.	Material Cost	Unit	Material Extension	Labor Hours	Unit	Labor Extension
Receptacles							
NEMA 5-20R Hospital Grade	9	*	*	*	19.00	C	1.71
NEMA 5-15 Floor Outlet	4	*	*	*	1.00	E	4.00
NEMA 15-20R (Symbol A)	5	*	*	*	19.00	C	0.95
NEMA 14-30R (Symbol B)	1	*	*	*	25.00	E	25.00
NEMA 15-50 R (Symbol C)	1	*	*	*	25.00	E	25.00
Panelboards							
100 Ampere, 3-Phase, 4-Wire	3	*	*	*	1.00	E	3.00
225 Ampere, 3-Phase, 4-Wire	3	*	*	*	2.50	E	7.50
Service Switchboard (2500 lb)							
Installation Labor In pounds	2,500	*	*	*	1.00	E	25.00
Conductor Terminations	135	*	*	*	0.15	E	20.25
3"PVC	30	$99.00	C	$29.70	7.00	C	2.10
3" PVC Elbow	6	$1,147.00	C	$68.82	32.00	C	1.92
3" PVC Coupling	15	$316.00	C	$47.40	7.00	C	1.05
3" PVC Bell End	3	$357.00	C	$10.71	32.00	C	0.96
3" PVC Male Adapter	3	$173.00	C	$5.19	23.00	C	0.69
3" Locknut	3	$250.00	C	$7.50	7.00	C	0.21
500 kcmil Wire	99	$3,044.00	M	$301.36	29.00	M	2.87
No. 1/0 Wire	33	$670.00	M	$22.11	13.50	M	0.45
Panel A (House Panel)							
100 Ampere, 3-Phase, 4-Wire	3	*	*	*	1.00	E	3.00
3/4" EMT	10	$20.00	C	$2.00	2.85	C	0.29
3/4" EMT Coupling	1	$40.00	C	$0.40	3.00	C	0.03
3/4" EMT Connector	2	$41.00	C	$0.82	3.00	C	0.06
No. 6 Wire	44	$250.00	M	$11.00	7.00	M	0.31
Panel B (Bakery)							
225 Ampere, 3-Phase, 4-Wire	3	*	*	*	2.50	E	7.50
1 1/2" EMT	30	$71.00	C	$21.30	4.25	C	1.28
1 1/2" EMT Coupling	3	$117.00	C	$3.51	8.00	C	0.24
1 1/2" EMT Connector	2	$112.00	C	$2.24	8.00	C	0.16
1 1/2" Plastic Bushing	2	$104.00	C	$2.08	11.00	C	0.22
No. 2/0 Wire	66	$816.00	M	$53.86	15.25	M	1.01
No. 4 Wire	33	$350.00	M	$11.55	7.25	M	0.24
Disconnect 60A, 3 Phase, NEMA	1	*	*	*	1.50	E	1.50
Panel C (Drug Store)							
100 Ampere, 3-Phase, 4-Wire	3	*	*	*	1.00	E	3.00
1 1/4" EMT	20	$50.00	C	$10.00	4.00	C	0.80
1 1/4" EMT Coupling	2	$99.00	C	$1.98	7.00	C	0.14
1 1/4" EMT Connector	2	$108.00	C	$2.16	7.00	C	0.14
1 1/4" Plastic Bushing	2	$89.00	C	$1.78	9.00	C	0.18
No. 2 Wire	66	$450.00	M	$29.70	7.75	M	0.51
No. 4 Wire	22	$350.00	M	$7.70	7.25	M	0.16
Disconnect 60A, 3 Phase, NEMA 3R	1	*	*	*	1.50	E	1.50
Total				$654.87			144.93

BILL-OF-MATERIAL (4 OF 4) Figure 7-41							
	Qnty.	Material Cost	Unit	Material Extension	Labor Hours	Unit	Labor Extension
Panel D (Beauty Salon)							
100 Ampere, 3-Phase, 4-Wire	3	*	*	*	1	E	3
1¼" EMT	50	$50.00	C	$25.00	4	C	2
1¼" EMT Coupling	5	$99.00	C	$4.95	7	C	0.35
1¼" EMT Connector	2	$108.00	C	$2.16	7	C	0.14
1¼" Plastic Bushing	2	$89.00	C	$1.78	9	C	0.18
No. 3 Wire	165	$375.00	M	$61.88	7.5	M	1.24
No. 6 Wire	55	$250.00	M	$13.75	7	M	0.39
Disconnect 60A, 3 Phase, NEMA 3R	1	*	*	*	1.5	E	1.5
Panel E (Doctor's Office)							
225 Ampere, 3-Phase, 4-Wire	3	*	*	*	2.5	E	7.5
1¼" EMT	70	$50.00	C	$35.00	4	C	2.8
1¼" EMT Coupling	7	$99.00	C	$6.93	7	C	0.49
1¼" EMT Connector	2	$108.00	C	$2.16	7	C	0.14
1¼" Plastic Bushing	2	$89.00	C	$1.78	9	C	0.18
No. 2 Wire	231	$450.00	M	$103.95	7.75	M	1.79
No. 4 Wire	77	$350.00	M	$26.95	7.25	M	0.56
Disconnect 60A, 1 Phase, NEMA 3R	1	*	*	*	1	E	1
Panel F (Insurance Office)							
225 Ampere, 3-Phase, 4-Wire	3	*	*	*	2.5	E	7.5
1½" EMT	40	$71.00	C	$28.40	4.25	C	1.70
1½" EMT Coupling	4	$117.00	C	$4.68	8	C	0.32
1½" EMT Connector	2	$112.00	C	$2.24	8	C	0.16
1½" Plastic Bushing	2	$104.00	C	$2.08	11	C	0.22
No. 1 Wire	155	$591.00	M	$91.61	8	M	1.24
No. 2 Wire	40	$450.00	M	$18.00	7.75	M	0.31
No. 8 Wire	40	$146.00	M	$5.84	6	M	0.24
Disconnect 60A, 3 Phase, NEMA 3R	1	*	*	*	1.5	E	1.5
Boiler Disconnect							
Boiler Disconnect	1	*	*	*	2	E	2
3" Nipple	2	$3.00	C	$0.06	0.4	C	0.01
3" Locknut	4	$250.00	C	$10.00	7	C	0.28
3" Plastic Bushing	4	$300.00	C	$12.00	19	C	0.76
500 kcmil Wire	60	$3,044.00	M	$182.64	29	M	1.74
No. 1/0 Wire	30	$670.00	M	$20.10	13.5	M	0.41
Total				$663.94			41.65

7.08 ESTIMATE AND BID SUMMARY

Use the information contained on the Specification Check List, Figure 7–18, and the Estimate and Bid Notes, Figure 7–19, to complete the Estimate/Bid Summary worksheet.

Step A – Total Labor-Hours

Labor Hour Calculation	
Labor-Unit Estimated Hours	660.75
Labor-Unit Adjustment, + 11%	72.68
Additional Labor-Hours	60.00
Total Hours (Rounded)	793.43

Step B – Labor Cost

Labor Cost = Labor-Hours × Labor Rate
Labor Cost = 793.43 hours × $10.65
Labor Cost = $8,450.03

Note. Don't apply a labor burden factor, because labor burden costs will be recovered when overhead is applied.

Step C – Adjusted Material Cost

See Chapter 5 for details.

Adjusted Material Cost Calculation	
Price Sheet Total Cost	$3,611.42
Miscellaneous Items, + 10%	$361.14
Waste and Theft, + 5%	$180.57
Small Tool Allowance, + 3%	$108.34
Total Adjusted Material Cost (Rounded)	$4,261.47

Step D – Quotes

Gear Cost = $8,000
Fixtures supplied by owner.

Step E – Sales Tax

Calculate sales tax at 7% of total taxable material cost.

Sales Tax Calculation	
Total Adjusted Material Cost	$4,261.47
Quote – Gear	$8,000
Total Taxable Material Cost	$12,261.47
Sales Tax, + 7%	$858.30
Total Cost of Material Including Tax	$13,119.77

Step F – Direct Cost

Total Direct Cost = $5,000

Step G – Estimated Prime Cost

Estimated Prime Cost Calculation	
Labor Cost at $10.65 per hour	$8,450.03
Total Material Cost, Including Tax	$13,119.77
Direct Cost	$5,000
Total Prime Cost	$26,569.80

Step H – Overhead Expenses

Overhead cost should be the lesser of 40% of prime cost or $13 per labor man-hour.

Percentage Method = $26,569.80 × 40% = $10,627.92

Hour Method = 793.43 hours × $13 = $10,314.59

Step I – Break Even Cost

Break Even Cost Calculation	
Total Prime Cost	$26,569.80
Overhead At $13 Per Man-hour	$10,314.59
Estimated Break Even Cost	$36,884.39

Step J – Profit

For this job, calculate profit to be 15% of break even cost.

Estimated break even cost = $36,884.39
Profit at 15% of $36,884.39 = $5,532.66

Step K – Other Final Costs (None)

Step L – Bid Price

The bid price is the sum of the estimated break even cost, plus profit, plus other final cost.

Bid Price Calculation	
Estimated Break Even Cost	$36,884.39
Profit, + 15% (Rounded)	$5,532.66
Other Cost: None	$0
Bid Price	$42,417.05

Figure 7–42 demonstrates the Estimate and Bid Summary worksheet for the commercial project.

ESTIMATE AND BID SUMMARY WORKSHEET Figure 7-42			
	Description		
Labor Calculations		Hours	
	Labor Worksheet Page 1 of 4	247.97	
	Labor Worksheet Page 2 of 4	226.20	
	Labor Worksheet Page 3 of 4	144.93	
	Labor Worksheet Page 4 of 4	41.65	
	Price/labor Worksheet Labor-Hours	660.75	
	Labor Unit Adjustment, Plus 11%	72.68	
	Additional Labor Hours	60.00	
Step A	Total Final Adjusted Hours	793.43	
Step B	Labor Cost - $10.65 per hour × 793.43 Hours		$8,450.03
Material Cost		Dollars	
	Price Worksheet Page 1 of 4	$701.25	
	Price Worksheet Page 2 of 4	$1,591.36	
	Price Worksheet Page 3 of 4	$654.87	
	Price Worksheet Page 4 of 4	$663.94	
	Pricing Worksheet Material Cost	$3,611.42	
	Miscellaneous Material Items, Plus 10%	$361.14	
	Waste And Theft, Plus 5%	$180.57	
	Small Tool Allowance, Plus 3%	$108.34	
Step C	Total Adjusted Material Cost	$4,261.47	
Step D	Quote - Gear	$8,000.00	
	Total Taxable Material Cost	$12,261.47	
Step E	Sales Tax At 7%	$858.30	
	Total Material Cost		$13,119.77
Step F	Direct Costs		$5,000.00
Step G	Total Prime Cost		$26,569.80
Step H	Overhead At $13 Per Man-hour		$10,314.59
Step I	Break Even Cost		$36,884.39
Step J	Profit At 15%		$5,532.66
Step K	Other Final Cost: None		$0.00
Step L	Bid Selling Price		$42,417.05

7.09 BID ACCURACY AND ANALYSIS

Make sure you have not made any of the following mistakes:
- Errors in multiplication or addition.
- Failing to include outside or underground work.
- Failure to visit site to determine job conditions.
- Forgetting a major item, such as gear quotes.
- Forgetting to include subcontract cost or equipment rental requirements.
- Forgetting to include the changes to the original specifications or blueprints.
- Hurrying and rushing the bid.
- Improper estimating forms.
- Not double checking all figures.
- Not transferring totals to the bid summary worksheet properly, transposing numbers.
- Omitting a section of the bid.
- Using supplier take-offs for quotes or depending on verbal quotes.
- Wrong extensions.
- Wrong scale on reduced blueprints.
- Wrong unit for labor-unit.
- Wrong unit for material cost.

Bid Analysis
Cost Distribution
Labor
$8,450.03/$42,417.05 =   19.92%
Material
$13,119.77/$42,417.05 =   30.93%
Direct Cost
$5,000/$42,417.05 =   11.79%
Overhead
$10,314.59/$42,417.05 =   24.32%
Profit
$5,532.66/$42,417.05 =   13.04%
Cost Per Square Foot
Cost Per Square Foot = $5.48
Cost Per Sq. Ft = $42,417.05/7,740 sq. ft.
Labor Analysis
Total Hours =   793.43 hours
Total 8 Hr. Days (4 men)
(794 hours/32 hours) =   24.79 Days
Total 5 Day Weeks
(24.81 days/5 days) =   4.96 weeks

7.10 THE PROPOSAL

Since this bid is based on a manual estimate, use a preprinted wiring contract and apply the necessary corrections. Be sure to review your Bid and Specification Notes. See Figure 7-19 when you prepare your proposal.

PART C – COMPUTER ASSISTED ESTIMATE AND BID

This Part explains the computer assisted estimating process for commercial wiring and gives you the estimated time to complete each phase of the bid. The total amount of time required from start to finish is about $2^{1}/_{2}$ to 3 hours.

7.11 BID PREPARATION AND TAKE-OFF

The bid preparation and the take-off using a computer is about the same as with the manual estimate method. With experience and proper take-off forms you should be able to review the blueprints and complete the take-off in about two to three hours.

7.12 INPUT TAKE-OFF INTO COMPUTER

The next step is to input the take-off quantities into the computer. Once you're up to speed, this should take less than 30 minutes.

7.13 BILL-OF-MATERIAL, PRICING, LABORING, EXTENDING, AND TOTALING

The computer automatically determines the bill-of-material, it prices and labors each material item, determines the extended cost and labor-hours for each material item, as well as the totals in about 20 seconds.

7.14 ESTIMATE AND BID SUMMARY

It only takes about 3 minutes to input the estimate summary adjustments in the computer to produce the bid price.

7.15 BID ACCURACY AND ANALYSIS

Before you submit your bid you must verify its accuracy and perform a bid analysis. The following is a description of typical computer analysis reports that can help you insure that you bid is correct.

Bid Summary Analysis – Report No. 1

This report permits you the opportunity to analyze the following:
1. Estimated hours, man-days and man-weeks.
2. Ratio of labor cost to selling price.
3. Material ratio to selling price.
4. Ratio of overhead to selling price.

Detailed Summary Analysis – Report No. 2

This report permits a detailed analysis of each take-off item, such as labor-hours, cost, sales tax, overhead and profit. If you made a mistake in your estimate, you'll see the error very quickly.

Labor Analysis – Report No. 3

This report analyzes which material items are most labor intensive and which are not. If you made a mistake on the quantity in your take-off, you will catch it here, or if you did not include sufficient labor or forgot to include the labor completely, you'll know about it.

Material Quantity Analysis – Report No. 4

You'll be able to analyze the quantity of material items estimated, their individual and extended cost as well as their unit and extended labor-hours.

Material Cost Analysis – Report No. 5

With this report you'll know which material items cost you the most. If you made a mistake on the quantity in your take-off, you will catch it here, or if you did not include sufficient cost or forgot to include the cost at all, you'll know about it.

Work Phase Analysis – Report No. 6

This reports analyzes the distribution of the material for the slab, the rough, the trim, etc. With this information you will be able to detect any material items that you might have missed.

Job Proposal – Report No. 7

You will be able to analyze the unit price for each take-off item, as well as the total cost for selective groups of the take-off, such as fixtures, switches, receptacles, special systems, equipment and service, etc.

Draw Schedule – Report No. 8

Before you submit your bid, you will want to compare the bid's cost per square foot against similar bids. In addition, you want to compare the bid cost distribution percentage for the slab, rough, trim, etc. This report gives you the information to help you manage your cash flow.

Chapter 7

Review Questions

Practice Estimate

The following questions are in reference to the residential blueprints marked CE-1, CE–2, and C–3 located in the *Estimating Workbook*. Before answering any of the following questions be sure to review these blueprints completely.

1. Determine the count for the following symbols in blueprint CE-1.

Lighting Fixture	Count
Type A	
Type B	
Type C	
Type D	
Type E	
Type F	
Type G	
Type H	
Type I	
Type J	
Type K	
Type L	
Sign Outlet	
Switch	
Single Pole Switch	
Switch 3 Way	
Switch (Dimmer)	
Switch 30A 2 Pole	
Box	
1 Gang Box With Ring	
2 Gang Box With Ring	
3 Gang Box With Ring	
Receptacle	
NEMA 5-15R	
NEMA 5-15R GFCI	
NEMA 5-20R	
NEMA 5-15R Floor Outlet	
NEMA 5-20R Hospital Grade	
NEMA 14-30R	

2. Determine the length of the branch circuits, including home runs according to the wiring listed in blueprint CE-2.

Home Runs	
Insurance Office	Length
½" EMT w/2 No. 12	
½" EMT w/3 No. 12	
½" EMT w/4 No. 12	
¾" EMT w/3 No. 8	
Beauty Salon	
½" EMT w/2 No. 12	
½" EMT w/3 No. 12	
½" EMT w/4 No. 12	
½" EMT w/2 No. 10	
½" EMT w/3 No. 10	
½" EMT w/5 No. 10	
Doctor's Office	
½" EMT w/2 No. 12	
½" EMT w/3 No. 12	
½" EMT w/4 No. 12	
½" EMT w/5 No. 12	
½" EMT w/2 No. 10	
¾" EMT w/4 No. 8	
House/Hall/Bath	
½" EMT w/2 No. 12	
½" EMT w/3 No. 12	
½" EMT w/4 No. 12	
½" EMT w/5 No. 12	

3. Determine the bill-of-material for the lighting fixtures.

Lighting Fixture	Quantity Fixture	Install Labor	Box 4" × 4"	Ring 1 Gang	Connector ½"
Type A	5				
Type B	18				
Type C	3				
Type D	2				
Type E	4				
Type F	11				
Type G	3				
Type H	9				
Type I	6				
Type J	16				
Type K	2				
Type L	4				
Sign	3				
Total		65	86	86	172

4. Determine the bill-of-material for the switches.

	Qnty.	Switch				Plate			Box		Ring			Conn.
		1 Pole	3 Way	Dim.	2 Pole	1G	2G	3G	4"x4"	3G	1G	2G	3G	1/2"
Single Pole Switch	21													
Switch 3 Way	2													
Switch (Dimmer)	3													
Switch 2 Pole 30 Amp	2													
Box														
1 Gang Box With Ring	21													
2 Gang Box With Ring	2													
3 Gang Box With Ring	1													
Total		21	2	3	2	28	2	1	30	1	28	2	1	62

5. Determine the bill-of-material for the receptacles.

	Qnty.	Receptacle				Plate			Box		Ring		Conn.
		5-15R	GFCI	5-20R	Special	1G	GFCI	3G	4"x4"	3G	1G	3G	1/2"
NEMA 5-15R	28												
NEMA 5-15R GFCI (Bath)	2												
NEMA 5-20R	5												
NEMA 5-20R Hospital Grade	9												
NEMA 5-15R Floor Outlet	4												
NEMA 14-30R	1												
Total		28	2	5	14	36	2	3	36	3	36	3	86

6. Determine the raceway and wire requirements for the branch circuits, including homeruns.

	Qnty.	EMT	Couplings	Wire (+ 10%)		
		1/2"	1/2"	No. 12	No. 10	No. 8
Branch Circuits						
Insurance Office						
1/2" EMT w/2 No. 12	333					
1/2" EMT w/3 No. 12	488					
1/2" EMT w/4 No. 12	76					
1/2" EMT w/3 No. 8	21					
Beauty Salon						
1/2" EMT w/2 No. 12	299					
1/2" EMT w/3 No. 12	66					
1/2" EMT w/4 No. 12	28					
1/2" EMT w/2 No. 10	15					
1/2" EMT w/3 No. 10	30					
1/2" EMT w/5 No. 10	32					
Doctor's Office						
1/2" EMT w/2 No. 12	226					
1/2" EMT w/3 No. 12	28					
1/2" EMT w/4 No. 12	81					
1/2" EMT w/5 No. 12	17					
1/2" EMT w/2 No. 10	2					
1/2" EMT w/4 No. 8	26					
House/Hall/Bath						
1/2" EMT w/2 No. 12	204					
1/2" EMT w/3 No. 12	84					
1/2" EMT w/4 No. 12	29					
1/2" EMT w/5 No. 12	13					
Total		2,098	211	5,376	312	183

7. Determine the total material cost and labor-hours for the following bill-of-material.

	Qnty.	Material			Labor		
		Cost	Unit	Extension	Hours	Unit	Extension
Lighting Fixtures							
Lighting Fixture Labor	1						
4" × 4" Box	86						
Ring - Square Round	86						
1/2" EMT Connector	172						
Switches							
Single Pole Switch	21						
3 Way Switch	2						
Dimmer	3						
Switch 2 Pole 30 Ampere	2						
Switch Plate 1 Gang	28						
Switch Plate 2 Gang	2						
Switch Plate 3 Gang	1						
4" × 4" Box Metal	30						
3 Gang Box	1						
1 Gang Ring	28						
2 Gang Ring	2						
3 Gang Ring	1						
1/2" EMT Connector	62						
Receptacles							
NEMA 5-15R (15A)	28						
NEMA 5-15R GFCI (15A)	2						
NEMA 5-20R (20A)	5						
Receptacle Plate 1 Gang	36						
Receptacle Plate GFCI	2						
Receptacle Plate 3 Gang	3						
4" × 4" Box	36						
3 Gang Box	3						
1 Gang Ring	36						
3 Gang Ring	3						
1/2" EMT Connector	86						
Branch Circuit And Home Run Wiring							
1/2" EMT	2,098						
1/2" EMT Coupling	211						
No. 12	5,376						
No. 10	312						
No. 8	183						
Total				$1,040.25			198.46

8. Determine the total material cost and labor-hours for the following bill-of-material.

	Qnty.	Material			Labor		
		Cost	Unit	Extension	Hours	Unit	Extension
Service Equipment							
600 Ampere Main Disconnect	1						
3" PVC	60						
3" PVC 90	4						
3" PVC Coupling	10						
3" PVC Bell End	2						
3" PVC Male Adapter	2						
3" Locknut	2						
3" Plastic Bushing	2						
Gutter 8" × 8" × 6'	1						
Grounding And Bonding							
Ground Rod - ½" Copper	2						
Ground Clamp - Direct Burial	2						
No. 6 Copper Wire (Bare)	15						
½" PVC	10						
½" PVC Male Adapter	2						
½" Locknut	2						
No. 2/0 Copper Wire	10						
1" Ground Clamp - Water Pipe	1						
Panel A (House Panel)							
Panel 60A 3 Phase, 4 Wire	3						
¾" EMT	30						
¾" EMT Coupling	3						
¾" EMT Connector	2						
No. 6 Wire	132						
Total				$264.63			20.63

9. Complete the Estimate/Bid Summary Worksheet and determine the selling price for blueprint CE-1, based on the following specifications.

	Qnty.	Material			Labor		
		Cost	Unit	Extension	Hours	Unit	Extension
Panel B (Beauty Salon)							
Panel 100A, 3 Ph, 4 Wire	1						
1¼" EMT	40						
1¼" EMT Coupling	4						
1¼" EMT Connector	2						
1¼" EMT Plastic Bushing	2						
No. 3 Wire	176						
Disconnect 30A, 3 Ph, NEMA 3R	1						
Panel C (Doctors Office)							
Panel 125A, 3 Phase, 4 Wire	1						
1¼" EMT	70						
1¼" EMT Coupling	7						
1¼" EMT Connector	2						
1¼" Plastic Bushing	2						
No. 2 Wire	231						
Disconnect 30A, 1 Ph, NEMA 3R	1						
Panel D (Insurance Office)							
Panel 150A 3 Ph, 4 Wire	3						
1½" EMT	20						
1½" EMT Coupling	2						
1½" EMT Connector	2						
1½" Plastic Bushing	2						
No. 1 Wire	66						
No. 2 Wire	22						
No. 8 Wire	22						
Disconnect 60A, 3 Ph, NEMA 3R	1						
Total				$316.70			26.14

10. Determine the total adjusted labor hours for the following.

Total Adjusted Labor Hours	
Labor-Hours - Price/Labor Worksheets	245.00
Labor-Unit Adjustment, +8%	
Total Adjusted Labor-Hours (Rounded)	

11. Determine the total labor cost based on question 10's total adjusted labor hours.

    Labor Cost = Labor Hours × Labor Rate

    Labor Cost = _____ hours × $10.65

    Labor Cost (Rounded) = $ _____

12. Determine the total material cost for the following.

Material Cost Calculation	
Material Cost - Price Worksheets	$1,622.00
Miscellaneous Materials, + 10%	$
Theft And Waste, + 5%	$
Small Tools, + 3%	$
Quotes	$2,000.00
Total Taxable Materials (Rounded)	$3,914.00
Sales Tax, + 7%	$
Total Material Cost (Rounded)	$

13. Determine the estimated prime cost based on questions 10 through 12.

Estimated Prime Cost	
Labor Cost - (265 hours)	$2,822
Material Cost Including Quotes And Tax	$
Other Direct Cost - Permit & Temporary	$1,000
Estimated Prime Cost	$

14. Determine the overhead not to exceed $13.00 per labor hour.

Overhead Calculation	
Estimated Prime Cost	$8,010
Overhead = $13.00 × 265 hours =	$
Break Even	$

15. Determine the bid price based on question 14.

Total Bid Price	
Estimated Break Even Cost	$11,455
Profit, + 15%	$
Other Cost: None	$0 0
Total Bid Price (Rounded)	$

16. Determine the percentage distribution for the following.

    Labor $2,822/$13,173 = _____ %
    Material $4,188/$13,173 = _____ %
    Direct Cost $1,000/$13,173 = _____ %
    Overhead $3,445/$13,173 = _____ %
    Profit $1,718/$13,173 = _____ %

17. Determine the cost per square foot, based on question 16.

    Cost per square foot = $13,173/2,580 sq. ft.

    Cost per square foot = $_____ per square foot

18. Complete the following labor analysis.

    Labor Analysis

    Total hours = _____ hours

    Total 8 hr. days (3 men)

    (265 hrs./24 hrs) = _____ days

    Total 5 day weeks

    (11.04 days/5 days) = _____ weeks

Chapter 8

Computer Estimating

INTRODUCTION

Today's computers come in a wide range of capabilities, speeds, sizes, and prices. Their use is common in all types of businesses for estimating, billing, word processing, accounting, and a wide variety of other purposes. Advances in personal computer design and software has improved the estimating process to permit increased bid accuracy, improved project management and the bottom line.

8.01 COMPUTERIZED ESTIMATING

A leading electrical contracting publication conducted a national survey of electrical contractors and learned that the number one function that computers were used for by electrical contractors was estimating and bid preparation. This was considered more important than general ledger, accounts payable, payroll, or job costing. It comes as no surprise that the second choice is material price updating.

Advantages And Benefits

Computer assisted estimating is more complete and more accurate than manual estimating and can be completed in less than one-fourth the time. This translates into additional time for other important functions, such as more time with the family, more time to better organize and manage your company, or time to estimate additional jobs for business expansion.

Estimating

A computer does thousands of mathematical computations in a fraction of a second, never makes an error, never gets tired or careless, and never forgets the information that is stored. Computer estimating will reduce your estimating time and cost because you no longer need to price, labor, extend, or total material or labor.

Reduce Material Cost

Once the take-off is completed and the take-off quantities have been entered into the computer, a report of all the materials required for the job can be printed and broken down by job phase. This list can be submitted to multiple suppliers in order to get competitive prices and fixed delivery dates. In addition you can reduce the storage requirements for material, waste and theft.

Labor Savings

A computer estimate permits the electricians on the job to have a labor budget to insure that the job gets completed in a timely fashion. Material will be on the job when needed and idle time and inventory handling will be reduced.

Reduces Overhead

Overhead will be decreased because of reduced estimating, inventory storage space, and financing costs. You'll also improve billing, expedite collections and increase cash flow.

Increase Business Volume

The saving in estimating time will permit you to estimate more jobs to expand your business volume.

Other Benefits

A computer estimate provides you with increased confidence and security that your bid price is correct, and you'll communicate a higher level of professionalism. This all adds up to improved competitiveness with increased profit margins.

Can I Afford It?

In today's world, it's highly unlikely that you can be competitive if you estimate without a computer. The question is not can you afford a computer, but, can you afford not to use a computer to assist in your estimates.

To determine the dollars required in sales to cover the purchase of computer estimating software, use the following formula:

Sales to cover overhead expenses = Estimating system/gross profit percent

> **Example.** Let's assume the following: You have a three-person shop and you're thinking of purchasing a computer estimating system. Cost including software and training is $6,000, and your accountant indicates that your gross profit is 30%.
>
> **Answer:** Sales = $6,000/.3 = $20,000

To pay for the software you only need to get $20,000 in additional sales over the life of the software!

8.02 FREQUENTLY ASKED QUESTIONS

Must I Be A Trained Computer Operator?

No. Most software vendors assume that you have no computer experience and design their software to be easy to learn and use. All you need to do is follow the commands on the screen to get the results you want. In fact, after a brief training period, a clerical employee can use the take-off and complete the estimate for you. This leaves you more free time to start another bid, or devote yourself to other responsibilities.

Must I Still Do The Take-Off?

Yes. However with a computer estimating system, you have two methods of performing the take-off. The manual method or the direct input method.

Manual Take-Off – This method requires two steps. The first step is the manual take-off where you pencil the information on to paper. The second step is the input of the take-off into the computer.

Direct Take-Off – With the proper computer estimating system, you can input the take-off directly into the computer. Saving this step permits the estimate to be completed more quickly and accurately.

How Much Time Can I Save?

It depends on the software, the estimator, and the complexity of the job. As the estimator becomes more familiar with the system, less time is required. The more complicated the job, the more time saved. As a general rule, you should be able to complete the estimate in one-fourth the time it would take you to estimate the job manually.

How Will My Bid Accuracy Be Improved?

A computer doesn't make mistakes when it is tired or overworked. It doesn't forget the data it has stored when distractions occur. It doesn't omit steps in calculation and it doesn't make errors in overlooking taxes, overhead, profit.

Will My Estimates Be As Complete?

Yes, with more information to increase your efficiency and impress your customers.

Must I Change My Methods Of Estimating?

No, the estimating system should be flexible; it adapts to your estimating style, but don't expect a computer estimating software program to fit all of your needs.

Can I Realistically Expect To Increase Profits?

Without question. You won't lose a job because of an error, nor will your bid be too low through error or omission. Getting more jobs, reducing estimating time and financing costs, and improved project management translate into a brighter profit picture.

How Big Must I Be To Derive Any Benefit?

Size doesn't matter. Any contractor who estimates needs to have a computer to increase business efficiency. Many small companies, where the owner or manager does the estimating, can use computerized estimating to free more time for administration, selling, and project management. The professional appearance of a computerized estimate gives the impression of significance in size and standing, and will help increase the percentage of successful bids.

How Long Will It Take Me To Learn How To Use A Computer For Estimating?

It depends on your experience with computers and estimating. Those that have lots of experience will be estimating in a matter of days or even hours, without attending a training class! Those with less experience require a training class and a few weeks to get comfortable with the process.

8.03 TRAINING AND SUPPORT

Even the most carefully designed software requires training, service, support and technical assistance. These requirements are even more important when the software is highly specialized, or requires special skills, experience, and background in the field for which the software was designed.

Specialized software systems should be purchased directly from the software developer, who is equipped to explain the product, provide the training and support if you run into difficulty. See Section 8.08.

Note. If you attempt to use your software without proper training, you may never learn all of the valuable features that are designed into the program.

8.05 PRICING SERVICES

Many companies have contracted with pricing services to save office time and to gain assurance they have current material prices. If you do decide to use a pricing service, be sure you select one whose prices are broken down by geographical area, so that they will more accurately reflect those where your job is located.

8.06 SOFTWARE SELECTION

Don't expect a computer estimating software program to fit all of your needs. However, a quality estimating system should not require you to make any major adjustments in your estimate style. As a matter of fact the computer assisted estimate should considerably improve your estimating technique.

Regardless of how well an estimating system is designed, don't expect optimum results without complete training. Along the same lines, find out what kind of technical support is available and the annual technical support cost. Also determine the cost of new versions and upgrades.

Before Purchasing

Some things to consider before purchasing:

Simple to Use

The software should be logical, intuitive, simple to use and easy to understand. The commands must be precise and there should not be any danger of losing data. There should be an on screen audit trail to review and modify the take-off at any time.

Portability

A software package must have the versatility for you to use it at the office, at home, or even in your car with a notebook or possibly a palmtop computer.

Flexibility

The software should have the ability to factor labor or material cost for every line of the take-off to reflect the diverse installation conditions. You should be able to view or change anything in the estimate at any point without the requirement of a printer.

Reports

The software should print a wide variety of reports that furnish information to the client, as well information for project management decisions.

8.07 HOW MUCH SHOULD IT COST?

The cost of computerizing a business is dependent upon whether or not you already have a computer. Make your decision on the value you receive for your investment. Be sure you can count on long-range service and close personal attention. A quality estimating and management software system generally cost between $4,000 and $8,000.

8.08 WHO SELLS ESTIMATING SOFTWARE?

You can find estimating software in computer stores. BUT, and this is a big but, it's not a good idea unless you are already familiar with the software in question. For the novice, or relative novice, what the computer store does not sell along with the software is the person with the technical knowledge necessary to install it, boot it up, and run it correctly and at its fullest capacity.

Think of the nightmares if you purchase this way. Say you have a problem with the software at ten o'clock at night—who do you call? A vendor is really the only way to go.

Find out about as much as you can about the software vendor. How long have they been in the business of selling software? How many customers do they have? What other software products do they sell? Don't get too excited, take your time to investigate the different vendors and make a selection based on facts, not opinions. If possible, see if you can have a trial period to try the software. Naturally this will cost you a few hundred dollars, so insist on a money back guarantee.

To locate those who sell estimating software, you can contact Mike Holt Enterprises, Inc. at 1-888-NEC CODE, search the Internet, or check out the advertisements in *Electrical Contractor* magazine. You can receive a free subscription by contacting the National Electrical Contractor's Association in Bethesda Maryland at 1-301-657-3110.

Chapter 8

Review Questions

Essay Questions

1. Explain the number one function that computers are used for by electrical contractors. _____.

2. In what way does computer estimating reduce estimating time and cost?

3. How can computer estimating reports be used to reduce material cost?

4. How can computer estimating reports be used to reduce labor cost?

5. How can computer estimating help in the reduction of overhead expenses?

6. How does computer estimating help in expanding business volume?

7. Name some other values that computer estimates and their associated reports can do to help improve profits.

8. Do you have to be a computer operator or programmer to operate computer estimating software?

9. Can computer estimating software eliminate the take-off?

10. How will my bid accuracy be improved?

11. Must I change my methods of estimating?

12. Can I realistically expect to increase profits?

13. How big must I be to derive any benefit?

14. How long will it take me to learn how to use a computer for estimating?

15. Why should estimating software be purchased directly from the software developer?

16. If you do decide to use a pricing service, why should you select one whose prices are broken down by geographical area?

17. What are some things that you should consider before purchasing an estimating and management system?

Multiple Choice Questions

1. A computer assisted estimating is more complete and more accurate than manual estimating and can be completed in less than _____ of the time as compared to a manual estimate.
    (a) ¼
    (b) ⅓
    (c) ½
    (d) none of these

2. Assume the following: You have a five-person shop and you're thinking of purchasing a computer estimating and management system. Cost including software and training is $4,500, and your accountant indicates that your gross profit is 35%. To pay for the software you need to get _____ in additional sales over the life of the software.
    (a) $10,000
    (b) $13,000
    (c) $17,000
    (d) $21,000

3. A quality estimating and management software system generally costs approximately _____.
    (a) between $500 – $1,000
    (b) between $1,500 – $3,000
    (c) between $3,500 – $6,000
    (d) none of these

Appendix

Estimating Forms And Worksheets

INTRODUCTION

This Appendix contains samples of typical estimating forms.

ESTIMATE RECORD SHEET, PAGE 255

The Estimate Record Sheet is used to list pertinent job information such as job name, job location and address, names of the owner, contractor, architect, engineer, and to whom the bid is to be submitted.

Other information that should be included is the telephone company and the electric utility contact names and phone numbers. Once you have completed the job information sheet, hang it up on the wall over your take-off bench for handy reference. This information will be handy when creating a bid proposal.

ESTIMATE AND BID NOTES, PAGE 256

Keep a notepad handy so you can keep track of any questions you have about the estimate. If there is anything you don't understand, get the answer as soon as possible. Don't wait until the last moment.

SPECIFICATION CHECK LIST, PAGE 257

In an effort to keep track of the blueprint and specification details, you will want to create a Specification Check List and hang it up over the take-off bench for quick reference while you do the estimate.

TAKE-OFF WORKSHEET, PAGE 258

The Take-Off Sheet is used to list the description and quantities of the symbols on the blueprint. A take-off is the action of counting and measuring all material required to install a complete electrical system.

BILL OF MATERIAL SPREADSHEETS, PAGES 259—260

This Appendix contains two Bill-Of-Material Spreadsheets, one seven columns wide and the other fourteen columns wide. Once you have completed the take-off, you need to determine the bill-of-material for the job. This is the process of converting the take-off symbols and quantities to determine the bill-of-material. If you are estimating without the use of a computer, you must determine the bill-of-material in the most efficient manner. Some parts of the take-off will lend themselves to the use of a spreadsheet, which helps you determine the quantity of the most common items. Use whichever of the two sheets (the seven or the fourteen) which best fits your needs.

PRICING/LABOR WORKSHEET, PAGE 261

Pricing consists of locating the material cost for each item in your price catalog, and pencilling the value to the price/labor sheet. In addition, you must be sure to also indicate the unit of measure that the cost reflects, such as E, C, or M. Laboring consists of locating the material labor-hours in your labor-unit manual for each item you will be using and transferring that value to the price/labor sheet.

Once you have penciled in the material cost and the labor-hour value for each item, you can then determine the material cost and labor-hour extension for each item. When you know the material cost and labor-hour extension for each item, you can determine the totals for each worksheet.

QUOTE WORKSHEET, PAGE 262

Once you have determined the quantities for each lighting fixture type, complete the Quote Worksheet and fax this, with a copy of the lighting fixture schedule, to at least three suppliers for pricing.

As you take off the circuits and service equipment, determine the items that require special pricing. Once you have determined the quantities, complete the Quote Worksheet and fax this to at least three suppliers for pricing.

ESTIMATE SUMMARY WORKSHEET, PAGE 263

Summarizing the cost to determine the selling price must be completely understood to insure that you provide your customer with an accurate bid. This phase of the estimating process requires you to make judgments on intangibles: job conditions, labor productivity, miscellaneous material requirements, consideration of small tools, direct job expenses, overhead, job risk, and what you anticipate the market will permit for profit.

ESTIMATE RECORD WORKSHEET

Job Information Detail   Estimate Job Number:

- Job Name: _____   Bid Due Date: _____
- Job Location: _____   City: _____
- Job Owner: _____   Phone: _____

Plans And Specifications Detail

Date Of Plans: _____   Blueprint Page Numbers: _____

Comments: _____

Contractor Information Detail

- Contractor: _____   Contact: _____
- Address: _____   City: _____
                                              E-mail: _____
- Phone: _____   Mobil: _____
- Beeper: _____   FAX: _____

Important Contacts

- Architect: _____   Phone: _____
- Engineer: _____   Phone: _____
- Telephone Utility: _____   Phone: _____
- Electric Utility: _____   Phone: _____
- Electrical Inspector: _____   Phone: _____

Notes: _____
_____
_____
_____
_____

Estimate And Bid Notes Page ___ of ___

Blueprint Questions or Comments:

Specification Questions or Comments:

Specification Check List

Labor-Unit Adjustment (See Section 3.09)
1. Building Conditions _____
2. Change Orders _____
3. Concealed And Exposed Wiring _____
4. Construction Schedule _____
5. Job Factors _____
6. Labor Skill (Efficiency) _____
7. Ladder and Scaffold _____
8. Management _____
9. Material (Miscellaneous) _____
10. Off Hours And Occupied _____
11. Overtime _____
12. Remodel (Old Work) _____
13. Repetitive Factor _____
14. Restrictive Working Conditions _____
15. Shift Work _____
16. Teamwork _____
17. Temperature _____
18. Weather And Humidity _____

Labor Adjustment

Additional Labor (See Section 5.01)
1. As-Built Plans _____
2. Demolition _____
3. Energized Parts _____
4. Environmental Hazardous _____
5. Excavation, Trenching And Backfill _____
6. Fire Seals _____
7. Job Location _____
8. Match-Up Of Existing Equipment _____
9. Miscellaneous Material Items _____
10. Mobilization (Startup) _____
11. Nonproductive Labor _____
12. OSHA Compliance _____
13. Plans and Specifications _____
14. Public Safety _____
15. Security _____
16. Shop Time _____
17. Site Conditions _____
18. Sub-Contract Supervision _____
19. Temporary, Stand-By Power _____

Hour Adder

Direct Job Expenses (See Section 5.05)
1. As-Built Plans $_____
2. Bus. & Occupational Fees $_____
3. Engineering Drawings $_____
4. Equipment Rental $_____
5. Field Office Expenses $_____
6. Fire Seals $_____
7. Guarantee $_____
8. Insurance – Special $_____
9. Miscellaneous, Labels $_____
10. Mobilization $_____
11. OSHA Compliance $_____
12. Out of Town Expenses $_____
13. Parking Fees $_____
14. Permits/Inspection Fees $_____
15. Public Safety $_____
16. Recycle Fees $_____
17. Storage & Handling $_____
18. Sub Contract: _____ $_____
19. Supervision Cost $_____
20. Temporary Wiring:
    Lighting $_____
    Power $_____
    Maintenance $_____
21. Testing $_____
22. Trash $_____
23. Utility Cost $_____

Total Direct Cost

Other Final Cost
1. Allowances $_____
2. Back-Charges $_____
3. Bond $_____
4. Completion Penalty $_____
5. Finance Cost $_____
6. Gross Receipts Or Net Profit Tax $_____
7. Inspector Problems $_____
8. Retainage $_____

Total Other Cost

Other Considerations
1. Conductor Size - Minimum? _____
2. Raceway Size - Minimum? _____
3. Control Wiring Responsibility? _____
4. Cutting Responsibility? _____
5. Demolition Responsibility? _____
6. Excavation/Back Fill Responsibility? _____
7. Painting Responsibility? _____
8. Patching Responsibility? _____
9. Special Equipment? _____
10. Specification Grade Devices Or Fittings? _____

TAKE-OFF WORKSHEET

Description				

BILL-OF-MATERIAL SPREADSHEET

Description	Quantity							
Totals								

BILL-OF-MATERIAL SPREADSHEET

Description	Quantity																											Totals

PRICING/LABORING WORKSHEET

Description	Quantity	Cost	Unit	Extension	Labor	Unit	Extension
Totals							

QUOTE WORKSHEET

Type	Description	Qnty.	Supplier-1		Supplier-2		Supplier-3	
			Unit	Extension	Unit	Extension	Unit	Extension
Totals								

Notes:

ESTIMATE SUMMARY WORKSHEET

	Description		
	Labor Calculations	Hours	
	Labor Worksheet Page 1 of 6		
	Labor Worksheet Page 2 of 6		
	Labor Worksheet Page 3 of 6		
	Labor Worksheet Page 4 of 6		
	Labor Worksheet Page 5 of 6		
	Labor Worksheet Page 6 of 6		
	Price/labor Worksheet Labor-Hours		
	Labor Unit Adjustment, _____		
	Total Adjusted Hours _____		
	Additional Labor Hours: _____		
Step A	Total Final Adjusted Hours _____		
Step B	Labor Cost - $_____ per hour × _____ Hours		$
	Material Cost	Dollars	
	Price Worksheet Page 1 of 6	$	
	Price Worksheet Page 2 of 6	$	
	Price Worksheet Page 3 of 6	$	
	Price Worksheet Page 4 of 6	$	
	Price Worksheet Page 5 of 6	$	
	Price Worksheet Page 6 of 6	$	
	Pricing Worksheet Material Cost	$	
	Miscellaneous Material Items, plus ___%	$	
	Waste and Theft, plus ____%	$	
	Small Tool Allowance, plus ____%	$	
Step C	Total Adjusted Material Cost	$	
Step D	Quote – Gear	$	
	Total Taxable Material Cost	$	
Step E	Sales Tax at ____%	$	
	Total Material Cost		$
Step F	Direct Cost:		$
Step G	Total Prime Cost		$
Step H	Overhead:		$
Step I	Break Even Cost		$
Step J	Profit at ____% of $_____		$
Step K	Other Final Cost:		$
Step L	Bid Selling Price		$

Index

A

Accurate, 8, 11, 15-16, 19-21, 23-25, 27-28, 53, 69, 73, 78-79, 87, 247, 252

Additional Labor, 68-69, 71, 85, 88

Adjusted Material Cost, 84, 139-141

Allowances, 79, 86

Analysis, 20-21, 23, 25, 35, 41, 78, 80, 87, 90, 142-143, 146, 188, 236, 246

As-Built Plans, 69, 73, 85-86

Average Run Length, 115-116, 220

B

Back-Charges, 79, 86

Ball-Park Price, 19, 25, 27

Bid Accuracy, 15-16, 20-21, 23-25, 79, 87, 142-143, 146, 236, 247-248, 251

Bid Accuracy And Analysis, 142-143, 236

Bid Analysis, 20-21, 23, 25, 35, 41, 78, 80, 87, 142-143, 146, 236

Bid Preparation, 143, 236, 247

Bid Preparation and Take-Off, 143, 236

Bid Price, 7-8, 19, 21, 25, 67-68, 72, 78-79, 140, 142-143, 188, 236, 246, 248

Bid Process, 7, 20, 25, 45, 67-87, 89-93

Bid Proposal, 20-21, 25, 48, 67, 72, 80

Bid Summary, 90, 139, 143, 146, 236, 244

Bid Summary Analysis, 146, 236

Bidding, 7, 9, 11, 13, 21, 27, 45, 47-48, 74, 76-78, 83, 87

Bill-Of-Material, 45, 48, 53, 56-57, 62-65, 67, 70, 72, 83, 87, 143, 181-182, 236, 239-240, 242-243

Billing software, 17

Bonds, 79, 140

Break Even Cost, 45, 67-68, 77-78, 84, 86, 88, 91-93, 140-141, 188, 246

Business And Occupational Fees, 73, 86

Buy Price, 9

C

Change Orders, 19-20, 28, 35-36, 69, 73, 80-83, 177-178

Commercial Wiring, 189-213, 236-237, 239-247

Competition, 7-11, 13, 78

Competitive, 8-11, 13, 16, 18, 23, 25, 27-28, 34-35, 47, 78, 247-248

Competitive Bid, 18, 25, 28

Competitors' Labor-Units, 29, 34, 41, 43

Completion Penalty, 79, 86

Computer, 15-18, 20-28, 34, 50, 55-57, 70, 72, 80, 87, 143, 146, 189, 209, 236, 247-252

Computer Assisted Estimate And Bid, 143, 236

Computer Assisted Method, 22-23

Computerized Estimating, 247, 249-250

Computers, 23, 34, 41, 247-252

Contractors' Organization, 10

Cost Of Material, 9, 45, 62, 72, 77, 140

Counting Sample, 53

Counting Symbols, 53, 62, 64

D

Demolition, 69, 85

Design Build, 18, 25, 28

Detailed Method, 20, 25

Detailed Summary Analysis, 146, 236

Determining The Bill-Of-Material, 56, 63, 67
Direct Cost, 20, 80, 83, 87-89, 140-141, 187-188, 245-246
Direct Job Expenses, 7, 19, 25, 67-68, 73, 75, 86, 88, 90, 139-140
Draw Schedule, 146, 237

E

E-Mail, 16-18, 23, 82
Electrical Contractor, 8-9, 11, 13, 18-19, 34, 46, 50, 74-76, 80-83, 177-178, 198, 200-202, 204, 206, 211-213, 250
Electrical Specifications, 189, 196, 198
Energized Parts, 69-70, 85
Environmental Hazardous Material, 69-70, 85
Equipment Rental, 67, 73, 86, 142, 146
Estimate, 7-8, 10-13, 15-16, 18-27, 29-30, 34-36, 40, 42, 45-65, 67-93, 139-143, 146, 179, 189, 213, 236, 238, 244, 247-250, 252
Estimate And Bid Notes, 50, 139
Estimate And Bid Process, 45
Estimate and Bid Summary, 139, 143, 236
Estimate Summary, 20-21, 23, 25, 45, 67-84, 86-93, 140, 213, 236
Estimated Prime Cost, 68, 75-78, 85-86, 139-140, 187, 245
Estimating, 7-65, 67-87, 89-93, 139-178, 180-213, 236-237, 239-252
Estimating And Bid Process, 20, 25
Estimating Forms And Worksheets, 50, 62, 64
Estimating Methods, 15, 19-20, 25
Estimating Service, 15, 22-24, 26-28
Estimating System, 8, 11, 13, 24, 55, 248-249
Estimating Tools, 16, 24, 27
Estimator, 8, 15-16, 22, 24, 27-28, 38, 248
Excavation, Trenching, And Backfill, 69-70, 85
Experience, 8-11, 18, 20-23, 25-30, 33-35, 37-38, 40-43, 69, 85, 143, 236, 248-249
Extending, 20-22, 25, 45, 57, 62, 67, 143, 236
Extending And Totaling, 20-22, 25, 45, 57, 62, 67, 143, 236

Extensions And Totals, 57, 61, 63

F

Field Office, 73, 86
Final Costs, 79, 86, 140
Finance Cost, 79, 86, 140
Financial Resources, 47
Fire Seals, 69-70, 73-74, 85-86
Fixed Fee Proposal, 19
Frequently Asked Questions, 248

G

Gross Receipts, 50, 79, 86
Guarantee, 73-74, 79, 86, 250
Gut Feelings, 47

H

How To Develop, 19, 29, 35, 41

I

Input Take-Off Into Computer, 143, 236
Insurance, 72-74, 80, 85-86, 89, 201, 207, 209, 212, 239, 241, 244
Internet, 16, 18, 23, 250
Internet E-Mail, 16

J

Job Average Labor Rate, 72, 89
Job Location, 37, 44, 48, 69-70, 85, 196
Job Proposal, 67, 74, 78, 146, 236
Job Risk, 9, 14
Job Selection, 46, 62

L

Labor Analysis, 80, 142, 146, 188, 236, 246
Labor Burden, 20, 32, 67, 72, 77, 85-86, 89, 139
Labor Cost, 7, 9, 11, 19-20, 25, 67-68, 71-73, 75, 77, 80, 84-86, 88-89, 91-93, 139-141, 146, 187, 236, 242-243, 245, 251
Labor Cost And Productivity, 9, 11
Labor-Hour Extension, 57, 60, 63, 65

Labor-Hours, 15, 30, 35-36, 40, 45, 60-63, 68-71, 76-77, 84-85, 88-89, 139, 141, 146, 183, 187, 236, 242-243, 245

Labor-Unit, 29-30, 32-35, 37, 40-44, 60, 69, 73, 80, 83, 85, 87-88, 90, 139, 142, 187, 245

Labor-Unit Competitor, 33

Labor-Unit Manuals, 33, 41

Labor-Units Are Expressed, 30, 40, 43

Labor-Units Do Not Include, 33, 41

Laboring, 20-22, 25, 45, 56-57, 60-63, 65, 67, 141, 143, 183-187, 236

Ladder And Scaffold, 35, 37, 69

Lighting Fixture And Switchgear Quotes, 23, 72

M

Management, 7-11, 14-16, 20, 22-24, 27-28, 30, 33, 35-38, 40, 43-44, 69, 73, 77-78, 86, 90, 247, 249-252

Manual Estimates, 22

Market, 7-10, 12-14, 16, 21, 78

Match-Up Of Existing Equipment, 69-70, 85

Material Cost Analysis, 146, 236

Material Cost Extension, 57

Material Quantity Analysis, 146, 236

Measuring Circuits, 53, 63

Measuring Sample, 55

Meeting The Lowest Bid, 19

Mike Holt Enterprises, 177, 250

Miscellaneous Material, 68-70, 72-74, 85, 89, 139-141, 245

Miscellaneous Material Items, 68-70, 72-74, 85, 89, 140-141

Mobilization, 69-70, 73-74, 85-86

N

Negotiated Work, 18-19, 25

Net Profit Tax, 79, 86

Nonproductive Labor, 32, 44, 69-71, 85

O

OSHA Compliance, 69-70, 73-74, 85-86

Other Final Costs, 79, 86, 140

Out Of Town Expenses, 73-74, 86

Overhead, 7, 9-11, 14-15, 19-21, 23, 25, 43, 45, 65, 67-68, 72-73, 75-78, 80, 83-93, 139-143, 146, 177, 187-188, 236, 245-246, 248, 251

Overhead Applying, 77

Overhead Calculation, 76, 140, 187, 245

Overtime, 35-36, 38, 44, 50, 69, 78, 82, 178

P

Parking Fees, 73-74, 86

Part G, 77, 86

Pay Scale, 9, 37

Percentage Method, 76-77, 86, 88, 140

Permit And Inspection Fees, 74

Plans, 39, 48, 50, 53, 69, 71, 73, 82, 85-86, 177, 189, 201-202, 204, 206, 211-212

Plans And Specifications, 69, 71, 82, 85, 177, 189

Pre-Qualified, 18, 28

Pricing, 17-22, 25, 27, 30, 45, 56-57, 60-63, 65, 67-68, 72-73, 83, 85, 87, 140-141, 143, 183-187, 236, 249, 251

Pricing And Laboring, 20-21, 25, 45, 56-57, 60-63, 65

Pricing Services, 249

Profit, 7-11, 13-15, 19-21, 23, 25, 45-47, 62, 64-65, 67-68, 71-73, 78-80, 83-84, 86-93, 140-143, 146, 177, 188, 236, 246, 248-249, 252

Profit How Much Is Reasonable, 78

Project Management, 7-8, 11, 15-16, 20, 22-24, 28, 247, 249-250

Project Manager, 8, 11, 39, 75

Proposal, 19-21, 25, 45, 48, 67, 72, 74-75, 78, 80-82, 87, 89-90, 142, 146, 177-178, 236

Public Safety, 69, 71, 73-74, 85-86

Purchasing Material, 15-16, 24

Q

Quotes, 9, 19, 23, 25, 56, 72-73, 75, 77, 80, 88, 140, 142, 145-146, 187, 245

R

Rate-Per-Hour Method, 77
Recycling Fees, 74
Repetitive Factor, 35, 39, 69
Report No. 1, 146, 236
Report No. 2, 146, 236
Report No. 3, 146, 236
Report No. 4, 146, 236
Report No. 5, 146, 236
Report No. 6, 146, 236
Report No. 7, 146, 236
Report No. 8, 146, 237
Retainage Cost, 79

S

Sales Tax, 19-20, 25, 45, 68, 72-73, 85, 88-89, 140-141, 146, 187, 236, 245
Schedule, 19, 24, 30, 35-36, 44, 69, 79, 82, 146, 177-178, 203-205, 237
Security, 44, 56, 69, 71, 74-75, 85, 248
Selling Price, 7, 11, 19-21, 25, 45, 67, 77-79, 83, 141, 146, 236, 244
Selling The Job At Your Price, 9-11
Shift Work, 35, 39, 69
Shop Average Labor Rate, 71, 89
Shop Time, 69, 71, 85
Shot-In-The-Dark, 19
Site Conditions, 30, 43, 48, 67, 69, 71, 79, 85, 198
Small Tools, 67-68, 72-73, 85, 89, 139, 187, 245
Software, 16-17, 23, 247-252
Software Selection, 249
Specficiations Footnotes, 213
Specification Check List, 50, 62, 64, 139
Specifications, 18-19, 22, 25, 28, 32, 45, 48, 50, 53, 62, 65, 69, 71-75, 79-82, 85-86, 90, 140, 142, 146, 177-178, 189, 196, 198, 203, 244
Spreadsheet, 17, 56-57
Square Foot Method, 19-20, 25
Step 1, 21, 48, 62
Step 2, 21, 50, 62
Step 3, 21, 56, 62-63
Step 4, 21, 56, 62-63
Step 5, 21, 57, 62-63
Step A, 68-69, 85, 139, 141
Step B, 68, 71, 85, 139, 141
Step C, 68, 72, 85, 139, 141
Step D, 68, 73, 85-86, 139-141
Step E, 68, 75, 85-86, 139-141
Street Map, 17
Street Price, 9
Sub-Contract Cost, 74-75, 142, 146
Sub-Contract Supervision, 69, 71, 85
Summary Worksheet, 68-69, 72, 79, 85, 139-142, 244
Supervision Cost, 73, 75, 86
Support, 204, 210, 212, 249

T

Take-Off, 15, 17, 21, 28, 45, 48, 50, 53, 55-56, 62-65, 67, 80, 89, 143-146, 236, 247-251
Take-Off Into Computer, 143, 236
Temperature, 35, 39, 44, 69
Temporary, Stand-By, And Emergency Power, 71
Temporary Wiring, 70, 73, 75, 82, 86, 145, 198
Testing And Certification Fees, 75
Time And Material, 17-20, 25, 27
Tools, 8, 15-16, 24, 27, 29, 33-35, 37-41, 43, 47, 55, 67-68, 72-74, 85, 89, 139, 187, 212, 245
Total Labor-Hours, 61, 63, 89, 139
Total Material Cost, 61, 63, 68, 72-73, 85, 89, 140-141, 183, 187, 242-243, 245
Totaling, 20-22, 25, 45, 57, 62, 67, 143, 236
Training, 9-10, 34, 37-38, 43, 70, 74, 248-249, 252
Trash Disposal, 73, 75, 86
Types Of Bids, 18, 25

U

Understanding Break Even Cost, 77
Understanding The Scope Of Work, 20, 25, 48, 62
Unit Price, 19-20, 25, 28, 83-84, 87, 90-93, 146, 236

Unit Price Example, 83
Unit Pricing, 18-19, 27, 67, 83, 87
Utility Charges And Fees, 75

V
Variables, 35, 42-44

W
Waste And Theft, 67, 72, 89, 139-141, 187, 245, 247
Weather And Humidity, 35, 39
Who Sells Estimating Software, 250
Word Processing, 17, 247
Work Experience, 30, 40
Work Phase Analysis, 146, 236
Workspace, 15-16, 24, 27

Notes:

Notes:

Notes:

Mike Holt's Illustrated Guide

Electrical Estimating

Workbook

neccode.com
888-NEC® CODE

Chapter 1

Price And Labor-Unit Catalog Introduction

PRICING

The material items in this workbook are priced based on South Florida as of September 1996. They are included only as a sample, and should not be used for estimating.

To properly price the material cost for each item, you need a current pricing catalog. Pricing catalogs are available in both book form as well as electronically from:

Electrical Resources, Inc.
7169 University Blvd.
Winter Park, Florida 32792
1-407-657-7001, Fax 1-407-657-0559

Trade Service Corporation
10996 Torreyanna Road
San Diego, California 92121-1192
1-800-854-1527

Note. Keeping your prices current takes a lot of energy and time, but it must be done. With a computer, it's just a matter of inserting a disk into the computer and updating your prices in a few seconds.

ABOUT LABOR-UNITS

Labor-units represent labor productivity under typical job site conditions with reasonable working conditions. A labor-unit represents the approximate time required to install an electrical product, component or piece of equipment. Labor-units are based on the assumption that a highly trained, skilled, and motivated electrician is completing the task under standard installation conditions with the proper tools. See Chapter 3 for details about labor-units.

A labor-unit is comprised of six major components, they include:

1. Installation	50%
2. Job layout	15%
3. Material handling and clean up	10%
4. Nonproductive time	5%
5. Supervision	10%
6. Tool handling	10%

LABOR-UNIT ADJUSTMENTS

There is no set of labor-units that can be applied to all jobs; therefore, the labor must be adjusted up or down to accommodate the varying job conditions. Some of these variables can be controlled and others must be accommodated. It generally makes more sense to adjust the total labor or sub-total labor rather than each individual labor-unit for the varying conditions.

With proper project management and the recording of historical data from past jobs, you will be able to develop adjustment factors to reflect your experiences. See Section 3.09 for details.

ABOUT THE LABOR-UNITS IN THIS CHAPTER

This chapter contains labor-units for residential, commercial, as well as some industrial installations. These labor-units were developed by Mike Holt and have been used successfully by thousands of electrical contractors since 1978. These labor-units are generally lower than those in most other labor-unit manuals, and it is recommended that you do not decrease these values until you have historical data.

Author's Comment. Not all electrical contractors can complete the task according to these labor-units. Some electrical contractors will be more efficient and organized and complete the tasks in less time, others will struggle to complete the task within 25% of the allotted time.

Table Of Contents
Price And Labor-Unit Catalog

Pricing 1
About Labor-Units 1
Labor-Unit Adjustments 1
About The Labor-Units In
This Chapter 2
Armored Cable 5
Box - Metal 5
Box - Plastic 6
Box - Weatherproof 6
Box - 4¹¹/₁₆" 6
Breaker 6
Breaker W/Enclosure 7
Bushing - Bonding 7
Bushing - Plastic 7
Cords 8
Dimmers 8
Disconnect - 1 Phase 8
Disconnect - 3 Phase 8
EMT 8
EMT Connector S.S. 9
EMT Connector Comp. 9
EMT Coupling S.S. 9
EMT Coupling Comp. 9
EMT Elbow 10
EMT Strap 10
ENT And Fitting 10
Excavation - Trenching 10
Fixture 11
Fixture - Fluorescent 11
Fixture - Miscellaneous 11
Fixture - Pole 12
Fixture - Track 12
Flexible Metal Conduit 12
Flex - Straight Connector 12

Flex - 90 Connector 12
Fuse 13
Fuse - Edison 13
Generator 13
Ground Fitting 14
Ground - Miscellaneous 14
Ground Rod 14
Gutter 14
Gutter Fittings 14
Hub 15
IMC 15
Junction Box 15
Knock Out Seal (KO) 16
Lamp (Bulb) 16
Locknut 16
Locknut Seal 17
Low Voltage 17
Lugs 17
Meter Equipment 18
Miscellaneous Items 18
Motor Control 18
Nipple 19
Panel (Load Center) 20
Plate 20
PVC - Schedule 40 21
PVC - Bell End 21
PVC - Coupling 21
PVC Elbow 22
PVC - Expansion Coup. 22
PVC - Female Adapter 22
PVC - L.B. & Cover 22
PVC Male Adapter 23
Receptacle 23

Reducing Bushing 23	SE Cable & Fitting 28
Reducing Washer 23	Surface Raceway 28
Rigid Conduit 24	Surface Raceway Fitting 28
Rigid - Connector SS 24	Switch 29
Rigid - Coupling SS 24	Television 29
Rigid Elbow 24	Telephone 29
Rigid Conduit - LB 25	Time Clock 29
Rigid - LB Cover 25	Transfer Switch 30
Rigid Strap 25	Transformer 30
Ring - Plaster 26	UF Cable 30
RX Fitting 26	U Channel And Fitting 30
RX – NM Cable 26	Weatherhead 31
Screw And Nut 26	Wire - Copper 31
Sealtight 27	Wire - Miscellaneous 32
Sealtight - Straight Connector 27	Wire Nut 32
Sealtight - 90 Connector 27	
SE Cable 27	

	#	Item Description	Cost	Labor	Unit
ARMORED CABLE	3	ARMORED CABLE 14/2	$244.00	9.60	M - Thousand
	4	ARMORED CABLE 14/3	$318.00	14.50	M - Thousand
	5	ARMORED CABLE 14/4	$518.00	16.00	M - Thousand
	6	ARMORED CABLE 12/2	$268.00	12.00	M - Thousand
	7	ARMORED CABLE 12/3	$394.00	16.00	M - Thousand
	8	ARMORED CABLE 12/4	$630.00	18.00	M - Thousand
	9	ARMORED CABLE 10/2	$530.00	14.50	M - Thousand
	10	ARMORED CABLE 10/3	$692.00	19.25	M - Thousand
	11	ARMORED CABLE 10/4	$1,040.00	26.50	M - Thousand
	12	ARMORED CABLE 8/2	$1,006.00	16.00	M - Thousand
	13	ARMORED CABLE 8/3	$1,245.00	25.50	M - Thousand
	14	ARMORED CABLE 8/4	$1,774.00	28.00	M - Thousand
	15	ARMORED CABLE 6/2	$1,374.00	20.00	M - Thousand
	16	ARMORED CABLE 6/3	$1,844.00	28.00	M - Thousand
	17	ARMORED CABLE CONN 3/8"	$20.00	4.50	C - Hundred
	18	ARMORED CABLE CONN 1/2"	$28.00	4.50	C - Hundred
	21	ARMORED CABLE BUSHING	$3.00	18.00	C - Hundred
	22	METAL CLAD CABLE 14/2	$297.00	9.60	C - Hundred
	23	METAL CLAD CABLE 12/2	$322.00	12.00	C - Hundred
BOX - METAL	25	BOX MET 2" × 4" SHALLOW	$65.00	18.00	C - Hundred
	26	BOX MET 2" × 4" REGULAR	$59.00	18.00	C - Hundred
	27	BOX MET 2" × 4" DEEP	$115.00	18.00	C - Hundred
	28	BOX MET 2" × 4" WORK REG	$117.00	18.00	C - Hundred
	29	BOX MET 2" × 4" WORK DP	$113.00	18.00	C - Hundred
	30	BOX MET 4" × 4" SHALLOW	$65.00	18.00	C - Hundred
	31	BOX MET 4" × 4" REGULAR	$59.00	18.00	C - Hundred
	32	BOX MET 4" × 4" DEEP	$162.00	18.00	C - Hundred
	33	BOX MET 3 GANG 1 1/2"	$516.00	20.00	C - Hundred
	34	BOX MET 4 GANG 1 1/2"	$668.00	24.00	C - Hundred
	35	BOX MET 5 GANG 1 1/2"	$849.00	32.00	C - Hundred
	36	BOX MET 6 GANG 1 1/2"	$1,581.00	40.00	C - Hundred
	37	BOX MET 3-4 OCTAGON	$94.00	18.00	C - Hundred
	38	BOX MET 3-4 PANCAKE	$166.00	18.00	C - Hundred
	41	BOX MET FLOOR	$15.00	1.00	E - Each
	42	BOX MET OCTAGON BAR	$361.00	23.00	C - Hundred
	43	BOX MET FLOOR COVER	$911.00	20.00	C - Hundred

	#	Item Description	Cost	Labor	Unit
BOX - METAL Cont'd	45	BOX MET 2" x 4" AC REG	$107.00	20.00	C - Hundred
	46	BOX MET 4" x 4" AC REG	$232.00	18.00	C - Hundred
	47	BOX MET FAN - PANCAKE	$199.00	18.00	C - Hundred
	48	BOX MET FAN - 1½" DEEP	$441.00	18.00	C - Hundred
BOX PLASTIC	50	BOX PLASTIC 1 GANG	$97.00	10.00	C - Hundred
	51	BOX PLASTIC 2 GANG	$167.00	12.00	C - Hundred
	52	BOX PLASTIC 3 GANG	$174.00	16.00	C - Hundred
	53	BOX PLASTIC 4 GANG	$225.00	20.00	C - Hundred
	54	BOX PLASTIC 5 GANG	$440.00	28.00	C - Hundred
	55	BOX PLASTIC OCTAGON	$89.00	10.00	C - Hundred
	56	BOX PLASTIC OCT/BAR	$179.00	20.00	C - Hundred
	57	BOX PLASTIC FLOOR	$1,095.00	48.00	C - Hundred
	58	BOX PLASTIC TRIM	$2,299.00	22.00	C - Hundred
	59	BOX PLASTIC FAN	$399.00	18.00	C - Hundred
BOX-WEATHERPROOF	65	BOX WP 1 GANG ½" KO	$224.00	31.25	C - Hundred
	66	BOX WP 2 GANG ½" KO	$604.00	36.00	C - Hundred
	67	BOX WP 3 GANG ½" KO	$1,270.00	45.00	C - Hundred
	68	BOX WP 3 GANG ¾" KO	$1,674.00	45.00	C - Hundred
	69	BOX WP 3"-4" ROUND	$427.00	31.25	C - Hundred
	70	BOX WP 1W/KO POOL JB	$15.00	0.50	E - Each
	71	BOX PVC 1 GANG ½" KO	$342.00	31.25	C - Hundred
	72	BOX PVC 1 GANG ¾" KO	$359.00	36.00	C - Hundred
BOX 4¹¹/₁₆"	80	BOX 4¹¹/₁₆" REGULAR	$222.00	22.50	C - Hundred
	81	BOX 4¹¹/₁₆" DEEP	$167.00	22.50	C - Hundred
	82	BOX 4¹¹/₁₆" EXT RING	$284.00	11.00	C - Hundred
	83	BOX 4¹¹/₁₆" RNG 1GAN	$123.00	5.00	C - Hundred
	84	BOX 4¹¹/₁₆" RNG 2GAN	$163.00	6.50	C - Hundred
	85	BOX 4¹¹/₁₆" SQ ROUND	$326.00	5.00	C - Hundred
	86	PLATE 4¹¹/₁₆" BLANK	$47.00	12.00	C - Hundred
	87	PLATE 4¹¹/₁₆" SW	$607.00	12.00	C - Hundred
	88	PLATE 4¹¹/₁₆" RECEPT	$532.00	12.00	C - Hundred
	89	PLATE 4¹¹/₁₆" DRYER	$619.00	12.00	C - Hundred
BREAKER	127	BREAKER 15-30 AMP 1 POLE	$5.00	0.10	E - Each
	128	BREAKER 15-60 AMP 2 POLE	$5.00	0.15	E - Each
	129	BREAKER 70-100 AMP 2 POLE	$36.00	0.20	E - Each
	130	BREAKER 125 AMP 2 POLE	$65.00	0.25	E - Each

Electrical Estimating

Chapter 1 – Price And Labor Unit Catalog

	#	Item Description	Cost	Labor	Unit
BREAKER Cont'd	131	BREAKER 150 AMP 2 POLE	$80.00	0.30	E - Each
	132	BREAKER 200 AMP 2 POLE	$83.00	0.35	E - Each
	133	BREAKER 15-60 AMP 3 POLE	$77.00	0.20	E - Each
	134	BREAKER 70 AMP 3 POLE	$101.00	0.30	E - Each
	135	BREAKER 70-100 AMP 3POLE	$76.00	0.30	E - Each
BREAKER W/ENCLOSURE	137	BRK ENCL 2PL 60 AMP SWITCH	$11.00	1.00	E - Each
	140	BRK ENCL 2PL 125 AMP	$30.00	2.00	E - Each
	141	BRK ENCL 3PL 125 AMP	$58.00	2.50	E - Each
	142	BRK ENCL 2PL 200 AMP	*	3.00	E - Each
	143	BRK ENCL 3PL 200 AMP	$58.00	2.50	E - Each
	144	BRK ENCL 2PL 300 AMP	*	7.00	E- Each
	145	BRK ENCL 3PL 300 AMP	*	8.00	E- Each
	146	BRK ENCL 2PL 400 AMP	*	8.00	E- Each
	147	BRK ENCL 3PL 400 AMP	*	9.00	E- Each
	148	BRK ENCL 2PL 600 AMP	*	9.00	E- Each
	149	BRK ENCL 3PL 600 AMP	*	12.00	E- Each
	150	BRK ENCL 2PL 800 AMP	*	11.00	E- Each
	151	BRK ENCL 3PL 800 AMP	*	12.00	E- Each
BUSHING - BONDING	180	BUSHING BONDING 1/2"	$299.00	7.50	C - Hundred
	181	BUSHING BONDING 3/4"	$379.00	8.00	C - Hundred
	182	BUSHING BONDING 1"	$429.00	8.00	C - Hundred
	183	BUSHING BONDING 1 1/4"	$599.00	12.00	C - Hundred
	184	BUSHING BONDING 1 1/2"	$591.00	13.00	C - Hundred
	185	BUSHING BONDING 2"	$656.00	15.00	C - Hundred
	186	BUSHING BONDING 2 1/2"	$597.00	22.00	C - Hundred
	187	BUSHING BONDING 3"	$816.00	27.00	C - Hundred
	188	BUSHING BONDING 3 1/2"	$882.00	30.00	C - Hundred
	189	BUSHING BONDING 4"	$1,148.00	37.00	C - Hundred
BUSHING - PLASTIC	190	BUSHING PLASTIC 1/2"	$67.00	3.00	C - Hundred
	191	BUSHING PLASTIC 3/4"	$71.00	5.00	C - Hundred
	192	BUSHING PLASTIC 1"	$77.00	7.00	C - Hundred
	193	BUSHING PLASTIC 1 1/4"	$89.00	9.00	C - Hundred
	194	BUSHING PLASTIC 1 1/2"	$104.00	11.00	C - Hundred
	195	BUSHING PLASTIC 2"	$139.00	13.00	C - Hundred
	196	BUSHING PLASTIC 2 1/2"	$69.00	15.00	C - Hundred
	197	BUSHING PLASTIC 3"	$79.00	19.00	C - Hundred
	198	BUSHING PLASTIC 3 1/2"	$89.00	21.00	C - Hundred
	199	BUSHING PLASTIC 4"	$104.00	25.00	C - Hundred

	#	Item Description	Cost	Labor	Unit
CORDS	221	CORD 15, 125 VOLT 6'	$4.00	0.25	E- Each
	223	CORD 15, 250 VOLT 6'	$4.00	0.25	E- Each
	225	CORD 20, 125 VOLT 6'	$7.00	0.25	E- Each
	227	CORD 20, 250 VOLT 6'	$7.00	0.25	E- Each
	228	CORD DRYER 30 AMP 4WIRE	$12.00	0.30	E- Each
	230	CORD RANGE 50 AMP 4WIRE	$12.00	0.30	E- Each
DIMMERS	235	DIM 600W 120 VOLT 1POLE (Push)	$6.10	0.25	E- Each
	236	DIM 600W 120 VOLT 3WAY	$13.00	0.35	E- Each
	237	DIM 1000W 120 VOLT	$25.00	0.35	E- Each
	238	DIM 1500W 120 VOLT	$54.00	0.35	E- Each
	239	DIM 2000W 120 VOLT	$74.00	0.40	E- Each
	240	DIM FLUOR 600W 120 VOLT	$45.00	0.35	E- Each
	241	DIM FLUOR 1,000W 120 VOLT	$55.00	0.35	E- Each
	242	DIM FLUOR 2,000W 120 VOLT	$75.00	0.35	E- Each
DISCONNECT - 1 PHASE	275	DISC. 30 AMP 1PHASE	$28.00	1.00	E- Each
	276	DISC. 60 AMP 1 PHASE	$47.00	1.00	E- Each
	277	DISC. 100 AMP 1PHASE	$97.00	1.50	E- Each
	278	DISC. 200 AMP 1PHASE	$202.00	2.50	E- Each
	279	DISC. 400 AMP 1PHASE	$543.00	4.00	E- Each
	280	DISC. 30 AMP 1P RAINTIGHT	$42.00	1.00	E- Each
	281	DISC. 60 AMP 1P RAINTIGHT	$74.00	1.00	E- Each
	282	DISC. 100 AMP 1PH RAINTIGHT	$110.00	1.50	E- Each
	283	DISC. 200 AMP 1PH RAINTIGHT	$275.00	2.00	E- Each
	284	DISC. 400 AMP 1PH RAINTIGHT	$844.00	4.00	E- Each
DISCONNECT - 3 PHASE	286	DISC. 30 AMP 3 PHASE	$43.00	1.50	E- Each
	287	DISC. 60 AMP 3 PHASE	$74.00	1.50	E- Each
	288	DISC. 100 AMP 3PHASE	$129.00	2.00	E- Each
	289	DISC. 200 AMP 3PHASE	$275.00	2.50	E- Each
	292	DISC. 400 AMP 3PHASE	$663.00	4.00	E- Each
	293	DISC. 600 AMP 3PHASE	$1,420.00	5.00	E- Each
EMT	305	EMT 1/2"	$13.00	2.25	C - Hundred
	306	EMT 3/4"	$20.00	2.85	C - Hundred
	307	EMT 1"	$31.00	3.75	C - Hundred
	308	EMT 1 1/4"	$50.00	4.00	C - Hundred
	309	EMT 1 1/2"	$71.00	4.25	C - Hundred
	310	EMT 2"	$91.00	4.75	C - Hundred
	311	EMT 2 1/2"	$194.00	6.75	C - Hundred
	312	EMT 3"	$247.00	7.60	C - Hundred

	#	Item Description	Cost	Labor	Unit
EMT Cont'd	313	EMT 3$\frac{1}{2}$"	$350.00	9.50	C - Hundred
	314	EMT 4"	$401.00	11.50	C - Hundred
EMT CONNECTOR SS	315	EMT CONN SS $\frac{1}{2}$"	$21.00	2.00	C - Hundred
	316	EMT CONN SS $\frac{3}{4}$"	$41.00	3.00	C - Hundred
	317	EMT CONN SS 1"	$64.00	5.00	C - Hundred
	318	EMT CONN SS 1$\frac{1}{4}$"	$108.00	7.00	C - Hundred
	319	EMT CONN SS 1$\frac{1}{2}$"	$112.00	8.00	C - Hundred
	320	EMT CONN SS 2"	$183.00	9.00	C - Hundred
	321	EMT CONN SS 2$\frac{1}{2}$"	$450.00	12.00	C - Hundred
	322	EMT CONN SS 3"	$600.00	14.00	C - Hundred
	323	EMT CONN SS 3$\frac{1}{2}$"	$750.00	16.00	C - Hundred
	324	EMT CONN SS 4"	$900.00	18.00	C - Hundred
EMT CONNECTOR COMP.	335	EMT CONN COMP $\frac{1}{2}$"	$43.00	3.00	C - Hundred
	336	EMT CONN COMP $\frac{3}{4}$"	$60.00	5.00	C - Hundred
	337	EMT CONN COMP 1"	$109.00	7.00	C - Hundred
	338	EMT CONN COMP 1$\frac{1}{4}$"	$181.00	9.00	C - Hundred
	339	EMT CONN COMP 1$\frac{1}{2}$"	$264.00	10.00	C - Hundred
	340	EMT CONN COMP 2"	$382.00	12.00	C - Hundred
	341	EMT CONN COMP 2$\frac{1}{2}$"	$1,182.00	14.00	C - Hundred
	342	EMT CONN COMP 3"	$1,657.00	18.00	C - Hundred
	343	EMT CONN COMP 3$\frac{1}{2}$"	$1,900.00	21.00	C - Hundred
	344	EMT CONN COMP 4"	$2,500.00	23.00	C - Hundred
EMT COUPLING SS	355	EMT COUP SS $\frac{1}{2}$"	$24.00	2.00	C - Hundred
	356	EMT COUP SS $\frac{3}{4}$"	$40.00	3.00	C - Hundred
	357	EMT COUP SS 1"	$62.00	5.00	C - Hundred
	358	EMT COUP SS 1$\frac{1}{4}$"	$99.00	7.00	C - Hundred
	359	EMT COUP SS 1$\frac{1}{2}$"	$117.00	8.00	C - Hundred
	360	EMT COUP SS 2"	$281.00	9.00	C - Hundred
	361	EMT COUP SS 2$\frac{1}{2}$"	$527.00	11.00	C - Hundred
	362	EMT COUP SS 3"	$634.00	14.00	C - Hundred
	363	EMT COUP SS 3$\frac{1}{2}$"	$732.00	16.00	C - Hundred
	364	EMT COUP SS 4"	$2,900.00	18.00	C - Hundred
EMT COUPLING COMP.	375	EMT COUP COMP $\frac{1}{2}$"	$52.00	3	C - Hundred
	376	EMT COUP COMP $\frac{3}{4}$"	$71.00	5	C - Hundred
	377	EMT COUP COMP 1"	$109.00	7	C - Hundred
	378	EMT COUP COMP 1$\frac{1}{4}$"	$198.00	9	C - Hundred
	379	EMT COUP COMP 1$\frac{1}{2}$"	$287.00	10	C - Hundred
	380	EMT COUP COMP 2"	$390.00	12	C - Hundred

	#	Item Description	Cost	Labor	Unit
EMT COUPLING COMP. Cont'd					
	381	EMT COUP COMP 2 1/2"	$1,589.00	14.00	C - Hundred
	382	EMT COUP COMP 3"	$2,123.00	18.00	C - Hundred
	383	EMT COUP COMP 3 1/2"	$5,576.00	20.00	C - Hundred
	384	EMT COUP COMP 4"	$5,883.00	24.00	C - Hundred
EMT ELL 90	395	EMT ELL 90 1/2"	$184.00	5.00	C - Hundred
	396	EMT ELL 90 3/4"	$175.00	5.00	C - Hundred
	397	EMT ELL 90 1"	$195.00	10.00	C - Hundred
	398	EMT ELL 90 1 1/4"	$311.00	15.00	C - Hundred
	399	EMT ELL 90 1 1/2"	$350.00	17.00	C - Hundred
	400	EMT ELL 90 2"	$406.00	18.00	C - Hundred
	401	EMT ELL 90 2 1/2"	$998.00	21.00	C - Hundred
	402	EMT ELL 90 3"	$1,472.00	23.00	C - Hundred
	403	EMT ELL 90 3 1/2"	$2,020.00	25.00	C - Hundred
	404	EMT ELL 90 4"	$2,322.00	27.00	C - Hundred
EMT STRAP	415	EMT STRAP 1H 1/2"	$6.00	2.50	C - Hundred
	416	EMT STRAP 1H 3/4"	$9.00	2.50	C - Hundred
	417	EMT STRAP 1H 1"	$17.00	2.70	C - Hundred
	418	EMT STRAP 1H 1 1/4"	$19.00	2.90	C - Hundred
	419	EMT STRAP 1H 1 1/2"	$22.00	3.00	C - Hundred
	420	EMT STRAP 1H 2"	$48.00	3.25	C - Hundred
	421	EMT STRAP 1H 2 1/2"	$58.00	3.50	C - Hundred
	422	EMT STRAP 1H 3"	$73.00	3.75	C - Hundred
	423	EMT STRAP 1H 3 1/2"	$88.00	4.00	C - Hundred
	424	EMT STRAP 1H 4"	$130.00	4.25	C - Hundred
ENT AND FITTING	436	ENT 1/2"	$18.00	1.50	C - Hundred
	437	ENT 3/4"	$24.00	2.50	C - Hundred
	438	ENT 1"	$46.00	5.00	C - Hundred
	441	ENT CONNECTOR 1/2"	$33.00	5.00	C - Hundred
	442	ENT CONNECTOR 3/4"	$64.00	6.00	C - Hundred
	443	ENT CONNECTOR 1"	$76.00	7.00	C - Hundred
	444	ENT COUPLING 1/2"	$26.00	4.00	C - Hundred
	445	ENT COUPLING 3/4"	$32.00	4.50	C - Hundred
	446	ENT COUPLING 1"	$45.00	5.00	C - Hundred
EXCAVATION - TRENCHING	464	TRENCH 6" × 12" DEEP	*	3.35	C - Hundred
	465	TRENCH 6" × 24" DEEP	*	6.25	C - Hundred
	466	TRENCH 6" × 36" DEEP	*	9.00	C - Hundred
	467	TRENCH 12" × 36" DEEP	*	11.00	C - Hundred

Chapter 1 – Price And Labor Unit Catalog

Category	#	Item Description	Cost	Labor	Unit
FIXTURE	473	FAN AND LIGHT	$30.00	0.75	E- Each
	474	FAN, EXHAUST	$20.00	1.00	E- Each
	475	FAN, PADDLE	$75.00	0.50	E- Each
	476	FAN, HOOD	*	0.50	E- Each
	477	HEAT LAMP	$55.00	0.75	E- Each
	478	RECESS FIXTURE	$25.00	1.00	E- Each
	479	BATTERY PACK - 2 LAMP	$50.00	1.00	E- Each
	480	PULL CHAIN FIXTURE	$3.00	0.25	E- Each
	482	FIX HI PRES 70-250W	$50.00	1.00	E- Each
	483	FIX MERC VAPOR	$50.00	1.50	E- Each
	484	FIX PAR HOLDER	$3.00	0.15	E- Each
	485	FIX QUARTZ 500 WATTS	$15.00	0.75	E- Each
	486	RECESSED FIXTURE VP	$35.00	1.00	E- Each
	488	2' UNDER COUNTER W/L	$8.00	1.00	E- Each
	491	OUTSIDE FIXTURE	$20.00	0.25	E- Each
	492	INSIDE FIXTURE	$25.00	0.25	E- Each
	500	SMOKE DETECTOR (125 VOLT)	$20.00	0.25	E- Each
	502	KEYLESS FIXTURE	$2.00	0.25	E- Each
	503	EXIT FIXTURE	$25.00	0.25	E- Each
	505	POST LIGHT	$50.00	1.00	E- Each
FIXTURE - FLUORESCENT	572	FLUO STRIP 1-4' 120 VOLT	$20.00	0.75	E- Each
	573	FLUO STRIP 2-4' 120 VOLT	$17.50	0.75	E - Each
	574	FLUO STRIP 1-6' 120 VOLT	$29.00	0.75	E- Each
	575	FLUO STRIP 2-6' 120 VOLT	$29.00	0.75	E- Each
	576	FLUO STRIP 1-8' 120 VOLT	$29.00	0.75	E- Each
	577	FLUO STRIP 2-8' 120 VOLT	$30.00	1.00	E- Each
	579	FLUO WRAP 2 × 2' 120 VOLT	$36.00	0.75	E- Each
	580	FLUO WRAP 2 × 4' 120 VOLT	$34.00	0.75	E- Each
	583	FLUO LAY-IN 2 × 4' 120 VOLT	$50.00	0.75	E- Each
	585	FLUO LAY-IN 2 × 2' 120 VOLT	$36.00	0.75	E- Each
FIXTURE - MISCELLANEOUS	588	WHIP FLUORESCENT	$3.50	0.15	E- Each
	589	JACK CHAIN STD	$30.00	2.50	C - Hundred
	594	BALLAST 4FT 120 VOLT	$15.00	1.00	E- Each
	595	BALLAST 4FT 277 VOLT	$15.00	1.00	E- Each
	596	BALLAST 8FT 120 VOLT	$23.00	1.00	E- Each
	597	BALLAST 8FT 277 VOLT	$30.005	1.00	E- Each
	598	BALLAST MERC 480 VOLT	$60.00	1.50	E- Each

Category	#	Item Description	Cost	Labor	Unit
FIXTURE - POLE	610	POLE FIXTURE	*	0.50	E- Each
	611	POLE ANC BASE AL 10'	*	3.00	E- Each
	612	POLE ANC BASE AL 12'	*	4.00	E- Each
	613	POLE ANC BASE AL 16'	*	5.00	E- Each
	614	POLE ANC BASE AL 20'	*	6.00	E- Each
FIXTURE - TRACK	620	TRACK 4'	$23.00	0.75	E- Each
	621	TRACK 8'	$44.00	1.00	E- Each
	622	TRACK 12'	$68.00	1.50	E- Each
	623	FIXTURE HEAD	$40.00	0.25	E- Each
	624	TRACK POWER OUTLET	$7.50	0.25	E- Each
	625	L - CONNECTOR	$12.00	0.25	E- Each
FLEXIBLE METAL CONDUIT	639	FLEX STEEL 3/8"	$32.00	3.25	C - Hundred
	640	FLEX STEEL 1/2"	$31.00	3.25	C - Hundred
	641	FLEX STEEL 3/4"	$32.00	3.75	C - Hundred
	642	FLEX STEEL 1"	$82.00	4.00	C - Hundred
	643	FLEX STEEL 1 1/4"	$73.00	4.50	C - Hundred
	644	FLEX STEEL 1 1/2"	$94.00	5.00	C - Hundred
	645	FLEX STEEL 2"	$136.00	6.25	C - Hundred
	646	FLEX STEEL 2 1/2"	$162.00	8.00	C - Hundred
	647	FLEX STEEL 3"	$201.00	10.00	C - Hundred
	648	FLEX STEEL 3 1/2"	$300.00	11.00	C - Hundred
	649	FLEX STEEL 4"	$375.00	13.00	C - Hundred
FLEX - CONNECTOR	650	FLEX ST CONN 3/8"	$22.00	6.25	C - Hundred
	651	FLEX ST CONN 1/2"	$55.00	6.25	C - Hundred
	652	FLEX ST CONN 3/4"	$62.00	7.50	C - Hundred
	653	FLEX ST CONN 1"	$95.00	9.00	C - Hundred
	654	FLEX ST CONN 1 1/4"	$282.00	9.50	C - Hundred
	655	FLEX ST CONN 1 1/2"	$452.00	12.00	C - Hundred
	656	FLEX ST CONN 2"	$540.00	13.50	C - Hundred
	657	FLEX ST CONN 2 1/2"	$1,382.00	16.00	C - Hundred
	658	FLEX ST CONN 3"	$2,004.00	18.00	C - Hundred
	659	FLEX ST CONN 3 1/2"	$2,500.00	20.00	C - Hundred
	660	FLEX ST CONN 4"	$3,500.00	26.00	C - Hundred
FLEX - 90 CONNECTOR	669	FLEX 90 CONN 3/8"	$54.00	7.50	C - Hundred
	670	FLEX 90 CONN 1/2"	$76.00	7.50	C - Hundred
	671	FLEX 90 CONN 3/4"	$154.00	8.50	C - Hundred
	672	FLEX 90 CONN 1"	$495.00	10.00	C - Hundred
	673	FLEX 90 CONN 1 1/4"	$641.00	10.00	C - Hundred

	#	Item Description	Cost	Labor	Unit
FLEX - 90 CONNECTOR Cont'd	674	FLEX 90 CONN 1½"	$1,220.00	11.25	C - Hundred
	675	FLEX 90 CONN 2"	$1,529.00	14.00	C - Hundred
	676	FLEX 90 CONN 2½"	$4,300.00	16.00	C - Hundred
	677	FLEX 90 CONN 3"	$5,803.00	19.00	C - Hundred
	678	FLEX 90 CONN 3½"	$7,020.00	20.50	C - Hundred
	679	FLEX 90 CONN 4"	$8,500.00	24.00	C - Hundred
FUSE	690	FUSE 10-30 A 250 VOLT ONE TIME	$61.00	2.75	C - Hundred
	695	FUSE 35-60 A 250 VOLT ONE TIME	$164.00	4.50	C - Hundred
	696	FUSE 70 AMP 250 VOLT ONE TIME	$680.00	4.50	C - Hundred
	697	FUSE 100 AMP 250 VOLT ONE TIME	$588.00	4.50	C - Hundred
	698	FUSE 110 AMP 250 VOLT ONE TIME	$1,657.00	6.25	C - Hundred
	699	FUSE 125 AMP 250 VOLT ONE TIME	$1,657.00	6.25	C - Hundred
	700	FUSE 200 AMP 250 VOLT ONE TIME	$1,421.00	6.25	C - Hundred
	701	FUSE 225 AMP 250 VOLT ONE TIME	$3,604.00	6.25	C - Hundred
	702	FUSE 250 AMP 250 VOLT ONE TIME	$3,604.00	7.50	C - Hundred
	703	FUSE 400 AMP 250 VOLT ONE TIME	$4,909.00	8.00	C - Hundred
	704	FUSE 450 AMP 250 VOLT ONE TIME	$4,909.00	8.50	C - Hundred
	705	FUSE 600 AMP 250 VOLT ONE TIME	$4,909.00	9.00	C - Hundred
	706	FUSE 10-30 AMP 600 VOLT ONE TIME	$481.00	2.75	C - Hundred
	710	FUSE 100 AMP 600 VOLT ONE TIME	$1,262.00	6.25	C - Hundred
	711	FUSE 200 AMP 600 VOLT ONE TIME	$3,006.00	6.25	C - Hundred
	712	FUSE 300 AMP 600 VOLT ONE TIME	$7,028.00	9.00	C - Hundred
	713	FUSE 400 AMP 600 VOLT ONE TIME	$7,028.00	9.00	C - Hundred
	714	FUSE 600 AMP 600 VOLT ONE TIME	$10,123.00	12.00	C - Hundred
FUSE - EDISON	715	FUSE 10-30 A 250 VOLT TD	$234.00	2.75	C - Hundred
	716	FUSE 35-60 A 250 VOLT TD	$414.00	4.50	C - Hundred
	717	FUSE EDISION 15-30 A	$208.00	4.50	C - Hundred
	718	FUSE NONTAMPER 15-30 A	$137.00	4.50	C - Hundred
GENERATOR	740	GENERATOR 50 kVA	*	17.50	E - Each
	741	GENERATOR 75 kVA	*	19.00	E - Each
	742	GENERATOR 100 kVA	*	20.00	E - Each
	743	GENERATOR 125 kVA	*	28.00	E - Each
	744	GENERATOR 150 kVA	*	32.00	E - Each
	745	GENERATOR 175 kVA	*	35.50	E - Each
	746	GENERATOR 200 kVA	*	41.00	E - Each
	747	GENERATOR 250 kVA	*	43.00	E - Each
	748	GENERATOR 300 kVA	*	47.00	E - Each

Category	#	Item Description	Cost	Labor	Unit
GENERATOR Cont'd	749	GENERATOR 350 kVA	*	52.00	E- Each
	750	GENERATOR 400 kVA	*	58.00	E- Each
	751	GENERATOR 500 kVA	*	67.00	E- Each
	752	GENERATOR 750 kVA	*	84.00	E- Each
GROUND FITTING	755	GROUND CLAMP BURIAL 1/2"	$174.00	13.50	C - Hundred
	756	GROUND CLAMP BURIAL 5/8"	$217.00	14.00	C - Hundred
	757	GROUND CLAMP 1/2" - 1"	$164.00	22.50	C - Hundred
	758	GROUND CLAMP 1 1/4" - 2"	$297.00	49.00	C - Hundred
GROUND-MISCELLANEOUS	759	GROUND SCREW 10/32"	$10.00	3.00	C - Hundred
	760	GROUND CLIP No. 14-12	$16.00	3.00	C - Hundred
	761	GROUND TAIL No. 12	$62.00	6.00	C - Hundred
GROUND ROD	762	GROUND ROD 1/2" × 8' CU	$905.00	50.00	C - Hundred
	764	GROUND ROD 5/8" × 8' GAL	$800.00	50.00	C - Hundred
	765	GROUND ROD 1/2" × 10' CU	$1,200.00	60.00	C - Hundred
	769	GROUND ROD 5/8" × 10' GAL	$1,000.00	60.00	C - Hundred
GUTTER	775	GUTTER 4" × 4" × 2'	$11.00	0.75	E- Each
	776	GUTTER 4" × 4" × 4'	$23.00	1.50	E- Each
	777	GUTTER 4" × 4" × 4' RT	$46.00	1.75	E- Each
	779	GUTTER 6" × 6" × 2'	$18.00	1.00	E- Each
	780	GUTTER 6" × 6" × 4'	$34.00	2.00	E- Each
	781	GUTTER 6" × 6" × 6'	$49.00	2.50	E- Each
	782	GUTTER 6" × 6" × 6' RT	$91.00	3.00	E- Each
	783	GUTTER 8" × 8" × 2'	$37.00	1.25	E- Each
	784	GUTTER 8" × 8" × 4'	$67.00	2.50	E- Each
	785	GUTTER 8" × 8" × 6'	$100.00	3.00	E- Each
	786	GUTTER 8" × 8" × 10'	$158.00	3.50	E- Each
	787	GUTTER 12" × 12" × 2'	$64.00	1.50	E- Each
	788	GUTTER 12" × 12" × 4'	$114.00	3.00	E - Each
GUTTER FITTINGS	791	GUTTER END 4" × 4"	$3.00	0.15	E- Each
	792	GUTTER END 6" × 6"	$4.00	0.15	E- Each
	793	GUTTER END 8" × 8"	$5.00	0.15	E- Each
	794	GUTTER END 12" × 12"	$11.00	0.20	E- Each
	795	GUTTER ELL 4"	$14.00	0.40	E- Each
	796	GUTTER ELL 6"	$17.00	0.50	E- Each
	797	GUTTER ELL 8"	$27.00	0.75	E- Each
	798	GUTTER ELL 12"	$62.00	1.00	E- Each

	#	Item Description	Cost	Labor	Unit
GUTTER FITTINGS Cont'd	799	GUTTER COUPLING 4"	$3.00	0.15	E- Each
	800	GUTTER COUPLING 6"	$4.00	0.15	E- Each
	801	GUTTER COUPLING 8"	$5.00	0.20	E- Each
	802	GUTTER COUPLING 12"	$11.00	0.20	E- Each
HUB	815	HUB 1/2"	$6.00	25.00	E- Each
	816	HUB 3/4"	$6.00	25.00	E- Each
	817	HUB 1"	$6.00	25.00	E- Each
	818	HUB 1 1/4"	$6.00	25.00	E- Each
	819	HUB 1 1/2"	$6.00	0.30	E- Each
	820	HUB 2"	$16.00	0.30	E- Each
	821	HUB 2 1/2"	$27.00	0.40	E- Each
	822	HUB 3"	$39.00	0.40	E- Each
	823	HUB 3 1/2"	$50.00	0.50	E- Each
	824	HUB 4"	$75.00	0.50	E- Each
	825	HUB CLOSING CAP	$3.00	0.20	E- Each
IMC	830	IMC 1/2"	$49.00	3.25	C - Hundred
	831	IMC 3/4"	$58.00	4.00	C - Hundred
	832	IMC 1"	$85.00	4.50	C - Hundred
	833	IMC 1 1/4"	$108.00	5.25	C - Hundred
	834	IMC 1 1/2"	$128.00	6.00	C - Hundred
	835	IMC 2"	$176.00	7.00	C - Hundred
	836	IMC 2 1/2"	$348.00	8.00	C - Hundred
	837	IMC 3"	$456.00	10.00	C - Hundred
	838	IMC 3 1/2"	$416.00	12.00	C - Hundred
	839	IMC 4"	$491.00	15.00	C - Hundred
JUNCTION BOX	840	JUNCTION BOX 4" × 4" × 4"	$5.00	1.00	E- Each
	841	JUNCTION BOX 4" × 4" × 4" RT	$9.00	1.00	E- Each
	842	JUNCTION BOX 6" × 6" × 4"	$7.00	1.00	E- Each
	843	JUNCTION BOX 6" × 6" × 4" RT	$11.00	1.00	E- Each
	844	JUNCTION BOX 6" × 6" × 6"	$8.00	1.00	E- Each
	845	JUNCTION BOX 6" × 6" × 6" RT	$13.00	1.00	E- Each
	848	JUNCTION BOX 8" × 8" × 6"	$11.00	1.50	E- Each
	849	JUNCTION BOX 8" × 8" × 6" RT	$17.00	1.50	E- Each
	850	JUNCTION BOX 12" × 12" × 4"	$16.00	1.50	E- Each
	851	JUNCTION BOX 12" × 12" × 4" RT	$24.00	1.50	E- Each
	852	JUNCTION BOX 12" × 12" × 6"	$19.00	2.00	E- Each
	853	JUNCTION BOX 12" × 12" × 6" RT	$28.00	2.00	E- Each
	854	JUNCTION BOX 15" × 15" × 4"	$22.00	2.00	E- Each
	855	JUNCTION BOX 15" × 15" × 4" RT	$28.00	2.00	E- Each

Chapter 1 – Price And Labor Unit Catalog

	#	Item Description	Cost	Labor	Unit
JUNCTION BOX Cont'd	856	JUNCTION BOX 15" × 15" × 6"	$25.00	2.50	E- Each
	857	JUNCTION BOX 18" × 12" × 4"	$21.00	2.50	E- Each
	858	JUNCTION BOX 18" × 12" × 6"	$25.00	3.00	E- Each
	860	JUNCTION BOX 18" × 18" × 6" RT	$55.00	3.50	E- Each
	861	JUNCTION BOX 18" × 18" × 6"	$33.00	4.00	E- Each
	862	JUNCTION BOX 24" × 18" × 6"	$42.00	4.00	E- Each
	863	JUNCTION BOX 24" × 24" × 6"	$67.00	4.00	E- Each
KNOCK OUT SEAL (KO)	880	KO SEAL 1/2"	$14.00	7.00	C - Hundred
	881	KO SEAL 3/4"	$16.00	9.00	C - Hundred
	882	KO SEAL 1"	$29.00	14.00	C - Hundred
	883	KO SEAL 1 1/4"	$34.00	17.00	C - Hundred
	884	KO SEAL 1 1/2"	$57.00	25.00	C - Hundred
	885	KO SEAL 2"	$68.00	32.00	C - Hundred
	886	KO SEAL 2 1/2"	$418.00	42.00	C - Hundred
	887	KO SEAL 3"	$480.00	48.00	C - Hundred
	888	KO SEAL 4"	$518.00	55.00	C - Hundred
LAMP (BULB)	890	LAMP, FLOOD, PAR 38	$7.00	0.10	E- Each
	891	LAMP, FLUO 2'	$3.00	0.10	E- Each
	892	LAMP, FLUO 4'	$2.00	0.10	E- Each
	893	LAMP, FLUO 6'	$6.00	0.10	E- Each
	894	LAMP, FLUO 8'	$6.00	0.10	E- Each
	895	LAMP, HEAT LAMP	$7.00	0.10	E- Each
	896	LAMP, HI PR SOD	$24.00	0.15	E- Each
	898	LAMP, MERC VAPOR	$46.00	0.15	E- Each
	900	LAMP F40 WATTMISER	$3.00	0.10	E- Each
	901	LAMP, QUARTZ 500W	$9.00	0.20	E- Each
	902	LAMP HI HAT 75 WATT	$5.00	0.10	E- Each
	907	EXIT LAMP 25/40 WATT	$5.00	0.20	E- Each
	908	U-TUBE FLUORESCENT	$11.00	0.20	E- Each
	909	LAMP, QUARTZ 1500W	$11.00	0.20	E- Each
LOCKNUT	910	LOCKNUT STEEL 1/2"	$15.00	4.00	C - Hundred
	911	LOCKNUT STEEL 3/4"	$17.00	4.25	C - Hundred
	912	LOCKNUT STEEL 1"	$23.00	4.50	C - Hundred
	913	LOCKNUT STEEL 1 1/4"	$38.00	5.00	C - Hundred
	914	LOCKNUT STEEL 1 1/2"	$43.00	5.50	C - Hundred
	915	LOCKNUT STEEL 2"	$69.00	6.00	C - Hundred
	916	LOCKNUT STEEL 2 1/2"	$200.00	6.50	C - Hundred
	917	LOCKNUT STEEL 3"	$250.00	7.00	C - Hundred

	#	Item Description	Cost	Labor	Unit
LOCKNUT Cont'd	918	LOCKNUT STEEL 3 1/2"	$350.00	7.50	C - Hundred
	919	LOCKNUT STEEL 4"	$450.00	8.00	C - Hundred
	920	LOCKNUT GND 1/2"	$78.00	4.50	C - Hundred
	921	LOCKNUT GND 3/4"	$98.00	6.00	C - Hundred
	922	LOCKNUT GND 1"	$136.00	6.50	C - Hundred
	923	LOCKNUT GND 1 1/4"	$179.00	7.00	C - Hundred
	924	LOCKNUT GND 1 1/2"	$232.00	7.25	C - Hundred
	925	LOCKNUT GND 2"	$242.00	7.75	C - Hundred
	926	LOCKNUT GND 2 1/2"	$471.00	8.00	C - Hundred
	927	LOCKNUT GND 3"	$594.00	8.50	C - Hundred
	928	LOCKNUT GND 3 1/2"	$973.00	9.00	C - Hundred
	929	LOCKNUT GND 4"	$1,200.00	11.50	C - Hundred
LOCKNUT SEAL	930	LOCKNUT SEAL 1/2"	$71.00	9.00	C - Hundred
	931	LOCKNUT SEAL 3/4"	$84.00	12.50	C - Hundred
	932	LOCKNUT SEAL 1"	$127.00	16.00	C - Hundred
	933	LOCKNUT SEAL 1 1/4"	$218.00	21.50	C - Hundred
	934	LOCKNUT SEAL 1 1/2"	$272.00	30.50	C - Hundred
	935	LOCKNUT SEAL 2"	$349.00	39.50	C - Hundred
LOW VOLTAGE	942	CHIME	$15.00	0.50	E - Each
	945	TRANSFORMER 8-16-24V 20W	$14.00	0.25	E - Each
	946	THERMOSTAT	*	0.50	E - Each
	947	THERMOSTAT 2 STAGE	*	1.00	E - Each
	948	PUSH-BUTTON	$3.00	0.25	E - Each
	949	REMOTE CONTROL 3 C CABLE	*	13.00	M - Thousand
LUGS	965	LUG CRIMP #4	$244.00	14.00	C - Hundred
	966	LUG CRIMP # 1/0	$419.00	29.50	C - Hundred
	967	LUG CRIMP # 4/0	$1,032.00	32.50	C - Hundred
	968	LUG CRIMP 250	$1,500.00	36.00	C - Hundred
	975	LUG, SLDRLS 18-8	$265.00	23.00	C - Hundred
	976	LUG, SLDRLS 14-6	$212.00	23.00	C - Hundred
	977	LUG, SLDRLS 8-2	$246.00	27.00	C - Hundred
	979	LUG, SLDRLS 4-3/0	$319.00	36.00	C - Hundred
	980	LUG, SET SCREW 350-1/0	$350.00	40.00	C - Hundred
	985	LUG, SPLIT BOLT BUG #6	$272.00	21.00	C - Hundred
	986	LUG, SPLIT BOLT BUG #3	$364.00	25.00	C - Hundred

Category	#	Item Description	Cost	Labor	Unit
LUG Cont'd	987	LUG, SPLIT BOLT BUG #2	$406.00	27.00	C - Hundred
	988	LUG, SPLIT BOLT BUG 2/0	$845.00	27.00	C - Hundred
	989	LUG, SPLIT BOLT BUG 4/0	$1,432.25	36.00	C - Hundred
	990	LUG, SPLIT BOLT BUG 250	$2,100.00	45.00	C - Hundred
	991	LUG, SPLIT BOLT BUG 350	$27.00	0.75	E- Each
	992	LUG, SPLIT BOLT BUG 500	$36.00	1.00	E- Each
	993	LUG, SPLIT BOLT BUG 1,000	$84.00	1.00	E- Each
METER EQUIPMENT	1005	METER/MAIN COMBO	$115.00	2.00	E- Each
	1006	METER CT 400-600AMP	*	4.00	E- Each
	1010	STACK MCB 800A 1 PH	*	9.00	E- Each
	1011	STACK MCB 1,000A 1 PH	*	10.00	E- Each
	1012	STACK MCB 1,200A 1 PH	*	11.00	E- Each
	1020	METER STACK 3U 150A	*	3.50	E- Each
	1021	METER STACK 4U 150A	*	4.00	E- Each
	1022	METER STACK 5U 150A	*	4.50	E- Each
	1023	METER STACK 6U 150A	*	5.00	E- Each
	1024	METER STACK 8U 150A	*	6.00	E- Each
MISCELLANEOUS ITEMS	1041	GREEN GROMMET 1"	$50.00	6.50	M Pcs.
	1042	KEEL	$5.00	*	E- Each
	1043	KO HOLE PUNCH $1/2$" - 1"	$119.00	0.25	E- Each
	1044	KO HOLE PUNCH $1/4$" - 2"	$236.00	0.30	E- Each
	1045	KO HOLE PUNCH $2^{1}/_{2}$" - $3^{1}/_{2}$"	$126.00	0.50	E- Each
	1046	KO HOLE PUNCH 4" - 6"	$180.00	0.75	E- Each
	1047	MADISON CLIPS/PAIR	$37.00	13.50	C - Hundred
	1050	PANEL FILLER	$2.50	0.20	E- Each
	1052	PIN, DRIVE $3/4$"	$7.00	1.50	C - Hundred
	1053	PIN, DRIVE 1"	$8.00	1.50	C - Hundred
	1054	PIN, DRIVE $1^{1}/_{2}$"	$12.00	1.50	C - Hundred
	1057	SHOTS - 22 CALIBER	$6.00	1.50	C - Hundred
	1058	SOAP PULL- QUART	$6.00	1.00	E- Each
	1060	TAPE, DUCT 2" x 66 YDS	$4.00	2.00	E- Each
	1062	TAPE, PHASE COLOR	$3.50	2.00	E- Each
	1063	TAPE, MASKING 2"	$2.00	2.00	E- Each
	1064	TAPE, RUBBER	$3.00	2.00	E- Each
	1065	EXT - BOX EXTENDER	$119.00	9.00	C - Hundred
	1074	TIE WRAP	$8.00	2.50	C - Hundred
MOTOR CONTROL	1100	MOTOR CONNECT 2-10 HP	*	1.00	E- Each
	1101	MOTOR CONNECT 1-30 HP	*	2.50	E- Each
	1102	MOTOR CONNECT 45-75 HP	*	4.00	E- Each

	#	Item Description	Cost	Labor	Unit
MOTOR CONTROL Cont'd	1105	STARTER 2 HP 3 POLE	$189.00	1.00	E- Each
	1106	STARTER 5 HP 3 POLE	$225.00	1.50	E- Each
	1107	STARTER 10 HP 3 POLE	$425.00	1.75	E- Each
	1108	STARTER 25 HP 3 POLE	$690.00	2.00	E- Each
	1109	STARTER 50 HP 3 POLE	$1,600.00	2.50	E- Each
	1110	STARTER 100 HP 3 POLE	$2,500.00	4.00	E- Each
	1115	START PUSH BUTTON 1U	$35.00	1.00	E- Each
	1116	START PUSH BUTTON 2U	$63.00	1.00	E- Each
	1117	START PUSH BUTTON 3U	$90.00	1.00	E- Each
	1120	STARTER REVERSE 2 HP	$220.00	1.50	E- Each
	1121	STARTER REVERSE 5 HP	$280.00	1.50	E- Each
	1122	STARTER REVERSE 10 HP	$310.00	2.00	E- Each
	1123	STARTER REVERSE 25 HP	$360.00	3.00	E- Each
	1124	STARTER REVERSE 50 HP	$450.00	3.00	E- Each
	1125	STARTER REVERSE 100 HP	$900.00	5.00	E- Each
NIPPLE	1140	NIPPLE, CHASE 1/2"	$38.00	4.00	C - Hundred
	1141	NIPPLE, CHASE 3/4"	$59.00	6.50	C - Hundred
	1142	NIPPLE, CHASE 1"	$114.00	8.00	C - Hundred
	1143	NIPPLE, CHASE 1 1/4"	$183.00	9.00	C - Hundred
	1144	NIPPLE, CHASE 1 1/2"	$205.00	10.00	C - Hundred
	1145	NIPPLE, CHASE 2"	$314.00	12.00	C - Hundred
	1146	NIPPLE, CHASE 2 1/2"	$674.00	13.50	C - Hundred
	1147	NIPPLE, CHASE 3"	$1,065.00	18.00	C - Hundred
	1148	NIPPLE, CHASE 3 1/2"	$1,996.00	22.00	C - Hundred
	1149	NIPPLE, CHASE 4"	$3,327.00	27.00	C - Hundred
	1150	NIPPLE, CLOSE 1/2"	$34.00	6.50	C - Hundred
	1151	NIPPLE, CLOSE 3/4"	$44.00	8.00	C - Hundred
	1152	NIPPLE, CLOSE 1"	$67.00	10.00	C - Hundred
	1153	NIPPLE, CLOSE 1 1/4"	$85.00	12.00	C - Hundred
	1154	NIPPLE, CLOSE 1 1/2"	$102.00	15.00	C - Hundred
	1155	NIPPLE, CLOSE 2"	$125.00	19.00	C - Hundred
	1156	NIPPLE, CLOSE 2 1/2"	$4.00	0.25	E- Each
	1157	NIPPLE, CLOSE 3"	$4.00	0.40	E- Each
	1158	NIPPLE, CLOSE 3 1/2"	$5.00	0.40	E- Each
	1159	NIPPLE, CLOSE 4"	$6.00	0.45	E- Each
	1160	NIPPLE, OFFSET 1/2"	$75.00	5.00	C - Hundred
	1161	NIPPLE, OFFSET 3/4"	$98.00	6.00	C - Hundred

Category	#	Item Description	Cost	Labor	Unit
NIPPLE Cont'd	1162	NIPPLE, OFFSET 1"	$118.00	7.50	C - Hundred
	1163	NIPPLE, OFFSET 1 1/4"	$186.00	9.00	C - Hundred
	1164	NIPPLE, OFFSET 1 1/2"	$229.00	10.50	C - Hundred
	1165	NIPPLE, OFFSET 2"	$338.00	12.50	C - Hundred
PANEL (LOAD CENTER)	1205	PANEL 60 AMP 1 PHASE	$28.00	1.00	E - Each
	1206	PANEL 100 AMP 1 PHASE	$37.00	1.00	E - Each
	1207	PANEL 200 AMP 1 PHASE	$72.00	2.50	E - Each
	1208	PANEL 60 AMP 3 PHASE	$110.00	1.50	E - Each
	1209	PANEL 100 AMP 3 PHASE	$110.00	1.50	E - Each
	1210	PANEL 200 AMP 3 PHASE	$380.00	3.00	E - Each
PLATE	1290	DECORATOR 1 GANG	$19.25	2.50	C - Hundred
	1291	DECORATOR 2 GANG	$48.00	4.00	C - Hundred
	1292	DECORATOR 3 GANG	$69.00	6.00	C - Hundred
	1293	DECORATOR 4 GANG	$386.00	7.50	C - Hundred
	1294	DECORATOR 5 GANG	$706.00	9.00	C - Hundred
	1260	PLATE PLASTIC 1G BLANK	$39.00	2.50	C - Hundred
	1261	PLATE PLASTIC 1G SWITCH	$47.00	2.50	C - Hundred
	1262	PLATE PLASTIC 1G SINGLE RE	$47.00	2.50	C - Hundred
	1263	PLATE PLASTIC 1G DUPLEX RE	$47.00	2.50	C - Hundred
	1264	PLATE PLASTIC 2G BLANK	$119.00	3.00	C - Hundred
	1265	PLATE PLASTIC 2G SWITCH	$40.00	4.00	C - Hundred
	1266	PLATE PLASTIC 30/50 AMP	$99.00	5.00	C - Hundred
	1267	PLATE PLASTIC 2G DB RECEPTACLE	$50.00	4.00	C - Hundred
	1268	PLATE PLASTIC 2G COMBO	$50.00	4.00	C - Hundred
	1269	PLATE PLASTIC 3G	$99.00	6.00	C - Hundred
	1270	PLATE PLASTIC 3G SWITCH	$99.00	6.00	C - Hundred
	1271	PLATE PLASTIC 4G	$201.00	7.00	C - Hundred
	1272	PLATE PUSH BUTTON	$33.00	4.00	C - Hundred
	1275	PLATE 2" × 4" BLANK	$35.00	4.00	C - Hundred
	1276	PLATE RAISED 2" × 4" SWITCH	$99.00	4.00	C - Hundred
	1277	PLATE RAISED 2" × 4" SINGLE RE	$99.00	4.00	C - Hundred
	1278	PLATE RAISED 2" × 4" DUPLEX RE	$99.00	4.00	C - Hundred
	1279	PLATE RAISED 4" × 4" SWITCH	$99.00	6.00	C - Hundred
	1280	PLATE RAISED 4" × 4" DB SW	$99.00	6.00	C - Hundred
	1281	PLATE RAISED 4" × 4" COMBO	$99.00	6.00	C - Hundred
	1282	PLATE RAISED 4" × 4" RECEPT	$99.00	6.00	C - Hundred
	1283	PLATE RAISED 4" × 4" SINGLE RECEPT	$99.00	6.00	C - Hundred
	1284	PLATE RAISED 4" × 4" DRYER	$99.00	6.00	C - Hundred

	#	Item Description	Cost	Labor	Unit
PLATE Cont'd	1285	PLATE BLANK 4" × 4"	$99.00	6.00	C - Hundred
	1286	PLATE RAISED GFCI RECEPT	$99.00	6.00	C - Hundred
	1301	PLATE WP 1G SWITCH	$199.00	4.00	C - Hundred
	1302	PLATE WP 1G SINGLE RECEPTACLE	$199.00	4.00	C - Hundred
	1303	PLATE WP 1G DUPLEX RECEPTACLE	$168.00	4.00	C - Hundred
	1304	PLATE WP 1G ROUND 4"	$287.00	8.00	C - Hundred
	1305	PLATE WP 2G BLANK	$183.00	6.00	C - Hundred
	1306	PLATE WP 2G DB SW	$370.33	6.00	C - Hundred
	1307	PLATE WP 2G COMBO	$418.43	6.00	C - Hundred
	1308	PLATE WP 2G DB DUPLEX	$715.00	6.00	C - Hundred
	1309	PLATE WP 2G DRYER	$343.29	8.00	C - Hundred
	1310	PLATE 4" ROUND BLANK	$300.00	6.00	C - Hundred
	1311	PLATE 2" × 4" FLOOD	$104.00	6.00	C - Hundred
	1312	PLATE WP GFI	$420.00	4.00	C - Hundred
PVC SCHEDULE 40	1350	PVC 1/2" SCHEDULE 40	$13.00	1.90	C - Hundred
	1351	PVC 3/4" SCHEDULE 40	$17.00	2.60	C - Hundred
	1352	PVC 1" SCHEDULE 40	$27.00	3.00	C - Hundred
	1353	PVC 1 1/4" SCHEDULE 40	$32.00	3.75	C - Hundred
	1354	PVC 1 1/2" SCHEDULE 40	$40.00	4.00	C - Hundred
	1355	PVC 2" SCHEDULE 40	$49.00	4.50	C - Hundred
	1356	PVC 2 1/2" SCHEDULE 40	$64.00	5.75	C - Hundred
	1357	PVC 3" SCHEDULE 40	$82.00	7.00	C - Hundred
	1358	PVC 3 1/2" SCHEDULE 40	$104.00	10.00	C - Hundred
	1359	PVC 4" SCHEDULE 40	$112.00	10.50	C - Hundred
PVC - BELL END	1360	PVC BELL END 2"	$305.00	22.50	C - Hundred
	1361	PVC BELL END 2 1/2"	$320.00	27.00	C - Hundred
	1362	PVC BELL END 3"	$357.00	32.00	C - Hundred
	1363	PVC BELL END 3 1/2"	$385.00	40.00	C - Hundred
	1364	PVC BELL END 4"	$426.00	45.00	C - Hundred
PVC - COUPLING	1370	PVC COUPLING 1/2"	$14.00	3.75	C - Hundred
	1371	PVC COUPLING 3/4"	$18.00	4.00	C - Hundred
	1372	PVC COUPLING 1"	$39.00	4.50	C - Hundred
	1373	PVC COUPLING 1 1/4"	$44.00	5.00	C - Hundred
	1374	PVC COUPLING 1 1/2"	$51.00	5.50	C - Hundred
	1375	PVC COUPLING 2"	$68.00	6.00	C - Hundred

	#	Item Description	Cost	Labor	Unit
PVC - COUPLING Cont'd	1376	PVC COUPLING 2 1/2"	$64.00	6.25	C - Hundred
	1377	PVC COUPLING 3"	$85.00	7.00	C - Hundred
	1378	PVC COUPLING 3 1/2"	$105.00	7.25	C - Hundred
	1379	PVC COUPLING 4"	$163.00	7.75	C - Hundred
PVC - EXPANSION COUP.	1380	PVC COUP EXPANSION 1/2"	$1,174.00	7.25	C - Hundred
	1381	PVC COUP EXPANSION 3/4"	$1,202.00	9.00	C - Hundred
	1382	PVC COUP EXPANSION 1"	$1,208.00	12.00	C - Hundred
	1383	PVC COUP EXPANSION 1 1/4"	$1,213.00	14.00	C - Hundred
	1384	PVC COUP EXPANSION 1 1/2"	$1,219.00	16.00	C - Hundred
	1385	PVC COUP EXPANSION 2"	$1,370.00	18.00	C - Hundred
	1386	PVC COUP EXPANSION 2 1/2"	$1,849.00	20.00	C - Hundred
	1387	PVC COUP EXPANSION 3"	$2,415.00	23.00	C - Hundred
	1388	PVC COUP EXPANSION 3 1/2"	$2,812.00	25.00	C - Hundred
	1389	PVC COUP EXPANSION 4"	$3,482.00	27.00	C - Hundred
PVC - ELBOW	1390	PVC ELBOW 90 1/2"	$42.00	7.00	C - Hundred
	1391	PVC ELBOW 90 3/4"	$45.00	8.00	C - Hundred
	1392	PVC ELBOW 90 1"	$79.00	9.00	C - Hundred
	1393	PVC ELBOW 90 1 1/4"	$99.00	14.00	C - Hundred
	1394	PVC ELBOW 90 1 1/2"	$101.00	16.00	C - Hundred
	1395	PVC ELBOW 90 2"	$107.00	18.00	C - Hundred
	1396	PVC ELBOW 90 2 1/2"	$276.00	22.00	C - Hundred
	1397	PVC ELBOW 90 3"	$483.00	32.00	C - Hundred
	1398	PVC ELBOW 90 3 1/2"	$667.00	36.00	C - Hundred
	1399	PVC ELBOW 90 4"	$836.00	40.00	C - Hundred
PVC - FEMALE ADAPTER	1400	PVC FEMALE ADAPT 1/2"	$22.00	4.50	C - Hundred
	1401	PVC FEMALE ADAPT 3/4"	$25.00	5.50	C - Hundred
	1402	PVC FEMALE ADAPT 1"	$41.00	6.25	C - Hundred
	1403	PVC FEMALE ADAPT 1 1/4"	$51.00	9.00	C - Hundred
	1404	PVC FEMALE ADAPT 1 1/2"	$61.00	11.00	C - Hundred
	1405	PVC FEMALE ADAPT 2"	$72.00	13.50	C - Hundred
	1406	PVC FEMALE ADAPT 2 1/2"	$97.00	18.00	C - Hundred
	1407	PVC FEMALE ADAPT 3"	$148.00	23.00	C - Hundred
	1408	PVC FEMALE ADAPT 3 1/2"	$194.00	26.00	C - Hundred
	1409	PVC FEMALE ADAPT 4"	$194.00	32.00	C - Hundred
PVC LB & COVER	1410	PVC LB 1/2"	$138.00	10.00	C - Hundred
	1411	PVC LB 3/4"	$187.00	11.00	C - Hundred
	1412	PVC LB 1"	$205.00	14.00	C - Hundred
	1413	PVC LB 1 1/4"	$292.00	19.00	C - Hundred

	#	Item Description	Cost	Labor	Unit
PVC LB & COVER Cont'd	1414	PVC LB 1 1/2"	$341.00	23.00	C - Hundred
	1415	PVC LB 2"	$466.00	27.00	C - Hundred
PVC - MALE ADAPTER	1420	PVC MALE ADAPT 1/2"	$22.00	4.50	C - Hundred
	1421	PVC MALE ADAPT 3/4"	$25.00	5.50	C - Hundred
	1422	PVC MALE ADAPT 1"	$41.00	6.25	C - Hundred
	1423	PVC MALE ADAPT 1 1/4"	$51.00	9.00	C - Hundred
	1424	PVC MALE ADAPT 1 1/2"	$61.00	11.00	C - Hundred
	1425	PVC MALE ADAPT 2"	$72.00	12.50	C - Hundred
	1426	PVC MALE ADAPT 2 1/2"	$91.00	18.00	C - Hundred
	1427	PVC MALE ADAPT 3"	$132.00	23.00	C - Hundred
	1428	PVC MALE ADAPT 3 1/2"	$173.00	26.00	C - Hundred
	1429	PVC MALE ADAPT 4"	$226.00	32.00	C - Hundred
RECEPTACLE	1490	RECEPTACLE DUPLEX 15A 120 VOLT	$52.00	18.00	C - Hundred
	1491	RECEPTACLE DUPLEX 20A 120 VOLT	$180.00	19.00	C - Hundred
	1492	RECEPTACLE GFCI 15A 120 VOLT	$12.00	0.30	E - Each
	1493	RECEPTACLE GFCI 20A 120 VOLT	$17.00	0.35	E - Each
	1494	RECEPTACLE DUPLEX 15A SELF GRD	$74.00	18.00	C - Hundred
	1495	RECEPTACLE SINGLE 15A 120 VOLT	$5.00	0.20	E - Each
	1496	RECEPTACLE SINGLE 20, 120 VOLT	$5.00	0.25	E - Each
	1497	RECEPTACLE SINGLE 20, 250 VOLT	$5.00	0.25	E - Each
	1498	RECEPTACLE DRYER 30A 4 WIRE	$7.00	0.25	E - Each
	1499	RECEPTACLE RANGE 50A 4 WIRE	$7.00	0.35	E - Each
	1504	CLOCK OUTLET & PLATE	$428.00	45.00	C - Hundred
	1505	RECEPTACLE 20A I.G. 120 VOLT	$1,100.00	25.00	C - Hundred
	1520	DECORATOR RECEPTACLE	$164.00	20.00	C - Hundred
	1521	MULTIOUTLET ASSEMBLY	*	1.00	E - Each
REDUCING BUSHING	1565	RED BUSH 3/4" × 1/2"	$75.00	5.00	C - Hundred
	1566	RED BUSH 1" × 3/4"	$90.00	6.00	C - Hundred
	1567	RED BUSH 1 1/4" × 1"	$242.00	7.00	C - Hundred
	1568	RED BUSH 1 1/2" × 1 1/4"	$302.00	8.00	C - Hundred
	1569	RED BUSH 2" × 1 1/2"	$502.00	9.50	C - Hundred
	1570	RED BUSH 2 1/2" × 2"	$852.00	11.00	C - Hundred
	1571	RED BUSH 3" × 2 1/2"	$1,094.00	14.00	C - Hundred
	1572	RED BUSH 3 1/2" × 3"	$1,409.00	16.00	C - Hundred
	1573	RED BUSH 4" × 3 1/2"	$2,122.00	18.00	C - Hundred
REDUCING WASHER	1574	RED/WASH 3/4" × 1/2"	$17.00	6.50	C - Hundred
	1575	RED/WASH 1" × 1/2"	$11.00	9.00	C - Hundred

Category	#	Item Description	Cost	Labor	Unit
REDUCING WASHER Cont'd	1576	RED/WASH 1" × 3/4"	$11.50	14.00	C - Hundred
	1577	RED/WASH 1 1/4" x 1"	$21.00	17.00	C - Hundred
	1578	RED/WASH 1 1/2" × 1 1/4"	$25.00	25.00	C - Hundred
	1579	RED/WASH 2" × 1 1/2"	$60.00	32.00	C - Hundred
	1580	RED/WASH 2 1/2" × 2"	$70.00	42.00	C - Hundred
RIGID CONDUIT	1590	RIGID 1/2"	$57.00	3.50	C - Hundred
	1591	RIGID 3/4"	$72.00	4.50	C - Hundred
	1592	RIGID 1"	$116.00	5.75	C - Hundred
	1593	RIGID 1 1/4"	$148.00	5.25	C - Hundred
	1594	RIGID 1 1/2"	$181.00	6.00	C - Hundred
	1595	RIGID 2"	$243.00	7.00	C - Hundred
	1596	RIGID 2 1/2"	$399.00	10.50	C - Hundred
	1597	RIGID 3"	$511.00	13.00	C - Hundred
	1598	RIGID 3 1/2"	$622.00	18.00	C - Hundred
	1599	RIGID 4"	$735.00	20.00	C - Hundred
RIGID - CONNECTOR SS	1600	RIGID CONN SS 1/2"	$92.00	6.50	C - Hundred
	1601	RIGID CONN SS 3/4"	$129.00	9.00	C - Hundred
	1602	RIGID CONN SS 1"	$201.00	12.00	C - Hundred
	1603	RIGID CONN SS 1 1/4"	$315.00	16.00	C - Hundred
	1604	RIGID CONN SS 1 1/2"	$452.00	19.00	C - Hundred
	1605	RIGID CONN SS 2"	$908.00	23.00	C - Hundred
	1606	RIGID CONN SS 2 1/2"	$2,511.00	40.00	C - Hundred
	1607	RIGID CONN SS 3"	$3,291.00	55.00	C - Hundred
	1608	RIGID CONN SS 3 1/2"	$10,649.00	70.00	C - Hundred
	1609	RIGID CONN SS 4"	$13,461.00	75.00	C - Hundred
RIGID - COUPLING SS	1620	RIGID COUP SS 1/2"	$137.00	5.00	C - Hundred
	1621	RIGID COUP SS 3/4"	$175.00	5.50	C - Hundred
	1622	RIGID COUP SS 1"	$286.00	6.50	C - Hundred
	1623	RIGID COUP SS 1 1/4"	$437.00	7.00	C - Hundred
	1624	RIGID COUP SS 1 1/2"	$563.00	8.00	C - Hundred
	1625	RIGID COUP SS 2"	$1,266.00	9.50	C - Hundred
	1626	RIGID COUP SS 2 1/2"	$2,918.00	11.00	C - Hundred
	1627	RIGID COUP SS 3"	$3,486.00	12.00	C - Hundred
	1628	RIGID COUP SS 3 1/2"	$11,310.00	13.00	C - Hundred
	1629	RIGID COUP SS 4"	$14,886.00	14.00	C - Hundred
RIGID ELBOW	1650	RIGID ELBOW 1/2"	$186.00	16.00	C - Hundred
	1651	RIGID ELBOW 3/4"	$223.00	21.00	C - Hundred
	1652	RIGID ELBOW 1"	$333.00	25.00	C - Hundred

	#	Item Description	Cost	Labor	Unit
RIGID ELBOW Cont'd	1653	RIGID ELBOW 1¼"	$476.00	32.00	C - Hundred
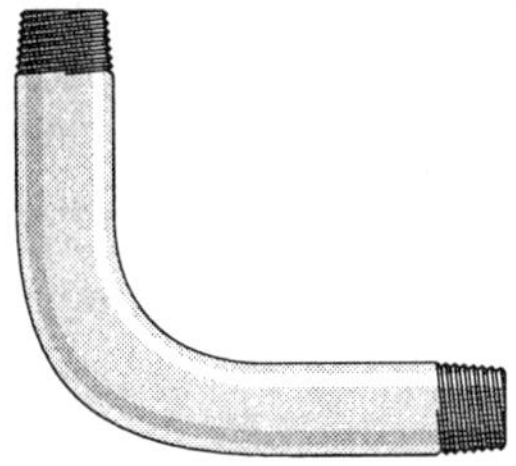	1654	RIGID ELBOW 1½"	$619.00	36.00	C - Hundred
	1655	RIGID ELBOW 2"	$878.00	40.00	C - Hundred
	1656	RIGID ELBOW 2½"	$1,604.00	44.00	C - Hundred
	1657	RIGID ELBOW 3"	$2,334.00	72.00	C - Hundred
	1658	RIGID ELBOW 3½"	$3,726.00	90.00	C - Hundred
	1659	RIGID ELBOW 4"	$4,186.00	115.00	C - Hundred
RIGID CONDUIT - LB	1660	RIGID LB ½"	$5.00	0.35	E - Each
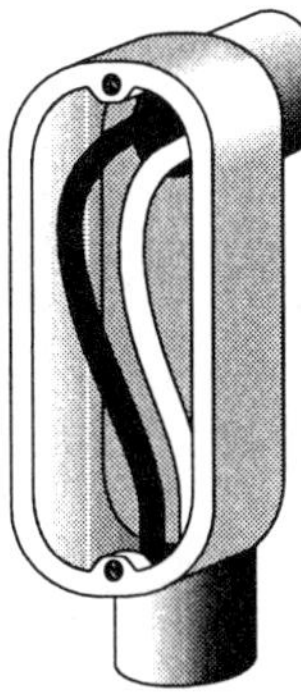	1661	RIGID LB ¾"	$7.00	0.40	E - Each
	1662	RIGID LB 1"	$9.00	0.50	E - Each
	1663	RIGID LB 1¼"	$16.00	0.60	E - Each
	1664	RIGID LB 1½"	$20.00	0.75	E - Each
	1665	RIGID LB 2"	$30.00	1.00	E - Each
	1666	RIGID LB 2½"	$60.00	1.25	E - Each
	1667	RIGID LB 3"	$70.00	1.25	E - Each
	1668	RIGID LB 3½"	$165.00	2.00	E - Each
	1669	RIGID LB 4"	$210.00	2.50	E - Each
RIGID STRAP	1680	RIGID STRAP 1H ½"	$16.00	2.25	C - Hundred
	1681	RIGID STRAP 1H ¾"	$20.00	2.50	C - Hundred
	1682	RIGID STRAP 1H 1"	$40.00	2.75	C - Hundred
	1683	RIGID STRAP 2H 1¼"	$18.00	3.00	C - Hundred
	1684	RIGID STRAP 2H 1½"	$21.00	3.00	C - Hundred
	1685	RIGID STRAP 2H 2"	$28.00	3.25	C - Hundred
	1686	RIGID STRAP 2H 2½"	$63.00	3.50	C - Hundred
	1687	RIGID STRAP 2H 3"	$92.00	4.00	C - Hundred
	1688	RIGID STRAP 2H 3½"	$119.00	4.00	C - Hundred
	1689	RIGID STRAP 2H 4"	$132.00	4.50	C - Hundred
RIGID - LB COVER	1690	CONDULET LB COVER ½"	$151.00	4.00	C - Hundred
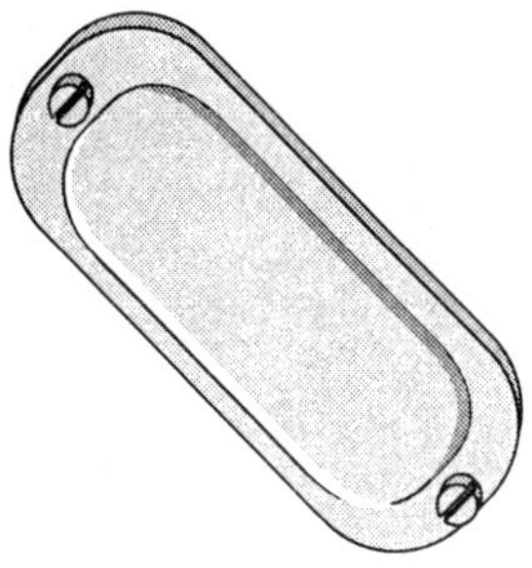	1691	CONDULET LB COVER ¾"	$184.00	4.00	C - Hundred
	1692	CONDULET LB COVER 1"	$251.00	4.00	C - Hundred
	1693	CONDULET LB COVER 1¼"	$18.40	4.00	C - Hundred
	1694	CONDULET LB COVER 1½"	$219.00	4.00	C - Hundred
	1695	CONDULET LB COVER 2"	$532.00	4.00	C - Hundred
	1696	CONDULET LB COVER 2½"	$607.00	4.00	C - Hundred
	1697	CONDULET LB COVER 3"	$607.00	4.00	C - Hundred
	1698	CONDULET LB COVER 3½"	$716.00	4.00	C - Hundred
	1699	CONDULET LB COVER 4"	$716.00	4.00	C - Hundred

Category	#	Item Description	Cost	Labor	Unit
RING - PLASTER	1710	RING SQUARE ROUND	$64.00	4.50	C - Hundred
	1711	RING 1 GANG	$39.00	4.50	C - Hundred
	1714	RING 2 GANG	$52.00	5.00	C - Hundred
	1715	RING 3 GANG	$100.00	9.00	C - Hundred
	1716	RING 4 GANG	$200.00	14.00	C - Hundred
	1719	BELL BOX EXTENSION	$499.00	9.00	C - Hundred
	1720	EXT RING 2" × 4"	$92.00	9.00	C - Hundred
	1721	EXT RING 4" × 4"	$135.00	9.00	C - Hundred
	1722	EXT RING 3-4" OCTAGON	$104.00	9.00	C - Hundred
RX – NM CABLE	1740	RX 14/2 COPPER	$108.00	10.00	M - Thousand
	1741	RX 14/3 COPPER	$186.00	12.00	M - Thousand
	1742	RX 12/2 COPPER	$148.00	12.00	M - Thousand
	1743	RX 12/3 COPPER	$267.00	14.00	M - Thousand
	1744	RX 10/2 COPPER	$216.00	14.00	M - Thousand
	1745	RX 10/3 COPPER	$397.00	16.00	M - Thousand
	1746	RX 8/2 COPPER	$583.00	16.00	M - Thousand
	1747	RX 8/3 COPPER	$601.00	20.00	M - Thousand
	1748	RX 6/2 COPPER	$820.00	20.00	M - Thousand
	1749	RX 6/3 COPPER	$1,014.00	22.00	M - Thousand
RX - FITTING	1750	RX CONN PLASTIC	$10.00	1.00	C - Hundred
	1751	RX CONN 1/2" METAL	$16.00	2.00	C - Hundred
	1752	RX CONN 3/4" METAL	$24.00	3.00	C - Hundred
	1753	RX CONN 1" METAL	$42.00	5.00	C - Hundred
	1760	RX STAPLE SMALL	$1.00	1.00	C - Hundred
	1761	RX STAPLE LARGE	$1.00	1.00	C - Hundred
	1762	KICK PLATES	$5.00	1.00	C - Hundred
SCREW AND NUT	1782	MACHINE SC 6/32 × 1"	$2.00	2.25	C - Hundred
	1785	MACHINE SC 8/32 × 1"	$3.00	2.25	C - Hundred
	1787	MACHINE SC 10/32 × 3/4"	$12.00	2.25	C - Hundred
	1790	MACHINE SC 1/4" × 20 × 1"	$6.00	2.25	C - Hundred
	1794	NUT 6/32"	$2.00	2.25	C - Hundred
	1795	NUT 8/32"	$2.00	2.25	C - Hundred
	1796	NUT 10/32"	$3.00	2.25	C - Hundred
	1797	NUT 1/4" × 20	$3.00	2.25	C - Hundred

Category	#	Item Description	Cost	Labor	Unit
SCREW AND NUT Cont'd	1798	SHEET METAL SCREW #8 × 3/4"	$2.00	2.25	C - Hundred
	1799	SHEET METAL SCREW #8 × 1 1/2"	$4.00	2.50	C - Hundred
	1800	SHEET METAL SCREW #10 × 3/4"	$3.00	2.25	C - Hundred
	1801	SHEET METAL SCREW #10 × 1 1/2"	$4.00	2.50	C - Hundred
SEALTIGHT	1890	SEALTIGHT 1/2"	$47.00	3.25	C - Hundred
	1891	SEALTIGHT 3/4"	$79.00	3.75	C - Hundred
	1892	SEALTIGHT 1"	$116.00	4.00	C - Hundred
	1893	SEALTIGHT 1 1/4"	$166.00	4.50	C - Hundred
	1894	SEALTIGHT 1 1/2"	$174.00	5.00	C - Hundred
	1895	SEALTIGHT 2"	$222.00	6.50	C - Hundred
	1896	SEALTIGHT 2 1/2"	$400.00	8.00	C - Hundred
	1897	SEALTIGHT 3"	$566.00	10.00	C - Hundred
	1898	SEALTIGHT 3 1/2"	$650.00	11.00	C - Hundred
	1899	SEALTIGHT 4"	$790.00	13.00	C - Hundred
SEALTIGHT - CONNECTOR	1900	SEAL CONN STRAIGHT 1/2"	$155.00	6.00	C - Hundred
	1901	SEAL CONN STRAIGHT 3/4"	$139.00	9.50	C - Hundred
	1902	SEAL CONN STRAIGHT 1"	$240.30	10.00	C - Hundred
	1903	SEAL CONN STRAIGHT 1 1/4"	$386.00	14.00	C - Hundred
	1904	SEAL CONN STRAIGHT 1 1/2"	$555.00	15.00	C - Hundred
	1905	SEAL CONN STRAIGHT 2"	$847.00	16.00	C - Hundred
	1906	SEAL CONN STRAIGHT 2 1/2"	$1,550.00	20.00	C - Hundred
	1907	SEAL CONN STRAIGHT 3"	$4,353.00	24.00	C - Hundred
	1908	SEAL CONN STRAIGHT 3 1/2"	$5,125.00	26.00	C - Hundred
	1909	SEAL CONN STRAIGHT 4"	$5,457.00	28.00	C - Hundred
SEALTIGHT - 90	1920	SEAL CONN 90 1/2"	$205.00	9.00	C - Hundred
	1921	SEAL CONN 90 3/4"	$219.00	11.00	C - Hundred
	1922	SEAL CONN 90 1"	$240.00	13.50	C - Hundred
	1923	SEAL CONN 90 1 1/4"	$679.00	16.00	C - Hundred
	1924	SEAL CONN 90 1 1/2"	$945.00	17.00	C - Hundred
	1925	SEAL CONN 90 2"	$1,430.00	18.00	C - Hundred
	1926	SEAL CONN 90 2 1/2"	$4,496.00	20.00	C - Hundred
	1927	SEAL CONN 90 3"	$6,090.00	25.00	C - Hundred
	1928	SEAL CONN 90 3 1/2"	$7,545.00	27.00	C - Hundred
	1929	SEAL CONN 90 4"	$9,025.00	30.00	C - Hundred
SE CABLE	1960	SER 8/3 COPPER	$1,062.00	20.00	M - Thousand
	1961	SER 6/3 COPPER	$1,445.00	25.00	M - Thousand
	1962	SER 4/3 COPPER	$2,128.00	26.00	M - Thousand
	1963	SER 2/3 COPPER	$3,553.00	32.00	M - Thousand

Category	#	Item Description	Cost	Labor	Unit
SE CABLE Cont'd	1964	SER 1/3 COPPER	$5,147.00	38.00	M - Thousand
	1965	SER 1/0-3 COPPER	$5,726.00	45.00	M - Thousand
	1966	SER 2/0-3 COPPER	$6,854.00	50.00	M - Thousand
	1967	SER 3/0-3 COPPER	$8,398.00	55.00	M - Thousand
	1968	SER 4/0-3 COPPER	$9,405.00	60.00	M - Thousand
SE CABLE & FITTING	1980	SE CONNECTOR 1/2"	$200.00	25.00	C - Hundred
	1981	SE CONNECTOR 3/4"	$300.00	30.00	C - Hundred
	1982	SE CONNECTOR 1"	$400.00	35.00	C - Hundred
	1983	SE CONNECTOR 1 1/4"	$400.00	40.00	C - Hundred
	1985	SE STRAP NO. 8-6	$100.00	20.00	C - Hundred
	1986	SE STRAP NO. 4-2	$200.00	25.00	C - Hundred
	1987	SE STRAP NO. 8-6	$300.00	30.00	C - Hundred
	1988	SE STRAP NO. 4-2	$400.00	35.00	C - Hundred
	1990	SE WEATHERHEAD 1/0-2/0	$740.00	50.00	C - Hundred
SURFACE RACEWAY	2040	SURFACE RACEWAY #200	$51.00	3.00	C - Hundred
	2041	SURFACE RACEWAY #500	$55.00	3.75	C - Hundred
	2042	BOX CEILING	$600.00	25.00	C - Hundred
	2043	BOX EXTENSION	$517.00	15.00	C - Hundred
	2044	BOX OUTLET	$700.00	25.00	C - Hundred
SURFACE RACEWAY FITTING	2045	COUPLINGS #200	$24.00	5.00	C - Hundred
	2046	COUPLING #500	$19.00	6.00	C - Hundred
	2047	ELBOWS FLAT #200	$225.00	10.00	C - Hundred
	2048	ELBOWS FLAT #500	$174.00	15.00	C - Hundred
	2049	ELBOWS INTERIOR #200	$325.00	10.00	C - Hundred
	2050	ELBOWS INTERIOR #500	$104.00	15.00	C - Hundred
	2051	ELBOWS EXTERIOR #200	$284.00	10.00	C - Hundred
	2052	ELBOWS EXTERIOR #500	$98.00	15.00	C - Hundred
	2053	JOINT CAP #200	$30.00	5.00	C - Hundred
	2054	JOINT CAP #500	$20.00	6.00	C - Hundred
	2055	STRAP #200	$32.00	5.00	C - Hundred
	2056	STRAP #500	$13.00	6.00	C - Hundred

	#	Item Description	Cost	Labor	Unit
SURFACE RACEWAY FITTING Cont'd	2057	TEE #200	$412.00	10.00	C - Hundred
	2058	TEE #500	$203.00	15.00	C - Hundred
SWITCH	2091	SW 15, 120 VOLT 1 POLE	$57.00	20.00	C - Hundred
	2092	SW 15, 120 VOLT 3 WAY	$127.00	25.00	C - Hundred
	2093	SW 15, 120 VOLT 4 WAY	$719.00	28.00	C - Hundred
	2094	SW 20, 120/277 VOLT 1 POLE	$258.00	20.00	C - Hundred
	2095	SW 20, 120/277 VOLT 3 WAY	$375.00	25.00	C - Hundred
	2096	SW 20, 120/277 VOLT 4 WAY	$1,200.00	28.00	C - Hundred
	2097	SW 20, 120 VOLT 2 POLE	$1,400.00	36.00	C - Hundred
	2098	SW 30, 250 VOLT 2 POLE	$1,200.00	32.00	C - Hundred
	2104	SW 15 AMP PILOT, 120 VOLT	$507.00	25.00	C - Hundred
	2106	SW DUPLEX 120 VOLT	$589.00	25.00	C - Hundred
	2107	SW/RECEPTACLE COMBO 120 VOLT	$627.00	28.00	C - Hundred
	2108	SW BATH/FAN/LIGHT	$17.00	0.50	E - Each
	2109	ATTIC FAN SWITCH	$15.00	0.50	E - Each
	2110	SW 15 120 VOLT FAN CONTROL	$930.00	25.00	C - Hundred
	2115	DECORATOR 1 POLE SWITCH	$264.00	25.00	C - Hundred
	2116	DECORATOR SWITCH 3 WAY	$420.00	28.00	C - Hundred
	2117	DECORATOR SWITCH 4 WAY	$650.00	30.00	C - Hundred
	2118	SW/RECEPTACLE COMBO 120 VOLT	$799.00	35.00	C - Hundred
TELEVISION	2215	ANTENNA WITH MAST	$90.00	3.00	E - Each
	2216	COUPLER (FEED THRU)	$95.00	0.10	E - Each
	2217	F CONNECTOR	$95.00	0.10	E - Each
	2220	SPLITTER 1-2	$3.00	0.10	E - Each
	2221	SPLITTER 1-3	$4.00	0.15	E - Each
	2222	TV RECEPTACLE	$3.00	0.25	E - Each
	2223	COAX TV CABLE	$180.00	12.00	M - Thousand
TELEPHONE	2224	TELEPHONE WIRE TWIST	$90.00	12.00	M - Thousand
	2225	PHONE JACK INDOOR	$3.00	0.25	E - Each
	2226	PHONE JACK WP	$8.00	0.25	E - Each
TIME CLOCK	2240	TIME CLOCK 40A, 120 VOLT T101	$33.00	1.00	E - Each
	2241	TIME CLOCK 40A, 120 VOLT RT	$52.00	1.00	E - Each
	2242	TIME CLOCK 40A, 250 VOLT T104	$47.00	1.00	E - Each
	2243	TIME CLOCK 40A, 250 VOLT RT	$65.00	1.00	E - Each
	2244	TIME CLOCK 40A, 120 VOLT 7 DAY	$93.00	1.25	E - Each

Category	#	Item Description	Cost	Labor	Unit
TRANSFER SWITCH	2256	TRANSFER SWITCH 100 AMP	$539.00	5.00	E- Each
	2257	TRANSFER SWITCH 200 AMP	$852.00	8.00	E- Each
	2258	TRANSFER SWITCH 400 AMP	$1,500.00	16.00	E- Each
	2259	TRANSFER SWITCH 600 AMP	$2,290.00	20.00	E- Each
	2260	TRANSFER SWITCH 800 AMP	$3,500.00	25.00	E- Each
TRANSFORMER	2269	TRANS 7.5 kVA 480/208 VOLT	*	5.00	E- Each
	2270	TRANS 9 kVA 480/208 VOLT	*	6.00	E- Each
	2271	TRANS 15 kVA 480/208 VOLT	$620.00	7.00	E- Each
	2272	TRANS 25 kVA 480/208 VOLT	*	7.50	E- Each
	2273	TRANS 30 kVA 480/208 VOLT	$742.00	8.00	E- Each
	2274	TRANS 37.5 kVA 480/208 VOLT	*	9.00	E- Each
	2275	TRANS 45 kVA 480/208 VOLT	$882.00	10.00	E- Each
	2276	TRANS 50 kVA 480/208 VOLT	*	11.00	E- Each
	2277	TRANS 75 kVA 480/208 VOLT	$1,313.00	13.00	E- Each
	2278	TRANS 100 kVA 480/208 VOLT	*	15.00	E- Each
	2279	TRANS 112.5 kVA 480/208 VOLT	$1,992.00	16.00	E- Each
	2280	TRANS 150 kVA 480/208 VOLT	$2,597.00	19.00	E- Each
UF CABLE	2290	UF CABLE 14/2 COPPER	$133.00	5.00	M - Thousand
	2291	UF CABLE 14/3 COPPER	$220.00	6.50	M - Thousand
	2292	UF CABLE 12/2 COPPER	$188.00	7.00	M - Thousand
	2293	UF CABLE 12/3 COPPER	$309.00	8.00	M - Thousand
	2294	UF CABLE 10/2 COPPER	$314.00	8.00	M - Thousand
	2295	UF CABLE 10/3 COPPER	$451.00	9.00	M - Thousand
	2296	UF CABLE 8/2 COPPER	$918.00	10.00	M - Thousand
	2297	UF CABLE 8/3 COPPER	$1,203.00	10.50	M - Thousand
	2298	UF CABLE 6/2 COPPER	$1,384.00	11.00	M - Thousand
	2299	UF CABLE 6/3 COPPER	$1,700.00	13.00	M - Thousand
U CHANNEL AND FITTING	2300	NUT 1/4"	$9.00	9.00	C - Hundred
	2301	NUT 3/8"	$13.00	9.00	C - Hundred
	2302	NUT 1/2"	$26.00	9.00	C - Hundred
	2304	ROD 1/4" × 6'	$2.00	0.10	E- Each
	2306	ROD 3/8" × 6'	$2.50	0.15	E- Each
	2308	ROD 1/2" × 6'	$5.00	0.25	E- Each
	2309	ROD COUPLING 1/4"	$95.00	9.00	C - Hundred
	2310	ROD COUPLING 3/8"	$130.00	9.00	C - Hundred
	2311	ROD COUPLING 1/2"	$194.00	9.00	C - Hundred
	2312	U CHANNEL 1 3/16" × 10'	$12.00	0.02	E- Each

	#	Item Description	Cost	Labor	Unit
U CHANNEL & FITTING Cont'd	2313	U CHANNEL 1⅝" × 10'	$15.00	0.30	E- Each
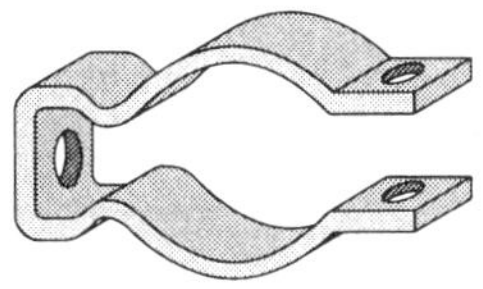 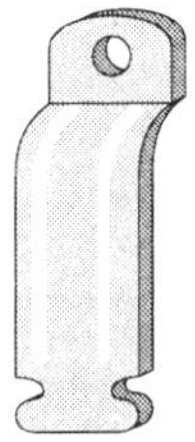	2314	U CHANNEL STRAP ½" RIG	$57.00	3.25	C - Hundred
	2315	U CHANNEL STRAP ¾" RIG	$64.00	4.50	C - Hundred
	2316	U CHANNEL STRAP 1" RIG	$70.00	5.50	C - Hundred
	2317	U CHANNEL STRAP 1¼" RIG	$106.00	7.25	C - Hundred
	2318	U CHANNEL STRAP 1½" RIG	$100.00	8.00	C - Hundred
	2319	U CHANNEL STRAP 2" RIG	$96.00	9.00	C - Hundred
	2323	U CHANNEL STRAP ½" EMT	$57.00	2.00	C - Hundred
	2324	U CHANNEL STRAP ¾" EMT	$64.00	4.00	C - Hundred
	2325	U CHANNEL STRAP 1" EMT	$70.00	5.00	C - Hundred
	2326	U CHANNEL STRAP 1¼" EMT	$106.00	6.25	C - Hundred
	2327	U CHANNEL STRAP 1½" EMT	$86.00	7.25	C - Hundred
	2328	U CHANNEL STRAP 2" EMT	$96.00	9.00	C - Hundred
WEATHERHEAD	2355	WEATHERHEAD ½"	$299.00	25.00	C - Hundred
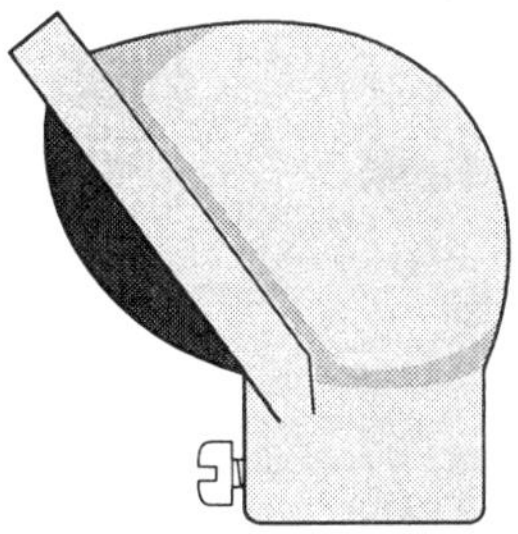	2356	WEATHERHEAD ¾"	$299.00	25.00	C - Hundred
	2357	WEATHERHEAD 1"	$299.00	30.00	C - Hundred
	2358	WEATHERHEAD 1¼"	$299.00	35.00	C - Hundred
	2359	WEATHERHEAD 1½"	$499.00	37.00	C - Hundred
	2360	WEATHERHEAD 2"	$7.00	0.40	E- Each
	2361	WEATHERHEAD 2½"	$20.00	0.45	E- Each
	2362	WEATHERHEAD 3"	$61.00	0.50	E- Each
	2363	WEATHERHEAD 3½"	$27.00	0.65	E- Each
	2364	WEATHERHEAD 4"	$24.00	0.75	E- Each
WIRE - COPPER	2390	WIRE #14 THHN CU 600V	$38.00	3.60	M - Thousand
	2391	WIRE #12 THHN CU 600V	$48.00	4.25	M - Thousand
	2392	WIRE #10 THHN CU 600V	$78.00	5.10	M - Thousand
	2393	WIRE #8 THHN CU 600V	$146.00	6.00	M - Thousand
	2394	WIRE #6 THHN CU 600V	$250.00	7.00	M - Thousand
	2395	WIRE #4 THHN CU 600V	$350.00	7.25	M - Thousand
	2396	WIRE #3 THHN CU 600V	$375.00	7.50	M - Thousand
	2397	WIRE #2 THHN CU 600V	$450.00	7.75	M - Thousand
	2398	WIRE #1 THHN CU 600V	$591.00	8.00	M - Thousand
	2404	WIRE #1/0 THHN CU 600V	$670.00	13.50	M - Thousand
	2405	WIRE #2/0 THHN CU 600V	$816.00	15.25	M - Thousand
	2406	WIRE #3/0 THHN CU 600V	$1,049.00	17.00	M - Thousand
	2407	WIRE #4/0 THHN CU 600V	$1,302.00	18.00	M - Thousand
	2408	WIRE 250 THHN CU 600V	$1,569.00	20.00	M - Thousand
	2409	WIRE 300 THHN CU 600V	$1,840.00	21.00	M - Thousand
	2410	WIRE 350 THHN CU 600V	$2,156.00	23.00	M - Thousand
	2411	WIRE 400 THHN CU 600V	$2,500.00	26.00	M - Thousand

	#	Item Description	Cost	Labor	Unit
WIRE - COPPER Cont'd	2412	WIRE 500 THHN CU 600V	$3,044.00	29.00	M - Thousand
	2419	WIRE GREEN #8 SOLID	$212.00	5.75	M - Thousand
	2420	WIRE BARE #8 COPPER	$146.00	6.00	M - Thousand
	2421	WIRE BARE #6 COPPER	$203.00	7.25	M - Thousand
	2422	WIRE BARE #4 COPPER	$303.00	9.00	M - Thousand
WIRE - MISCELLANEOUS	2423	WIRE FIX #18 TFF 600	$33.00	4.00	M - Thousand
	2424	WIRE FIX #16 TFF 600	$40.00	4.00	M - Thousand
	2425	THERMO WIRE 18/2	$70.00	13.00	M - Thousand
	2426	THERMO WIRE 18/5	$120.00	13.00	M - Thousand
	2427	PULL WIRE	$20.00	6.50	M - Thousand
	2432	PLENUM 18/2 TS	$232.00	12.00	M - Thousand
	2433	PLENUM 18/3 TS	$296.00	13.00	M - Thousand
	2434	PLENUM 18/4 TS	$355.00	14.00	M - Thousand
WIRE NUT	2440	WIRE NUT ORANGE	$4.00	1.00	C - Hundred
	2441	WIRE NUT YELLOW	$5.00	1.00	C - Hundred
	2442	WIRE NUT RED	$8.00	1.00	C - Hundred
	2443	WIRE NUT GRAY	$10.00	1.00	C - Hundred
	2444	WIRE NUT BLUE	$12.00	1.00	C - Hundred
	2445	WIRE NUT GREEN	$10.00	1.00	C - Hundred

Chapter 2

The blueprints listed in this chapter are used in conjunction with the *Electrical Estimating* book, authored by Charles Michael Holt. You will be instructed when to use the specific blueprints contained in this chapter. This chapter contains the following blueprints:

CHAPTER 6 - ESTIMATING RESIDENTIAL WIRING BLUEPRINTS (TOTAL 12)

Blueprint Name	Description	Area Served	Page
Blueprint - E–3	Lighting Symbols	First Floor	35
Blueprint - E–4	Lighting Symbols	Basement	36
Blueprint - E–5	Switch Symbols	First Floor	37
Blueprint - E–6	Switch Symbols	Basement	38
Blueprint - E–7	Convenience Receptacle Symbols	First Floor	39
Blueprint - E–8	Convenience Receptacle Symbols	Basement	40
Blueprint - E–9	Low Voltage Systems	First Floor	41
Blueprint - E–10	Low Voltage Systems	Basement	42
Blueprint - E–11	Circuit Wiring	First Floor	43
Blueprint - E–12	Circuit Wiring	Basement	44
Blueprint - E–13	Homeruns and Individual Circuits	First Floor	45
Blueprint - E–14	Homeruns and Individual Circuits	Basement	46

CHAPTER 6 - BLUEPRINTS FOR RESIDENTIAL SUMMARY QUESTIONS (TOTAL 4)

Blueprint Name	Description	Page
Blueprint - RE–1	Lighting, Switch, Receptacle, and Low Voltage Symbols	47
Blueprint - RE–2	Low Voltage System Wiring	48
Blueprint - RE–3	Circuit Wiring	49
Blueprint - RE–4	Homerun Wiring	50

CHAPTER 7 - ESTIMATING COMMERCIAL WIRING BLUEPRINTS (TOTAL 9)

Blueprint Name	Description	Area Served	Page
Blueprint - EC–4	Light, Switch, Receptacle, Low Voltage Symbols	Basement	51
Blueprint - EC–5	Light, Switch, Receptacle, Low Voltage Symbols	First Floor	52
Blueprint - EC–6	Light, Switch, Receptacle, Low Voltage Symbols	Second Floor	53
Blueprint - EC–7	Circuit Wiring	Basement	54
Blueprint - EC–8	Circuit Wiring	First Floor	55
Blueprint - EC–9	Circuit Wiring	Second Floor	56
Blueprint - EC–10	Homeruns and Individual Circuits	Basement	57
Blueprint - EC–11	Homeruns and Individual Circuits	First Floor	58
Blueprint - EC–12	Homeruns and Individual Circuits	Second Floor	59

CHAPTER 7 - BLUEPRINTS FOR COMMERCIAL SUMMARY QUESTIONS (TOTAL 3)

Blueprint Name	Description	Page
Blueprint - CE–1	Lighting, Switch, and Receptacle Symbols	60
Blueprint - CE–2	Circuit and Homerun Wiring	61
Blueprint - CE–3	Service and Feeder Riser	62

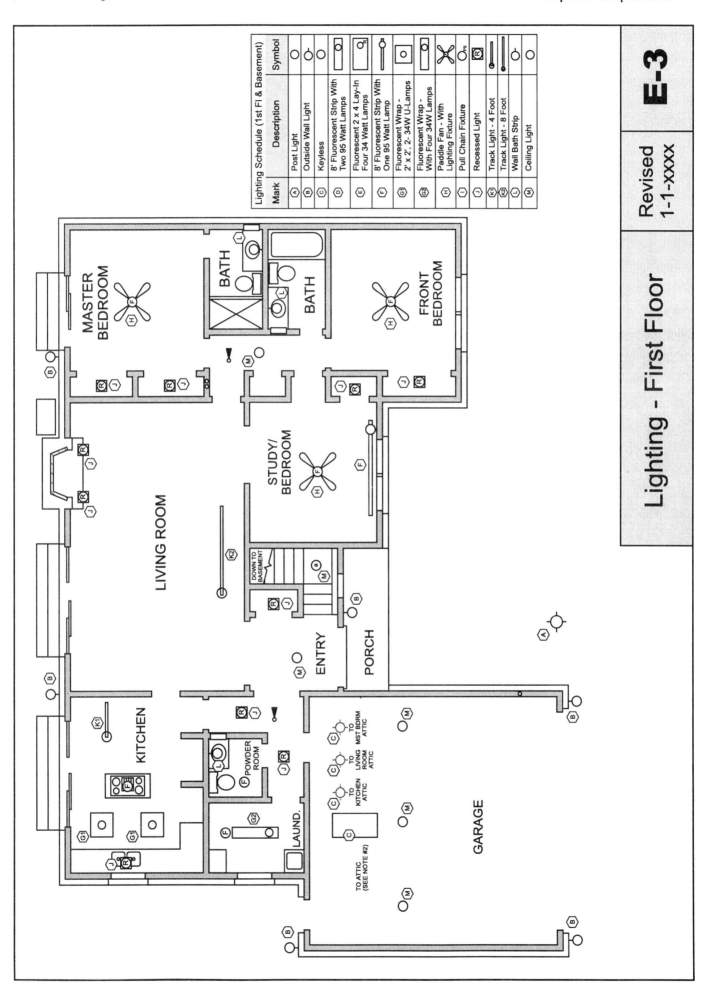

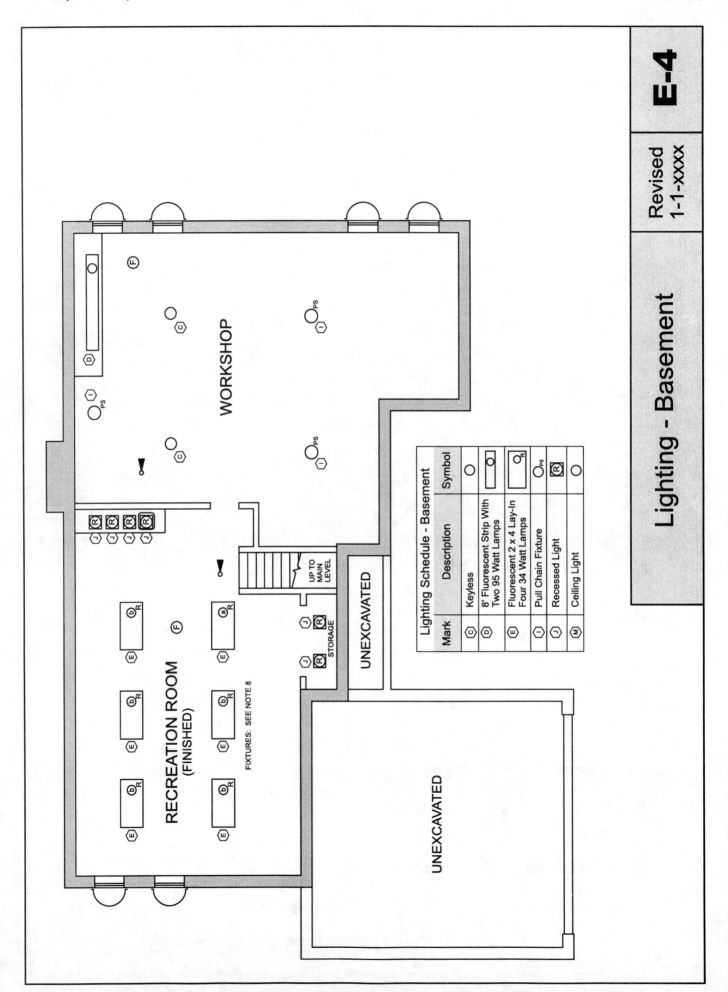

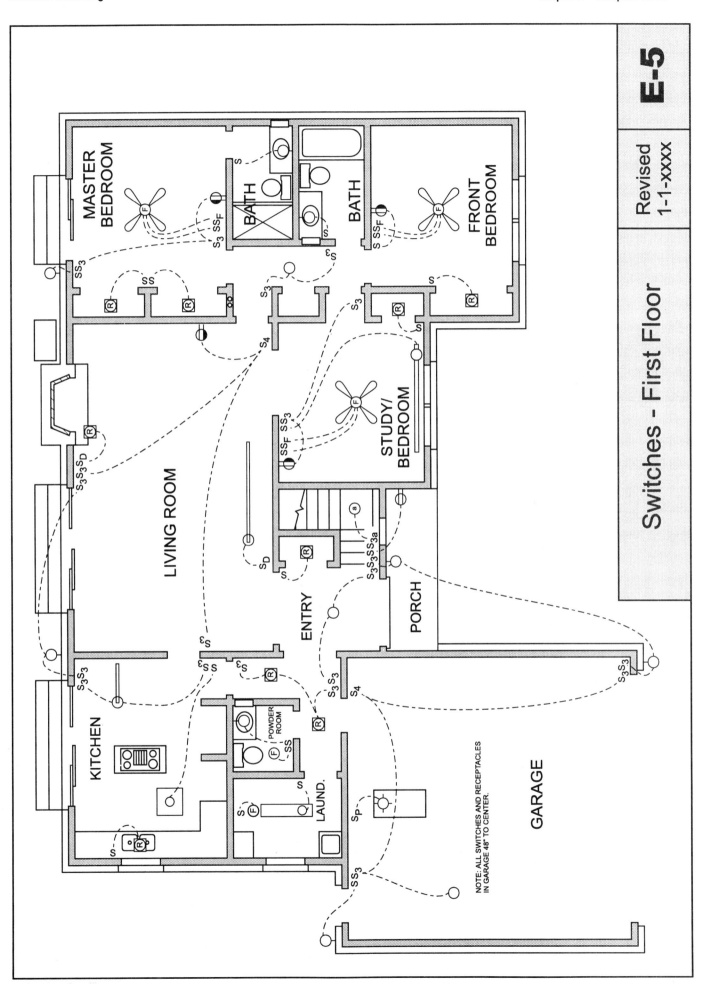

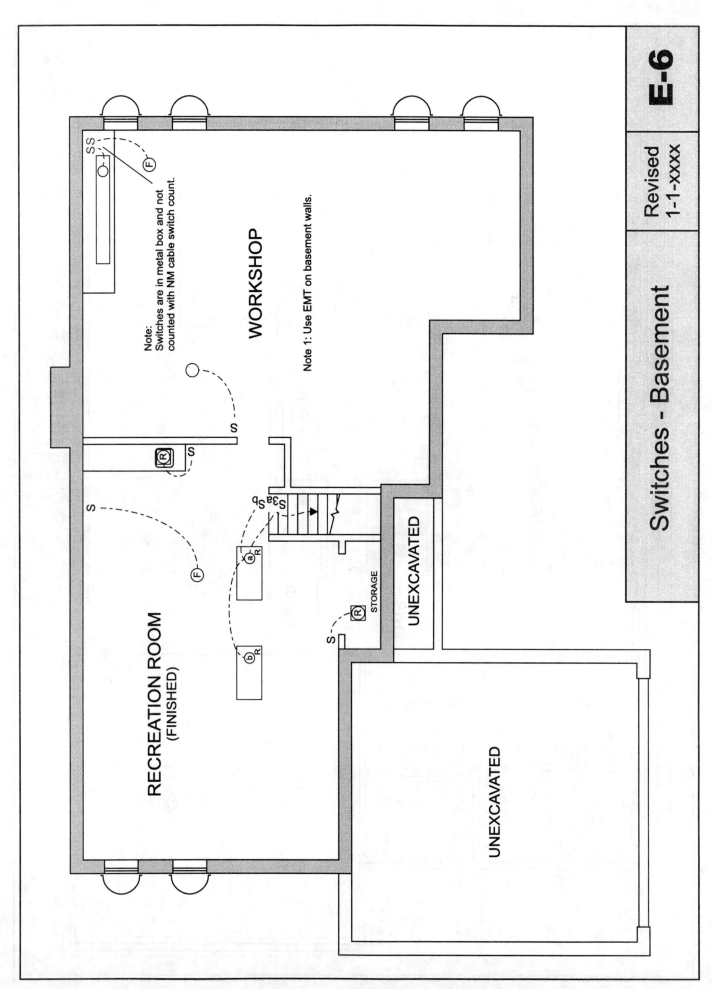

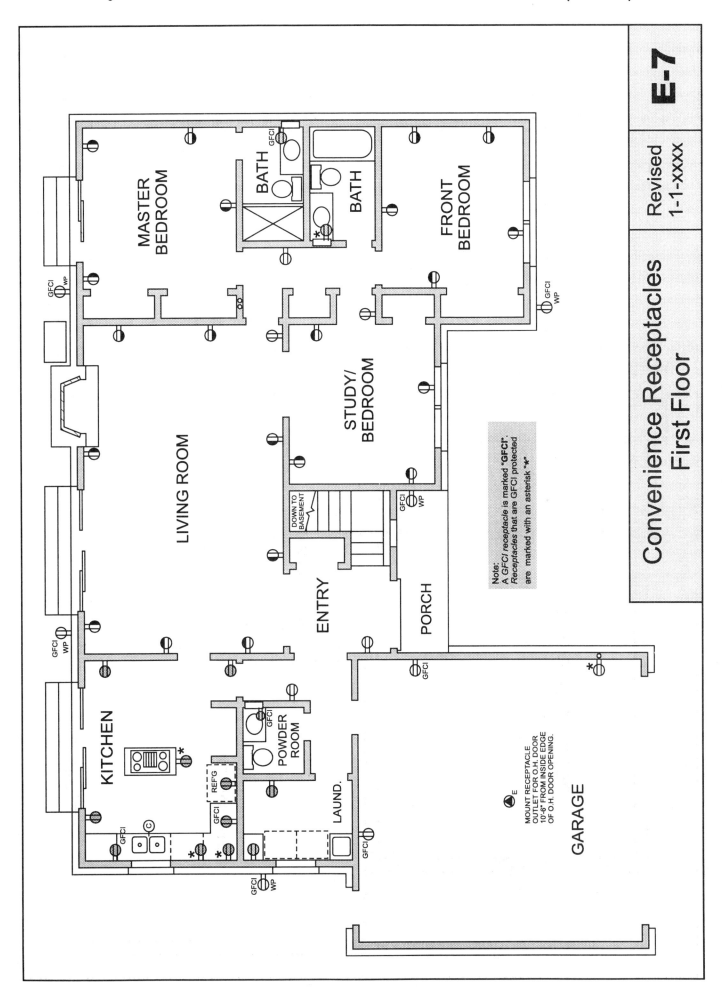

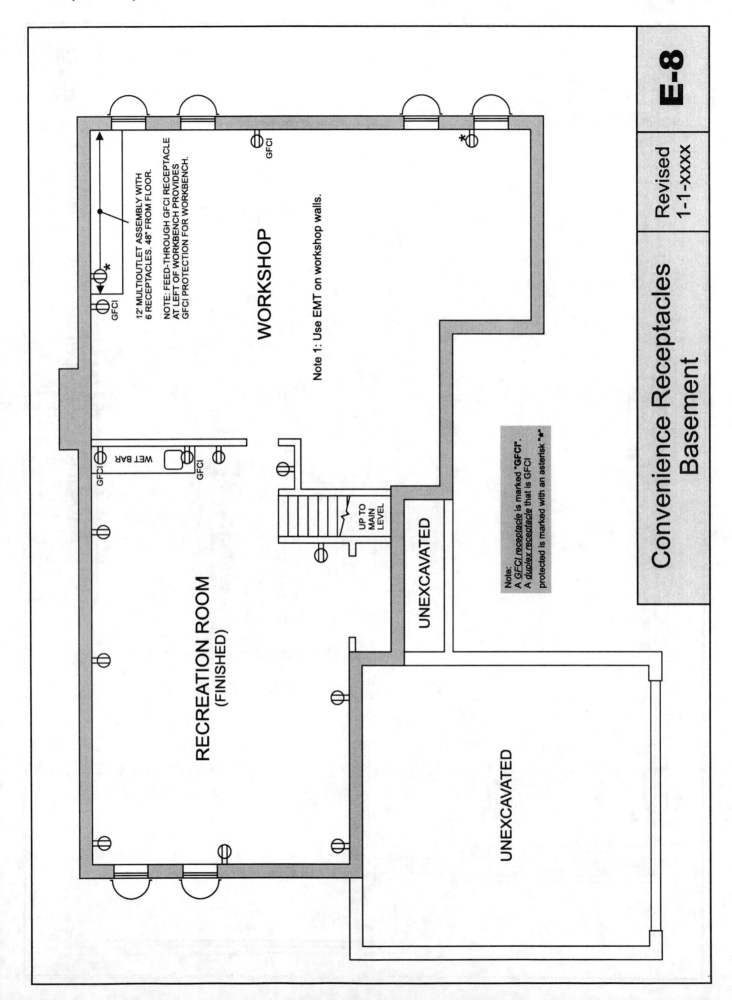

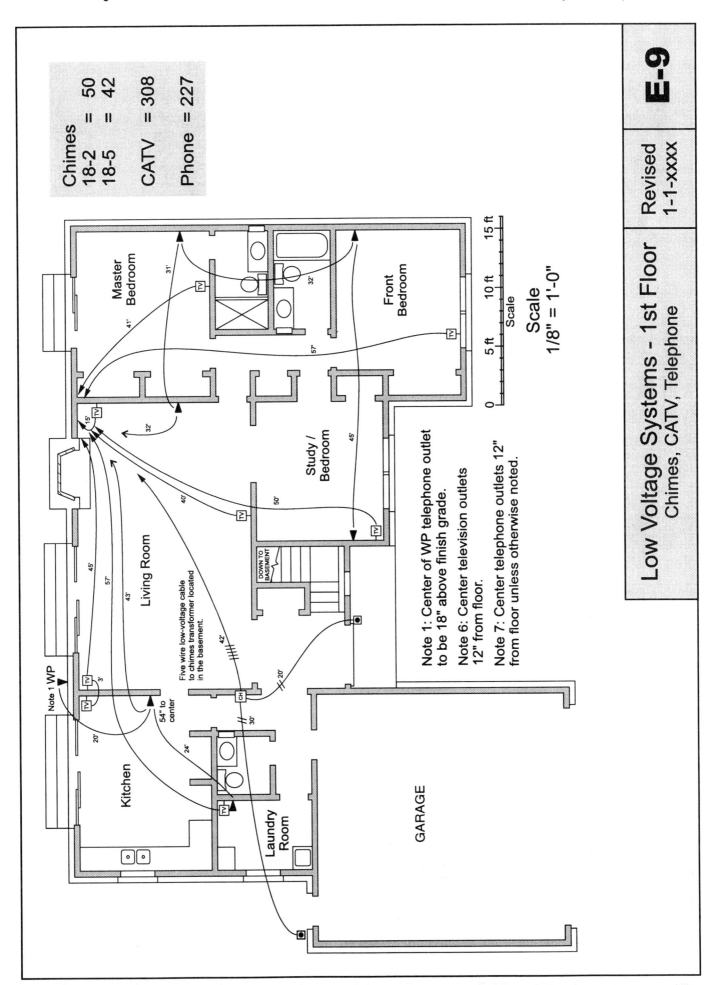

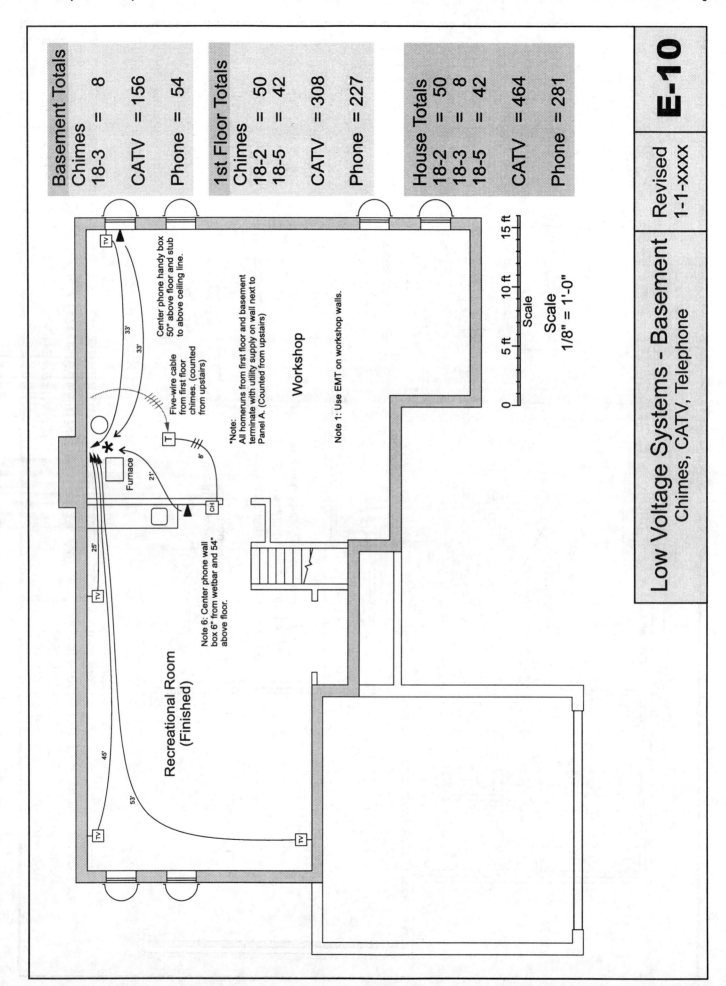

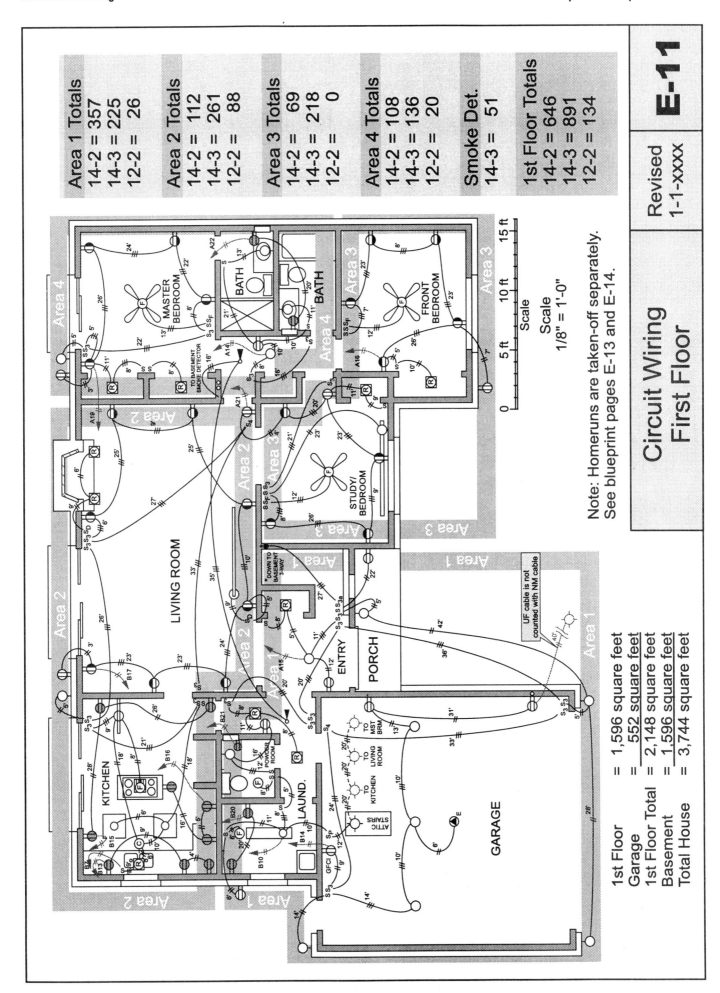

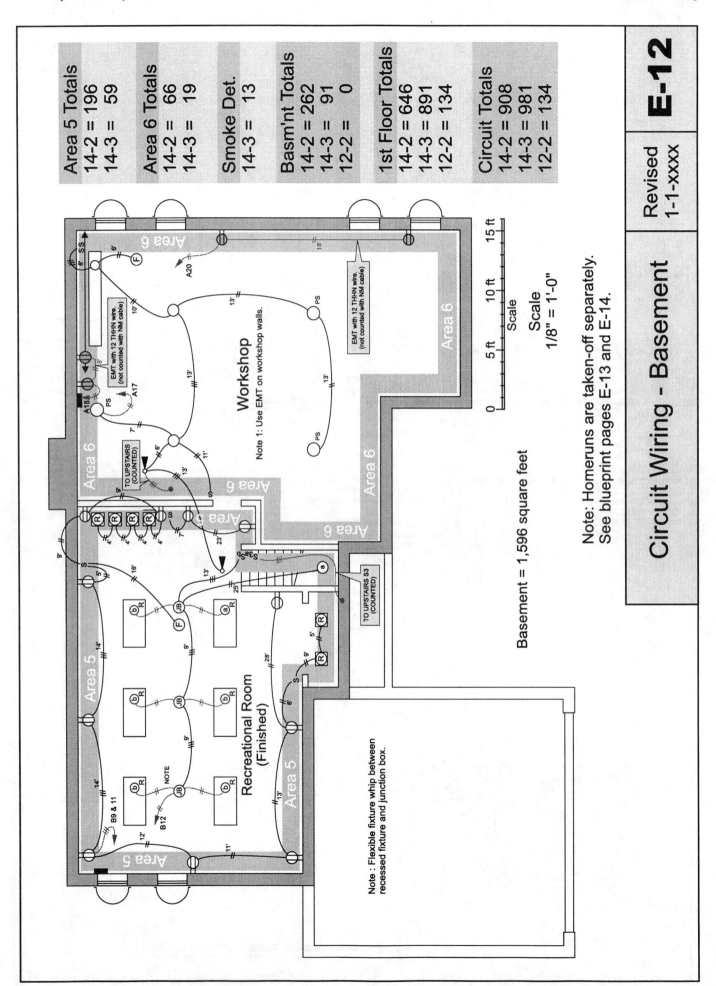

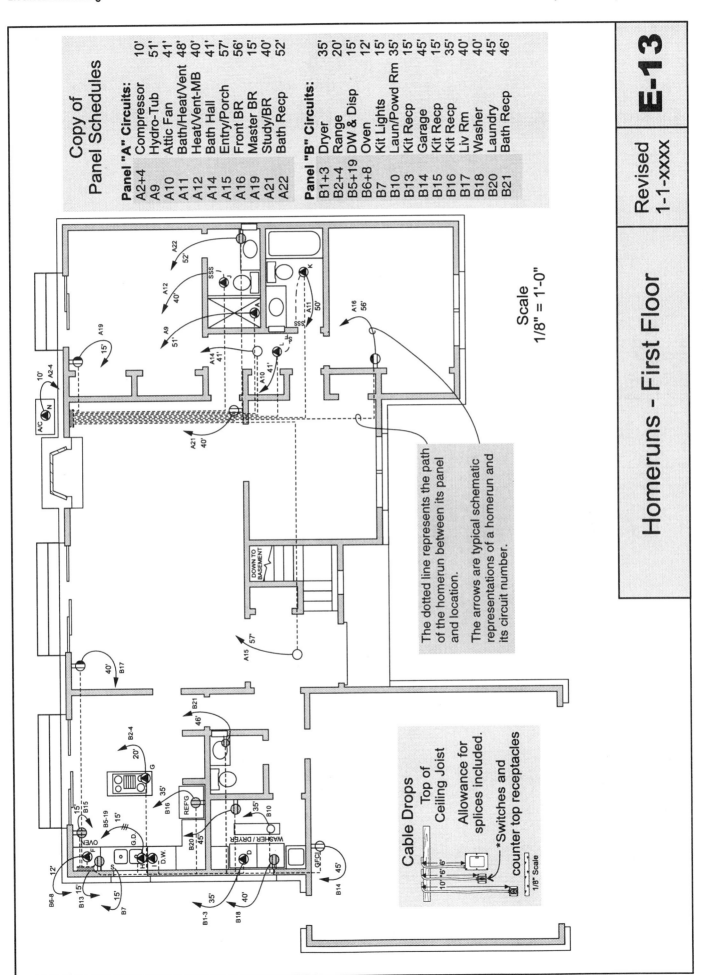

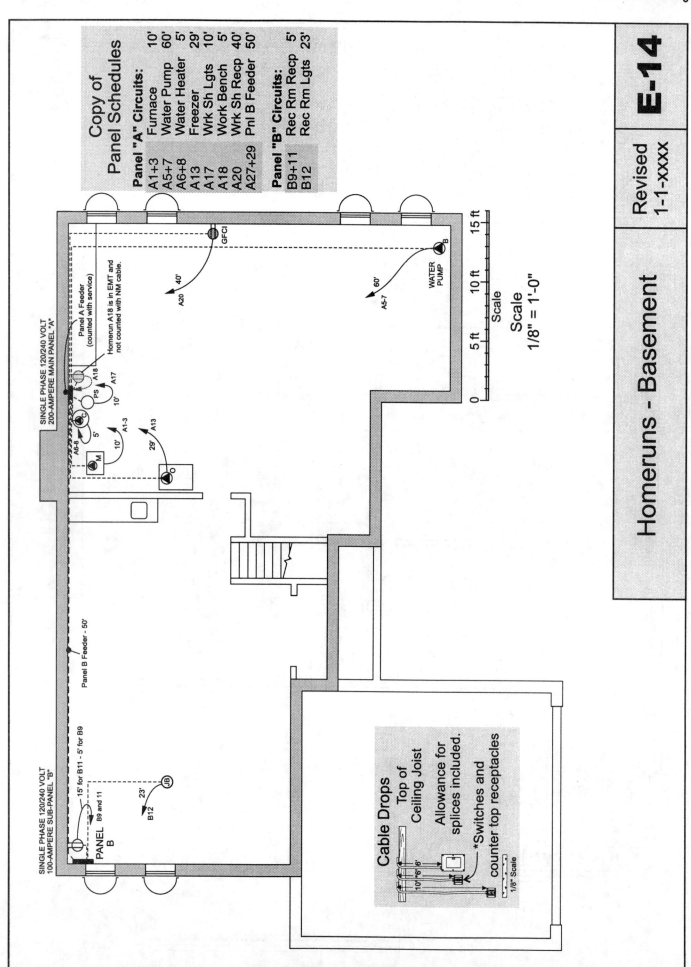

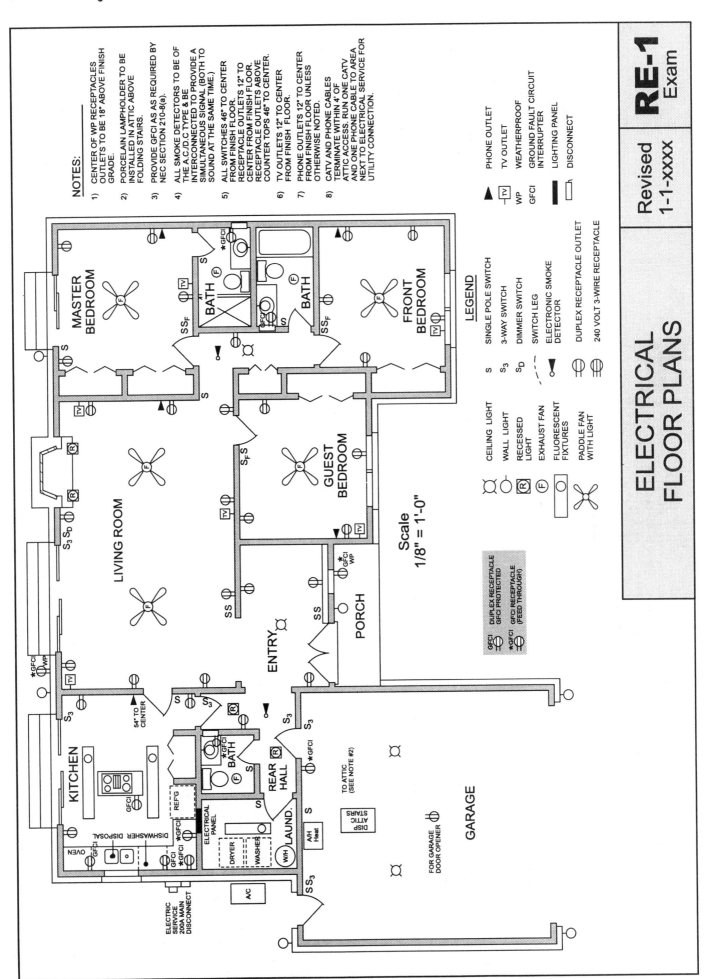

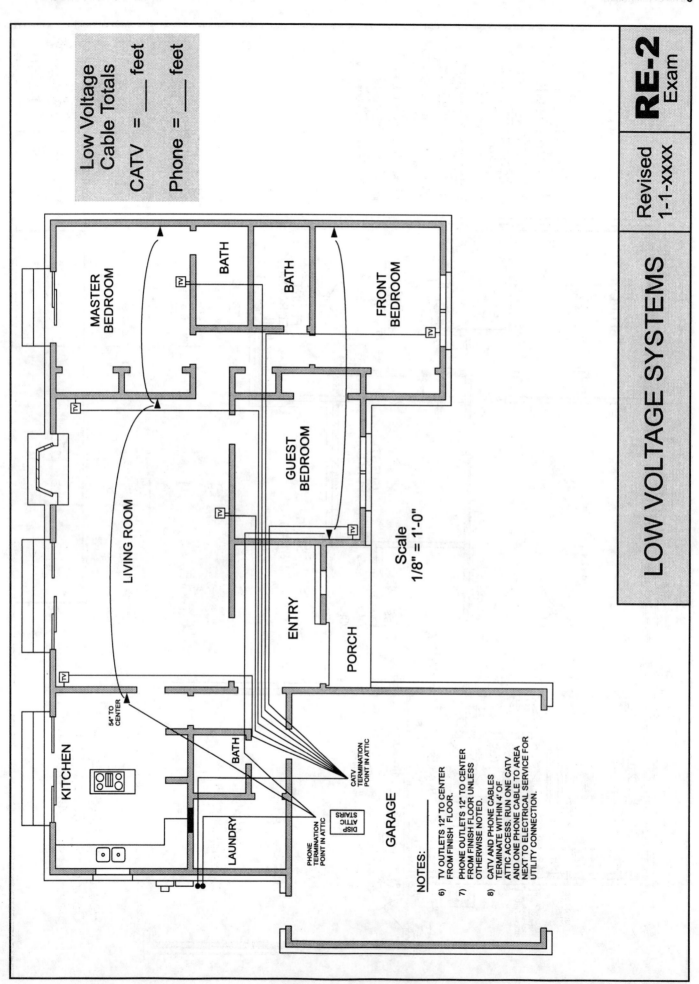

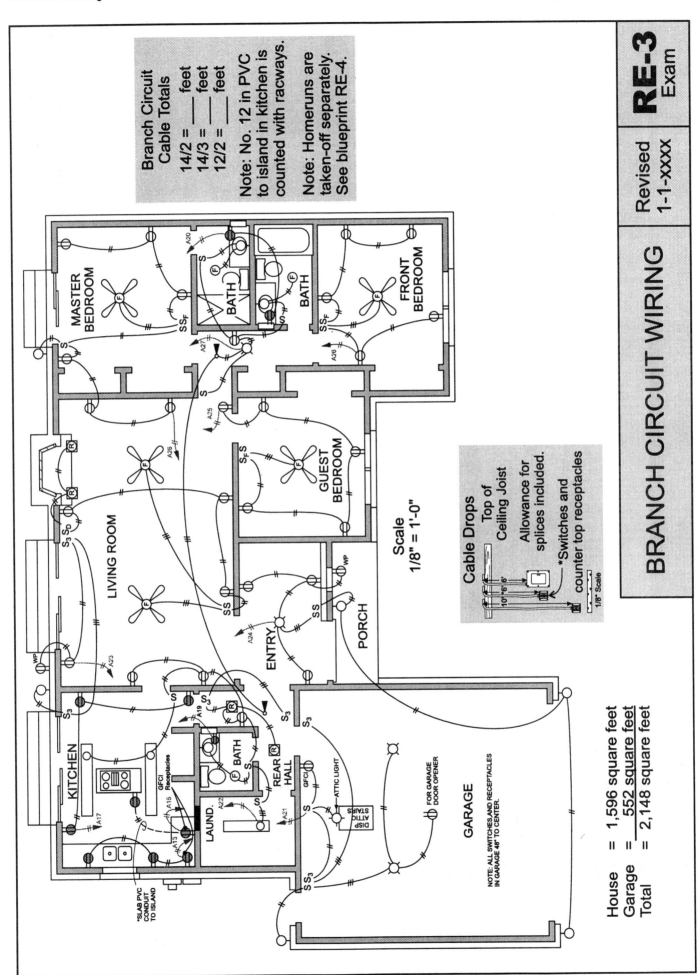

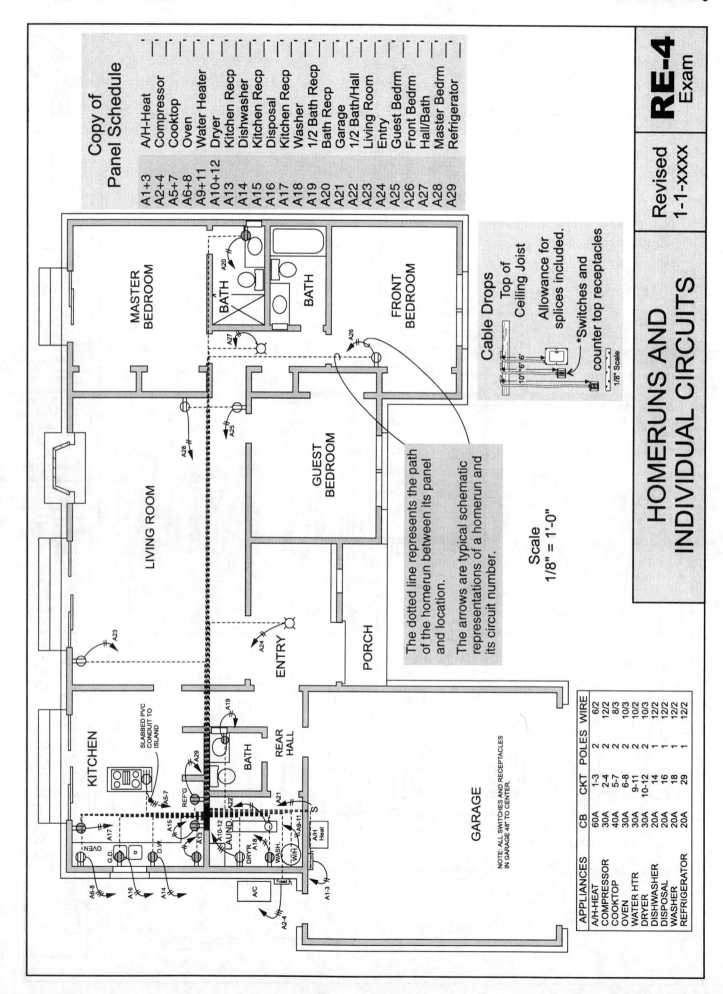

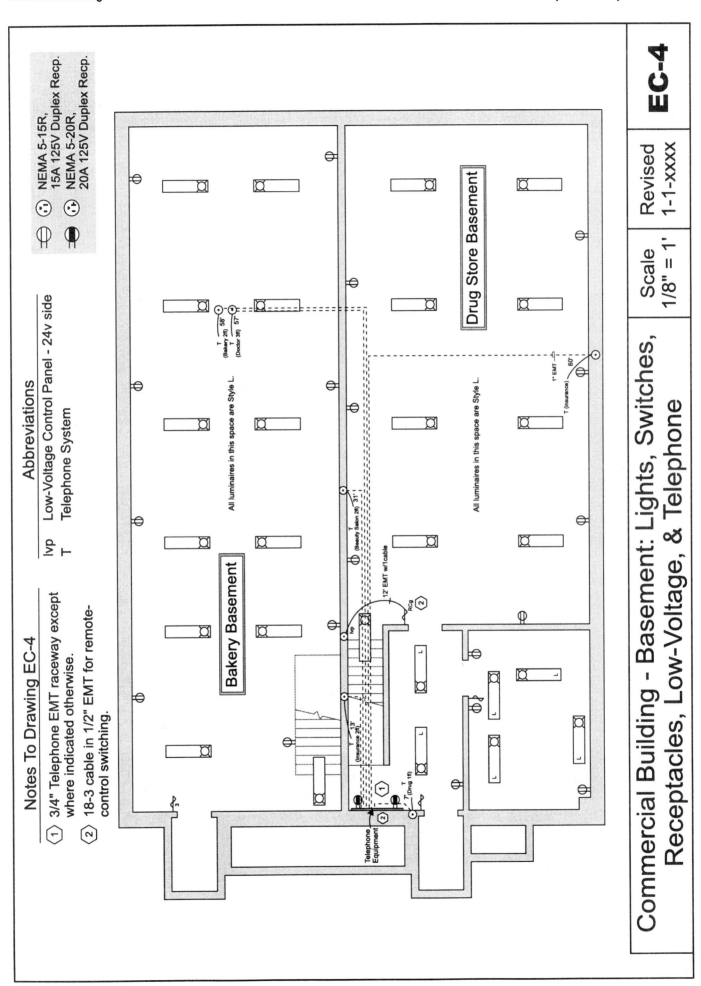

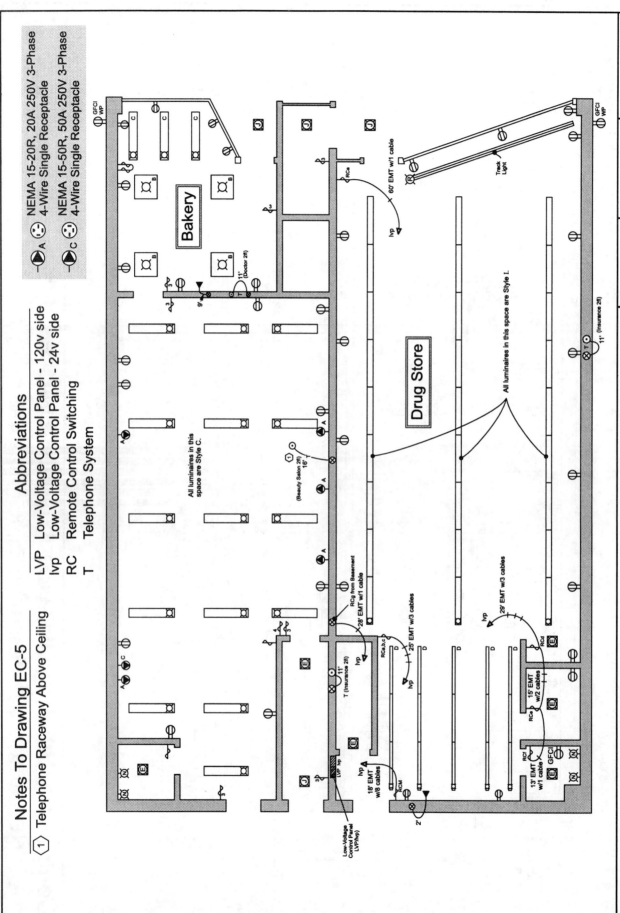

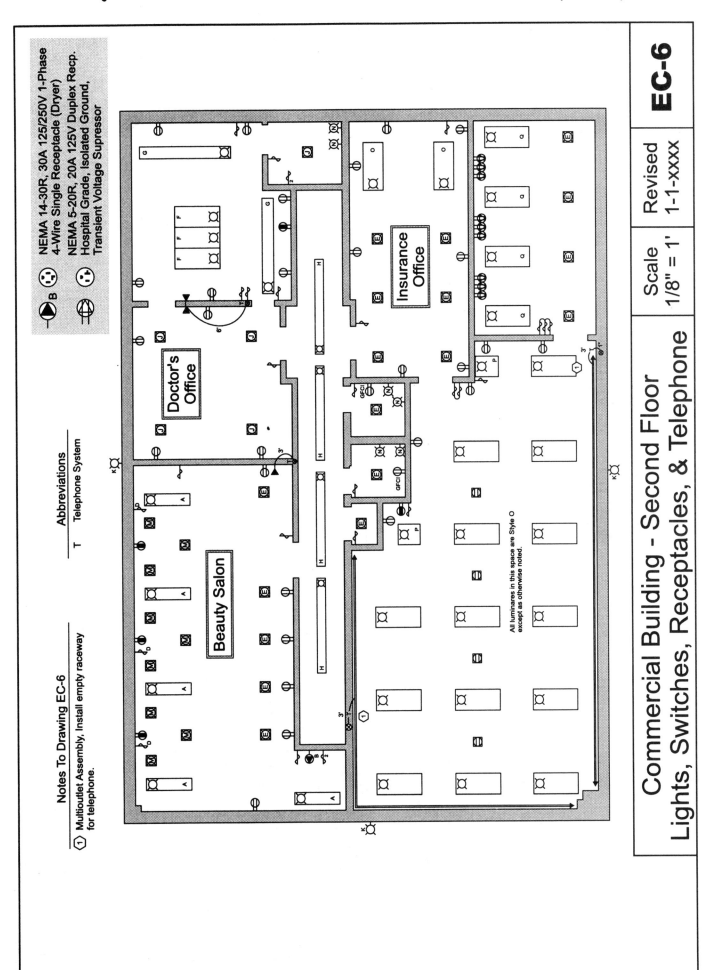

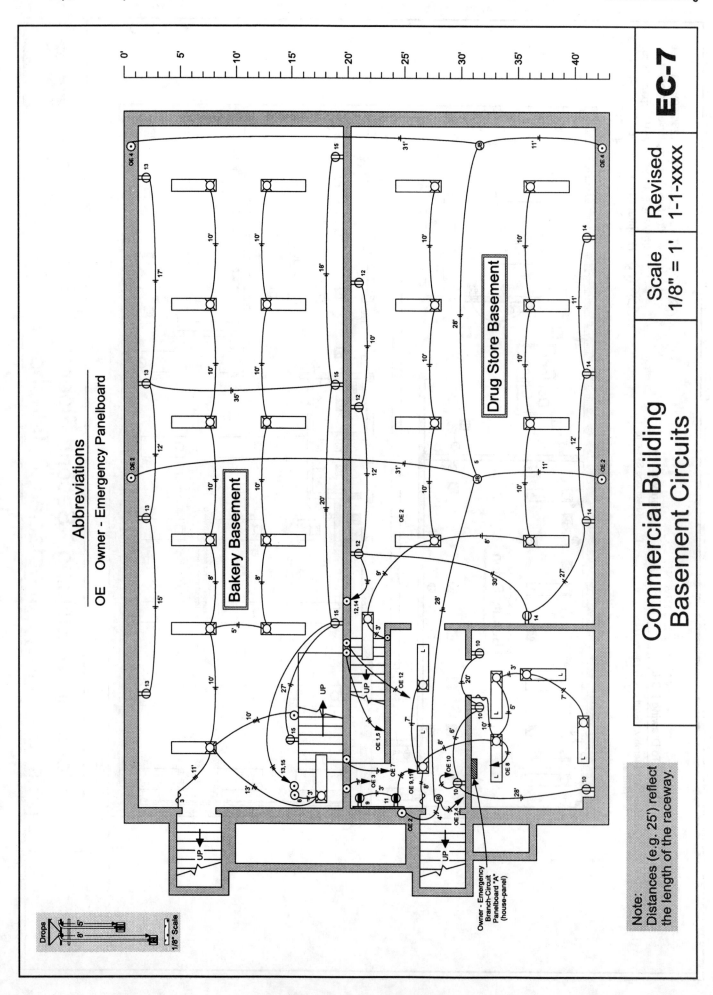

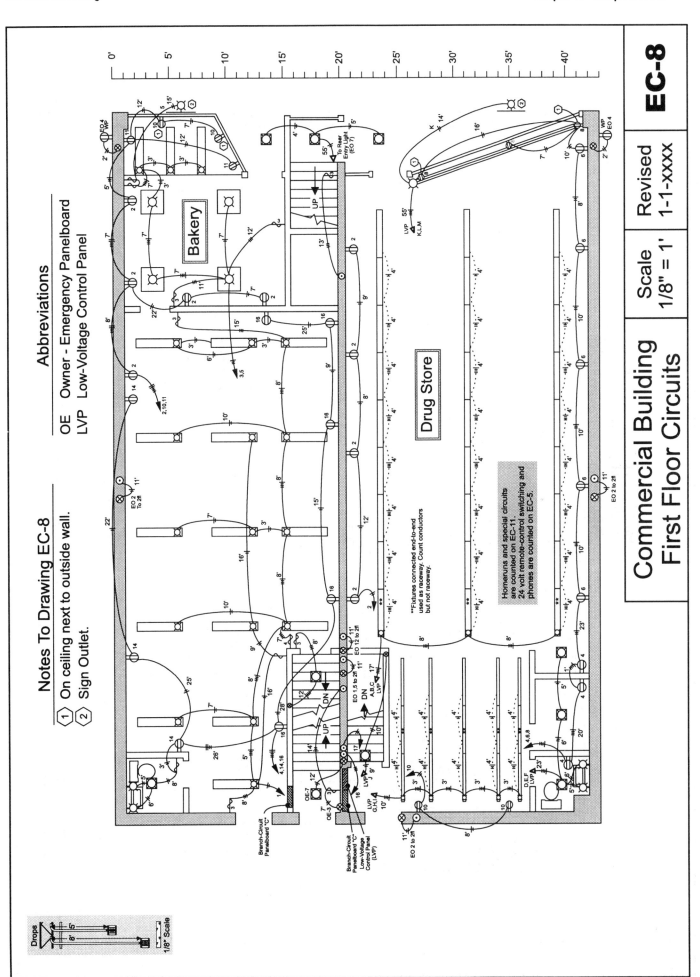

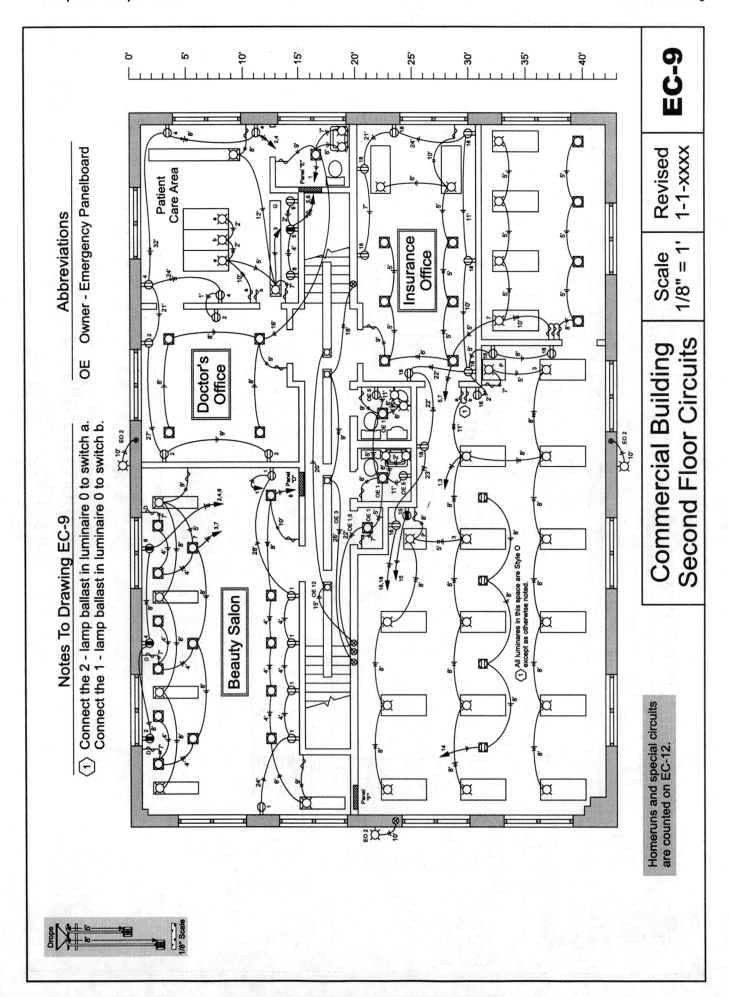

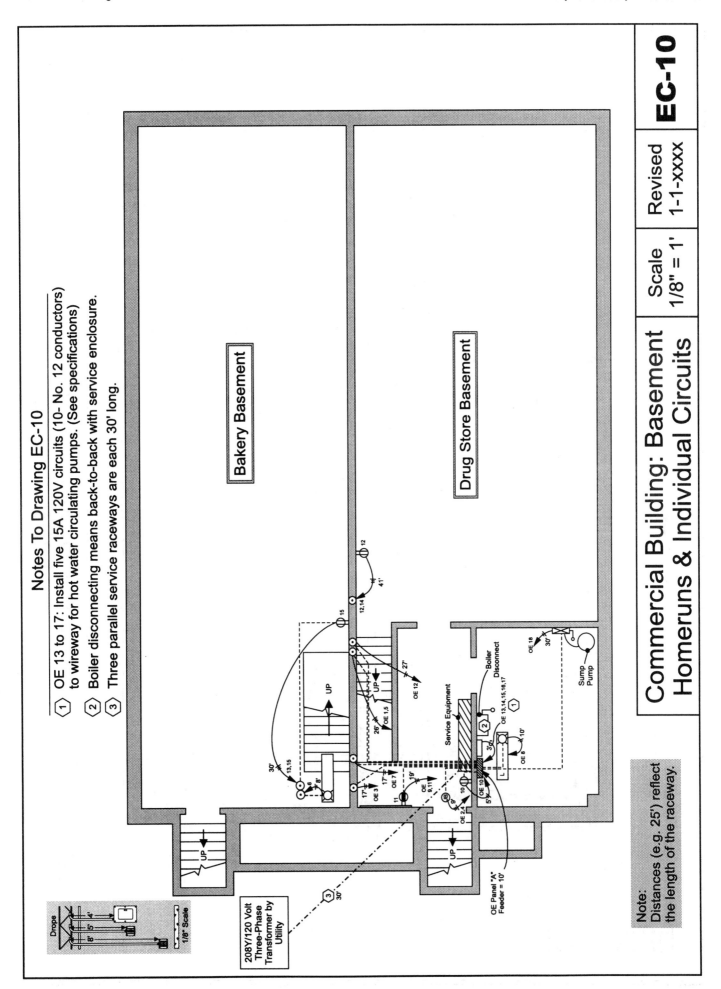

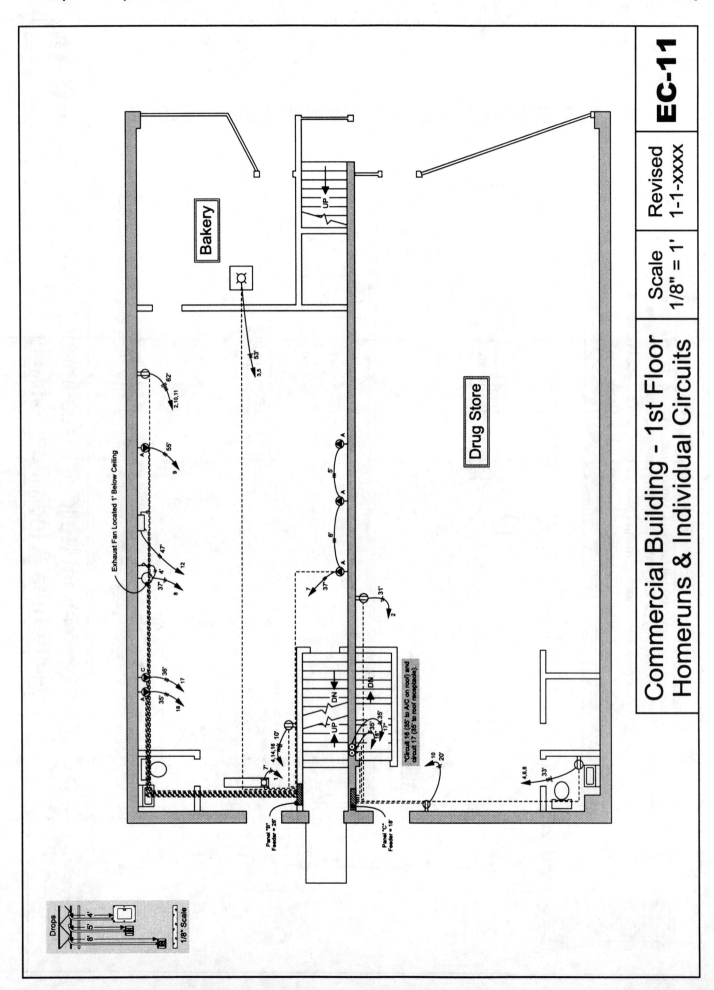

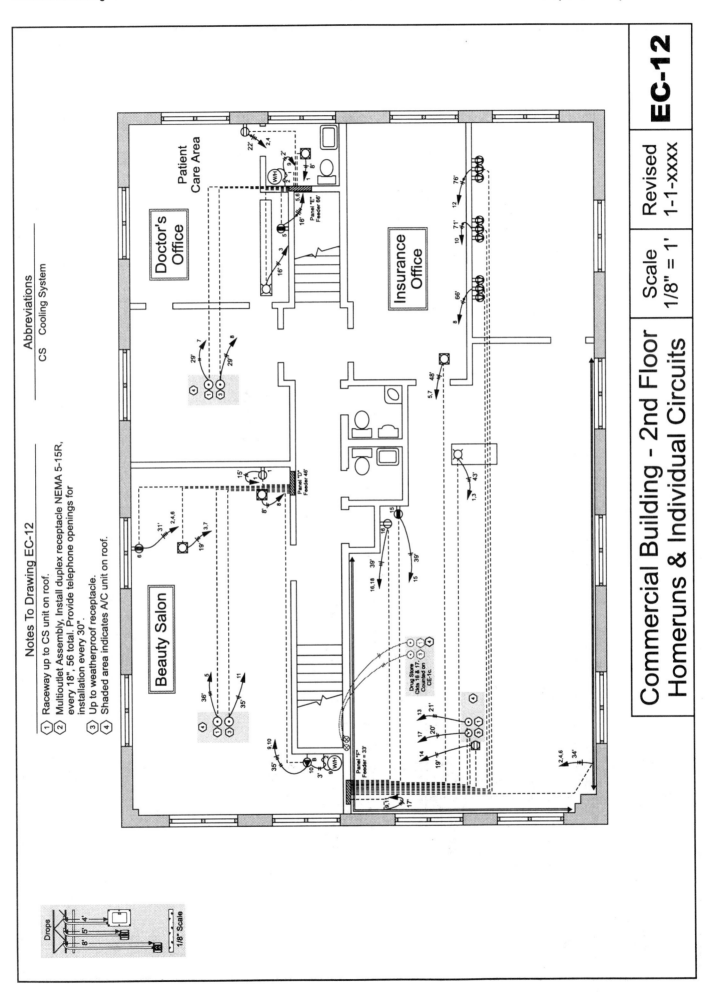

Chapter 2 – Blueprint Book

Notes To Drawing CE-1

1. Receptacles in Doctor's office Patient Care Area require insulated grounding conductor.
2. Multioutlet Assembly, Install duplex receptacle NEMA 5-15R, every 18", 56 total.
3. Connect the 2 - lamp ballast in luminaire J to switch a.
   Connect the 1 - lamp ballast in luminaire J to switch b.
4. Sign Outlet circuit to individual units.
5. Time clock controls entry and security lights, circuits A-2 & 4.
6. A/C unit on roof.
7. Rooftop weatherproof GFCI receptacle on each A/C unit.

Building is 2,580 sq. ft.

CE-1 Exam

Revised 1-1-xxxx

Scale 1/8" = 1'

Commercial Building

See CE-3 for Symbol Schedule, Fixture Schedule, and Service Riser.

Chapter 2 – Blueprint Book

Notes To Drawing CE-2

1. Receptacles in Doctor's office Patient Care Area require insulated grounding conductor.
2. Multioutlet Assembly, Install duplex receptacle NEMA 5-15R, every 18", 56 total.
3. Connect the 2 - lamp ballast in luminaire J to switch a. Connect the 1 - lamp ballast in luminaire J to switch b.
4. Sign Outlet circuit to individual units.
5. Time clock controls entry and security lights, circuits A-2 & 4.
6. A/C unit on roof.
7. Rooftop weatherproof GFCI receptacle on each A/C unit.

CE-2 Exam

Revised: 1-1-xxxx
Scale: 1/8" = 1'

Commercial Building Circuits, Homeruns, Feeders

See CE-3 for Symbol Schedule, Fixture Schedule, and Service Riser.

Luminaire - Lamp Schedule

Style	Nominal Size	Lamp	Ballast	Mounting	Symbol
A	18 inches by 4 feet	Two F40/Spec 30/RS	Energy Saving	Surface	
B	9 inches diameter	Two 26W Quad T4	Compatible	Recessed	
C	2 by 4 feet	Four F023/35K	Matching Electronic	Recessed	
D	1 by 8 feet	Four F023/35K	Matching Electronic	Recessed	
E	9 inches by 8 feet	Two F40/Spec 41/RS	Energy Saving	Surface	
F	8 inches diameter	One 150W, A21	NA	Recessed	
G	16 inches square	One 70W, HPS	Standard	Surface	
H	7 inches diameter	12V 50W NFL	Transformer	Surface	
I	20 inches long	One 60W, 120V	NA	Recessed	
J	4 feet by 20 inches	Three F40/Spec 41/RS	Two Ballast	Recessed	
K	2 feet by 20 inches	Two FB40/Spec 41/RS	One Ballast	Recessed	
L	4 feet by 20 inches	Three F40/Spec 41/RS	Two Ballast	Recessed	

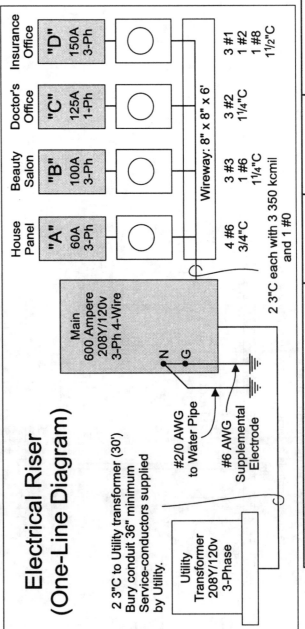

Electrical Riser (One-Line Diagram)

2 3"C to Utility transformer (30')
Bury conduit 36" minimum
Service-conductors supplied by Utility.

#2/0 AWG to Water Pipe
#6 AWG Supplemental Electrode

Utility Transformer 208Y/120v 3-Phase

Main 600 Ampere 208Y/120v 3-Ph 4-Wire

House Panel "A" 60A 3-Ph — 4 #6 3/4"C
Beauty Salon "B" 100A 3-Ph — 3 #3 1 #6 1¼"C
Doctor's Office "C" 125A 1-Ph — 3 #2 1¼"C
Insurance Office "D" 150A 3-Ph — 3 #1 1 #2 1 #8 1½"C

Wireway: 8" x 8" x 6'
2 3"C each with 3 350 kcmil and 1 #0

Commercial Building Details

Scale: None | Revised: 1-1-xxxx | CE-3 Exam

Electrical Symbol Schedule

- Surface Raceway
- Panelboard
- Lighting Outlet, Ceiling
- Lighting Outlet, Wall
- Lighting Outlet, Recessed
- Duplex receptacle outlet, NEMA 5-15R Floor mounted
- Duplex receptacle outlet, NEMA 5-15R Wall mounted
- Duplex receptacle outlet, NEMA 5-20R Wall mounted
- Duplex receptacle outlet, NEMA 5-20R Hospital Grade, Isolated Ground, Transient Voltage Suppressor
- Receptacle outlet, NEMA 14-30R Wall mounted
- Switch, Single Pole, Wall mounted
- Double Pole Switch, 30A 250V
- Switch, 3-way, Wall mounted
- Dimmer Switch
- Numbers Indicate Branch-Circuit Connection
- Luminaire
  - Uppercase Letter Indicates Style
  - Lowercase Letter Indicates Switching
  - Lower case letter indicates switching arrangement
- Branch-Circuit Wiring. Short lines indicate the number of ungrounded conductors, a long line indicates a grounded conductor, and a N indicates an insulated grounding conductor.
- Branch-circuit homerun with circuit numbers and panelboard label if required

NEMA Configurations of Receptacles

- NEMA 5-15R, 15A 125V Duplex Recp.
- NEMA 5-20R, 20A 125V Duplex Recp.
- NEMA 5-20R, 20A 125V Duplex Recp. Hospital Grade, Isolated Ground, Transient Voltage Supressor
- NEMA 14-30R, 30A 125/250V 1-Phase 4-Wire Single Receptacle (Dryer)